Sustainable Crop Plants Protection: Implications for Pest and Disease Control

Editors

Barlin Orlando Olivares Campos
Miguel Araya-Alman
Edgloris E. Marys

Basel • Beijing • Wuhan • Barcelona • Belgrade • Novi Sad • Cluj • Manchester

Editors

Barlin Orlando Olivares Campos
Biodiversity Management Research Group (GESBIO-UCO)
University of Cordoba
Córdoba, Spain

Miguel Araya-Alman
Department of Agrarian Sciences
Catholic University of Maule
Talca, Chile

Edgloris E. Marys
Centro de Microbiologia y Biología Celular
Instituto Venezolano de Investigaciones Cientificas
San Antonio de Los Altos, Venezuela

Editorial Office
MDPI
St. Alban-Anlage 66
4052 Basel, Switzerland

This is a reprint of articles from the Special Issue published online in the open access journal *Sustainability* (ISSN 2071-1050) (available at: https://www.mdpi.com/journal/sustainability/special_issues/Crop_Plants_Protection_Pest_Disease_Control).

For citation purposes, cite each article independently as indicated on the article page online and as indicated below:

Lastname, A.A.; Lastname, B.B. Article Title. *Journal Name* **Year**, *Volume Number*, Page Range.

ISBN 978-3-0365-9151-3 (Hbk)
ISBN 978-3-0365-9150-6 (PDF)
doi.org/10.3390/books978-3-0365-9150-6

Cover image courtesy of Barlin Orlando Olivares Campos

Contents

About the Editors

Barlin Orlando Olivares Campos

Dr. Olivares is a researcher in the Biodiversity Management Research Group (GESBIO-UCO) at the University of Cordoba (UCO) and completed his Ph.D. Cum Laude in Agricultural, Food, Forestry Engineering, and Sustainable Rural Development from the Agrifood Campus of International Excellence (CeiA3) of the University of Cordoba, Spain (2022) through a scholarship from the Iberoamerican Postgraduate University Association (AUIP). His outstanding scientific activity earned him the UCO Extraordinary Ph.D. Award (2022).

He is an Agricultural Engineer from the Central University of Venezuela (2009) and received his master's degree in Environmental Technology from the International University of Andalusia, Spain (2015) through a Carolina Foundation scholarship.

Dr. Olivares started in the scientific field in 2008, when he obtained a scholarship to stay at the Water Center for Arid and Semi-arid Zones of Latin America and the Caribbean (CAZALAC) based in Chile, within the framework of the UNESCO International Hydrological Program. Later, he was a young researcher at the National Institute of Agricultural Research (INIA) of Venezuela (September 2009 to January 2014), carrying out scientific activities and knowledge transfer to farmers. He was also a researcher at the Institute for Advanced Studies Foundation, Venezuela (IDEA) (March-August 2014), and finally a researcher at the National Institute of Meteorology and Hydrology, Venezuela (INAMEH) (January 2016 to May 2017).

Dr. Olivares is the author of an outstanding scientific production of journal papers and obtained various fellowships, accreditations, and recognition, as well as participation in projects for the promotion and dissemination of science in agricultural spaces. He has carried out six research stays in research institutes and universities of international prestige in Argentina, Chile, Costa Rica, Colombia, Spain, and Panama.

Miguel Araya-Alman

Dr. Araya-Alman obtained his Doctorate in precision agriculture from the prestigious University of Talca in Chile in 2020. This high-level training provided him with a solid foundation to address the challenges and opportunities that precision agriculture offers. Prior to his PhD, Dr. Araya-Alman had already demonstrated his commitment to agricultural research and development by obtaining a Master's Degree in Horticulture from the same University of Talca in 2014. His Master's Degree in Horticulture provided him with a deep understanding of plants and crops, essential knowledge for precision agriculture.

Dr. Araya-Alman began his academic journey by obtaining his Agricultural Engineering degree from the University of Talca in 2011, where he acquired the fundamental technical skills to address agricultural and forestry challenges. His academic experience spans two prominent educational institutions in Chile. For a decade, from 2009 to 2019, he played a fundamental role as a Researcher and Transfer Officer at the University of Talca, specifically in the field of Agrarian Sciences. During this period, he contributed significantly to the research and application of precision agriculture techniques in Chilean agriculture.

Since 2019, Dr. Araya-Alman has continued his academic career as a full-time Professor and Researcher at Universidad Católica del Maule (Chile), where he works in the field of Agrarian and Forestry Sciences. His commitment to teaching and research at this institution demonstrates his passion for training future generations of agricultural and forestry professionals and his desire to further drive sustainable agricultural development.

Edgloris E. Marys

Edgloris Elena Marys Saravia earned her B.S. degree in Science Education from Andrés Bello Catholic University (UCAB, Caracas, Venezuela), M.S degree in Microbiology from the Venezuelan Institute for Scientific Research (IVIC, Caracas, Venezuela), and her Ph.D. in Microbiology from IVIC. She pursued postdoctoral research at the University of California, Davis, and in 2002, she was appointed to the role of Assistant Researcher at IVIC. She was named Head of the Biotechnology and Plant Virology Laboratory at IVIC in February of 2012.

Much of Dr. Marys' work has been on plant pathology, specializing in the diagnostics and characterization of plant viruses and other phytopathogens of relevance to agricultural biosecurity. Her work is applicable to plant pathogens that can be intercepted at borders or detected by the general surveillance of field settings. Her current research includes pathogen detection, discrimination, and diagnostics. She is also interested in the identification of molecular landmarks and signatures, and the implications of genetic data for elucidating taxonomic relationships, host-pathogen associations, and pathogen detection. She has described new species of begomovirus, potyvirus, and tenuivirus infecting horticultural, cereal, and fruit crops, as well as the etiological agent causing citrus huanglongbing (HLB) in Venezuela. Her research has been published in leading international journals for rapid reporting on new, emerging, and established plant diseases. In 2019, she co-founded the Emerging Plant Disease Cluster in Venezuela. Recently, served as an ad hoc Science Advisor at INSAI, the Venezuelan National Plant Protection Organization, developing methods for the surveillance, prevention, and management of the fungus Foc TR4, which she diagnosed in bananas. Dr. Marys has supervised three B.S, four Masters, and two Ph.D. theses and one postdoctoral fellow.

Preface

This reprint, titled "Sustainable Crop Plants Protection: Implications for Pest and Disease Control", has become of scientific interest again for a reason: to continue its mission of advancing the science and practice of sustainable agriculture. This Special Issue delves into the multifaceted world of crop protection, exploring innovative approaches, technologies, and solutions that are pivotal for the future of agriculture. Our objectives remain as relevant today as they were when the first call went out for the Special Issue of *Sustainability*, to shed light on sustainable pest and disease control strategies, foster scientific collaboration, and provide a platform for sharing knowledge that can transform agriculture for the better.

The motivation behind the creation of this Special Issue was, and continues to be, an unswerving commitment to safeguarding our global food supply and the ecosystems that support it. The challenges facing agriculture are complex and ever-evolving, from the relentless march of pests and diseases to the imperative of maintaining ecological balance. In response to these challenges, a group of passionate scientists set out to compile and disseminate research that explores sustainable solutions, fostering a harmonious coexistence between humanity and the environment. The pages within represent a collective effort to address these critical issues head-on and drive sustainable practices forward.

Sustainability has emerged as a central theme in agriculture in recent decades, driven by the imperative to feed a growing global population while preserving our planet's delicate ecosystems. The original publication of this reprint was a testament to the collective efforts of researchers, scientists, and experts who have dedicated their careers to finding innovative ways to protect crop plants against pests and diseases without compromising the environment or jeopardizing human health.

This reprint is intended for a diverse audience, including scientists, researchers, agronomists, policymakers, students, and anyone with an interest in the future of agriculture. We recognize that addressing the challenges of crop protection requires a multidisciplinary approach, and as such, we welcome readers from various backgrounds who share a common goal: the advancement of sustainable agriculture practices. We hope that the knowledge contained herein will inspire collaboration, spark innovative ideas, and ultimately lead to a more resilient and sustainable food system.

The contributions within this reprint are the result of the dedication and expertise of a wide array of authors from Argentina, Burundi, China, Egypt, Kenya, Lebanon, Malaysia, Martinique, Panama, Peru, Saudi Arabia, Switzerland, Taiwan, and Venezuela, who have poured their knowledge, insights, and passion into their respective studies. We extend our heartfelt gratitude to all the authors who have made this work possible.

Barlin Orlando Olivares Campos, Miguel Araya-Alman, and Edgloris E. Marys
Editors

 *sustainability*

Article

Relationship of Microbial Activity with Soil Properties in Banana Plantations in Venezuela

Barlin O. Olivares [1],*, Juan C. Rey [2,3], Guillermo Perichi [3] and Deyanira Lobo [3]

[1] Programa de Doctorado en Ingeniería Agraria, Alimentaria, Forestal y del Desarrollo Rural Sostenible, Campus Rabanales, Universidad de Córdoba, Carretera Nacional IV, km 396, 14014 Cordoba, Spain

[2] Unidad de Recursos Agroecológicos, Instituto Nacional de Investigaciones Agrícolas (INIA-CENIAP), Av. Universidad vía El Limón, Maracay 02105, Venezuela

[3] Facultad de Agronomía, Universidad Central de Venezuela, Av. Universidad, Maracay 02105, Venezuela

* Correspondence: barlinolivares@gmail.com or ep2olcab@uco.es

Abstract: The present work aims to analyze the relationship of microbial activity with the physicochemical properties of the soil in banana plantations in Venezuela. Six agricultural fields located in two of the main banana production areas of Venezuela were selected. The experimental sites were differentiated with two levels of productivity (high and low) of the "Gran Nain" banana. Ten variables were selected: total free-living nematodes (FLN), bacteriophages, predators, omnivores, Phytonematodes, saturated hydraulic conductivity, total organic carbon, nitrate (NO_3), microbial respiration and the variable other fungi. Subsequently, machine learning algorithms were used. First, the Partial Least Squares-Discriminant Analysis (PLS-DA) was applied to find the soil properties that could distinguish the banana productivity levels. Second, the Debiased Sparse Partial Correlation (DSPC) algorithm was applied to obtain the correlation network of the most important variables. The variable free-living nematode predators had a degree of 3 and a betweenness of 4 in the correlation network, followed by NO_3. The network shows positive correlations between FLN predators and microbial respiration (r = 1.00; p = 0.014), and NO_3 (r = 1.00; p = 0.032). The selected variables are proposed to characterize the soil productivity in bananas and could be used for the management of soil diseases affecting bananas.

Keywords: soil microbes; microbial respiration; organic carbon; Phytonematodes; soil; machine learning

Citation: Olivares, B.O.; Rey, J.C.; Perichi, G.; Lobo, D. Relationship of Microbial Activity with Soil Properties in Banana Plantations in Venezuela. *Sustainability* **2022**, *14*, 13531. https://doi.org/10.3390/su142013531

Academic Editor: Chenggang Gu

Received: 11 September 2022
Accepted: 18 October 2022
Published: 19 October 2022

1. Introduction

Bananas are the most consumed fruits in the world, and one of the most dynamic crops in international trade, considered among the most exported fruits [1]. As a staple food, bananas, including plantains and other types of cooking bananas, contribute to the food security of millions of people in much of the developing world. The banana is rich in carbohydrates (20%), and an important source of Potassium, Vitamin C and Iron [2–4]. According to FAOSTAT [5] Venezuela ranked as the ninth largest banana-producing country in the Americas in 2019 with a production of 650,051 t of bananas from a harvested area of 41,708 ha.

The most recent studies in Venezuela are those of Hernández et al. [6], Delgado et al. [7], González-Pedraza et al. [8], Olivares et al. [9], Martínez et al. [1], Olivares et al. [10,11], González-García et al. [12,13]. The latest research focuses on the properties of the banana soils of Aragua and Trujillo states in Venezuela and emphasizes the importance of microbiological properties in banana productivity.

In agroecosystems, soil organisms are divided into several groups: micro, meso, macro, and megafauna. These organisms are represented by unicellular and microscopic forms, nematodes, insect larvae, earthworms, and arthropods,. They perform the function of transforming all organic and inorganic compounds in the soil, thus having a direct or indirect influence on properties such as soil porosity, water transport, and the union of particles for the formation of stable soil aggregates [14,15].

In the studies carried out by Delgado et al. [7] and González-García et al. [13] to determine a quality index of banana soils in Venezuela, it was concluded that the biological properties of the soil represented the highest scores in the construction of the quality index. These results establish the importance of considering these variables in research that addresses the quality of banana soils.

In the last decade, the productivity of banana plantations in Venezuela has been affected mainly by the application of high-cost inputs including huge amounts of agro-chemicals, and by soil conservation practices, generating a deterioration of the physical, chemical, and biological properties of soil [7,10]. Hence, the importance of evaluating these soil properties is linked to the decomposition of organic matter and the by-products that are generated by their determining role in the soil [14,16].

The characterization, measurement, and analysis of biological diversity in agroecosystems are performed to provide a new and solid knowledge of the different organisms that make their lives in the soil, as well as to obtain indicators that would allow making decisions or recommendations based on the conservation of taxa or areas where the conservation of the agroecosystem is threatened [17–19].

The link to microbial activity is indirectly estimated by determining basal respiration (O_2 consumption in the medium or CO_2 concentration released) and current productivity; this is fundamental in the search for higher productivity and sustainability in bananas. As noted, most studies rely heavily on soil quality using traditional statistical methods, while the potential of machine learning algorithms such as Debiased Sparse Partial Correlation (DSPC), Partial Least Squares-Discriminant Analysis (PLS-DA), and Random Forest, has been only moderately explored. Various algorithms have been studied; however, to the best of our knowledge, they have not been widely used in banana fields. This is a novel element in our study, in which soil biological properties were explored as promising new soil indicators to assess banana productivity in Venezuelan soils. Our study is a pioneer in the application of algorithms such as DSPC, which is completely new in banana soils. However, beyond its application in the case of a specific crop (banana) and geography (Venezuela), we have developed a scientific logic that is easily transferable to other areas, not only in agriculture but in soil science in general.

In this sense, and motivated by the scarcity of information on the biological quality of soil in the banana areas of Venezuela and its relationship with banana productivity, the objective of this research is to analyze the relationship of microbial activity with the physicochemical properties of the soil in banana plantations in Venezuela and identify the main soil variables responsible for the differentiation between high and low productivity sites, as well as the differences between banana farms. The results of this microbial characterization would reflect the microbiological processes of the organisms from the soil and would be a potential indicator of soil quality.

2. Materials and Methods

2.1. Description of the Study Areas and Banana Farms

The study area corresponds to six banana farms (Table 1), of which four are in the Depression of the Valencia Lake Basin of Aragua state (PL, SM, PZ, and CH), whose climate is tropical savanna (Aw) with an average annual rainfall of 980.0 mm [20]. The rainfall is seasonal, with five to six rainy months (May-October period). The average annual temperature is 26.2 °C and the average annual relative humidity is 70.0% [21]. The slope ranges from 0–2%. The PL farm is located on the fourth level of the lake terrace, while the SM, PZ, and CH farms are found on alluvial soils, whose texture classes are medium to silty. The predominant soil orders are Mollisols and Inceptisols with characteristics of moderate to good drainage, a neutral to alkaline pH and generally good fertility and organic matter contents above 4.0%.

Table 1. Location (latitude, longitude, height), states and bananas planted area of six sites evaluated in Venezuela.

Farm Code	Latitude	Longitude	Height (masl)	State	Planted Area (ha)	Average Yield (t ha^{-1} year^{-1})
PL	10°12′20″ N	67°30′10″ W	435	Aragua	135	74.9
BA	09°29′14″ N	70°57′05″ W	16	Trujillo	300	69.6
KA	09°28′31″ N	70°55′46″ W	17	Trujillo	270	64.9
SM	10°12′55″ N	67°23′42″ W	502	Aragua	11	11.1
PZ	10°11′30″ N	67°31′04″ W	514	Aragua	20	11.4
CH	10°11′34″ N	67°31′34″ W	498	Aragua	9	12.3

The other banana sites (BA and KA) are in the southeast Region of Maracaibo Lake in the Trujillo state, whose climate is tropical savanna (Aw), with annual rainfall amounts of 1094.0 mm with two precipitation peaks, one occurring in April or May and the other in October [22]. The annual average temperature is 27.5 °C and there is an average relative humidity of 78.0%. According to Martinez et al. [1], the terrain is flat with predominantly Entisols soil order, whose drainage is moderate to poor, pH neutral to alkaline, with average organic matter contents of 2.75%; that is, they are soils of moderate fertility.

2.2. Soil Sampling

The delimitation of the sampling areas was carried out according to Rosales et al. [23]. To establish the productivity levels in each site, two plots (4.0 ha for PL, BA and KA with four replicated) were identified, and the productivity index (PI) of the "Gran Nain" banana variety was calculated according to the procedure described in Olivares et al. [9]. On the large plantations ($\geq$50 ha, PL, BA and KA), the average yield of high productivity plots was 69.8 ± 5.0 t ha^{-1} yr^{-1} and in low productivity plots, it was 59.7 ± 5.3 t ha^{-1} yr^{-1}. On the other sites (SM, PZ, and CH), the two levels of productivity were identified in lots of 1 ha with two replicated plots in each lot. On the small plantations (<25 ha, SM, PZ and CH) the average yield on small plantations was 11.5 ± 0.7 t ha^{-1} yr^{-1} for high productivity plots and 1.6 t ha^{-1} yr^{-1} for those with low productivity. A total of 36 disturbed samples of 2.0 kg of soil were obtained (24 samples for PL, KA, and BA and 12 samples for PZ, CH, and SM) for the determination of the chemical and biological properties of soils by means of soil pits in the first 60 cm depth, following the guidelines suggested in Rosales et al. [23]. The following variables were obtained: total count of populations of bacteria, fungi, and actinomycetes (colony-forming units. g^{-1} soil); microbial respiration (mg 100 g^{-1}.10 days^{-1}); microbial biomass (mg) and carbon associated with microbial biomass (mg); plant-parasitic nematodes: *Radopholus similis* and *Helicotylenchus multicinctus* (Logarithm (N + 1)); total free-living nematodes (FLN) (number of nematodes in 100 g soil) which includes FLN bacteriophages, predators, and omnivores, Phytonematodes (PN); endophytic fungi in banana root system, namely *Trichoderma*, *Fusarium*; and other fungi. The chemical variables such as nitrate (NO$_3$) included in the analysis were determined according to Rosales et al. [23]. This study considers the same number of undisturbed samples (36) obtained with an Uhland-type sampler (mean depth 29.6 ± 17.6 cm) to obtain the saturated hydraulic conductivity (Ks, cm.h^{-1}).

2.3. Statistical Analysis

Before data analysis, we checked the data integrity. There were no missing values or negative values. The normalization of the soil and microbial variables was carried out using the statistical package in R software version 4.0.2 based on the geometric mean, and a generalized logarithmic transformation using "glog" function in R was performed to make the variables more comparable [24].

Before statistical analysis, a Principal Component Analysis (PCA) was performed as an exploratory analysis to check the presence of outliers and identify patterns in the data. The objective of this application was to summarize the information of many variables (29 in our study) in a few latent variables, trying to avoid overfitting as new components are added. The PCA allowed us to select only ten variables in our study: total free-living nematodes (FLN); bacteriophages; predators; omnivores; Phytonematodes (PN); saturated hydraulic conductivity (Ks); total organic carbon (C tot); nitrate (NO_3); microbial respiration (MR); variable other fungi (which includes non-pathogenic strains of endophytic fungi).

For determining the relative importance of different nematode orders, genera, and microbial variables in the global areas, we used the R circlize package, which provides an implementation of circular layout generation in R as an efficient way for the visualization of microbial information [25].

The Partial Least Squares-Discriminant Analysis (PLS-DA) was applied for the identification of the most important variables in the discrimination of banana productivity sites, which is a supervised method that uses multivariate regression techniques to extract via a linear combination of original variables (X) the information that can predict the class membership (Y). The PLS regression is performed using the plsr function provided by the R pls package [26]. The classification and cross-validation are performed using the corresponding wrapper function offered by the caret package.

The number of optimal components was estimated from cross-validation methods, which calculate the predictive power of the analysis. These methods are based on removing groups of data, calculating a new model from the remaining data, and predicting the values of the removed data. Many iterations of the same process are performed and, from the predicted and observed data, the predictive residual sum of squares (PRESS) is calculated. From the results of all the iterations, the new corresponding R^2y is calculated, and the Q^2 that represents the predictive capacity of the model [27]. This parameter is a good indicator to decide which components to include in the model.

The variable importance in projection (VIP) > 1.0, the weighted sum of the PLS-regression coefficient, and corresponding |loading values| > 0.2 in the model were used to identify the key variables [28]. Additionally, the non-parametric Kruskal-Wallis test was applied to obtain the significant variables. Subsequently, the correction was made for the Benjamini-Hochberg multi-test (FDR, false discovery rate) whose established significance level was 95% ($p < 0.05$).

Finally, we applied the Debiased Sparse Partial Correlation algorithm (DSPC), which is based on the recently proposed de-sparsified graphical lasso modeling procedure [29]. Through the graphical result, it is possible to visualize correlation networks where the nodes represent variables under study and the edges represent correlations between them.

3. Results and Discussion

3.1. Proportion of Free-living Nematodes (FLN) and Phytonematodes (PN) in Banana Sites

According to our results, Phytonematodes predominate in the banana system, with eleven important genera (Figure 1A), such as *Rotylenchulus* (28.5%), *Meloidogyne* (27.6%) and *Helicotylenchus* (24.5%), having the highest proportion. Within the group of FLN, only seven orders were identified, with Rhabditida (60.9%) and Dorylaimida (28.0%) having the highest proportion. Figure 1B shows that no significant differences were found for these nematode variables (data not shown); however, the proportion of Phytonematodes (1.60–3.10 log10) was higher in most of the evaluated banana farms compared to the FLN (1.41–2.48 log10), regardless of the productivity level (Table A2 in Appendix A).

Figure 2A shows the logarithmic representation of bacteria and fungi according to banana productivity levels: high (H) and low (L) in the six evaluated banana sites, indicating that there were no significant differences in representation between the productivity levels; the average representation of bacteria was slightly higher (5.95 ± 0.16 log10) than that of fungi (4.69 ± 0.23 log10) (Table A1 in Appendix A). On the other hand, Figure 2B presents the endophytic fungal populations that colonize the internal tissues of a plant without causing any pathogenic processes. In our case, emphasis was placed on the populations of mutual endophytic fungi such as *Trichoderma*, *Fusarium*, and other fungi because these microorganisms confer protection against pathogen attack and the colonization of plant organs and tissues. The non-pathogenic strain of *Fusarium* presented an average of 52.71 ± 20.46 CFU·g^{-1} soil. It is worth mentioning the high population of *Trichoderma* with 45.00 ± 21.21CFU·g^{-1} soil at the high productivity level of the CH farm, distinguishing it from the rest of the farms. Similarly, the variable "other fungi" was higher in PA (L) (Table A1).

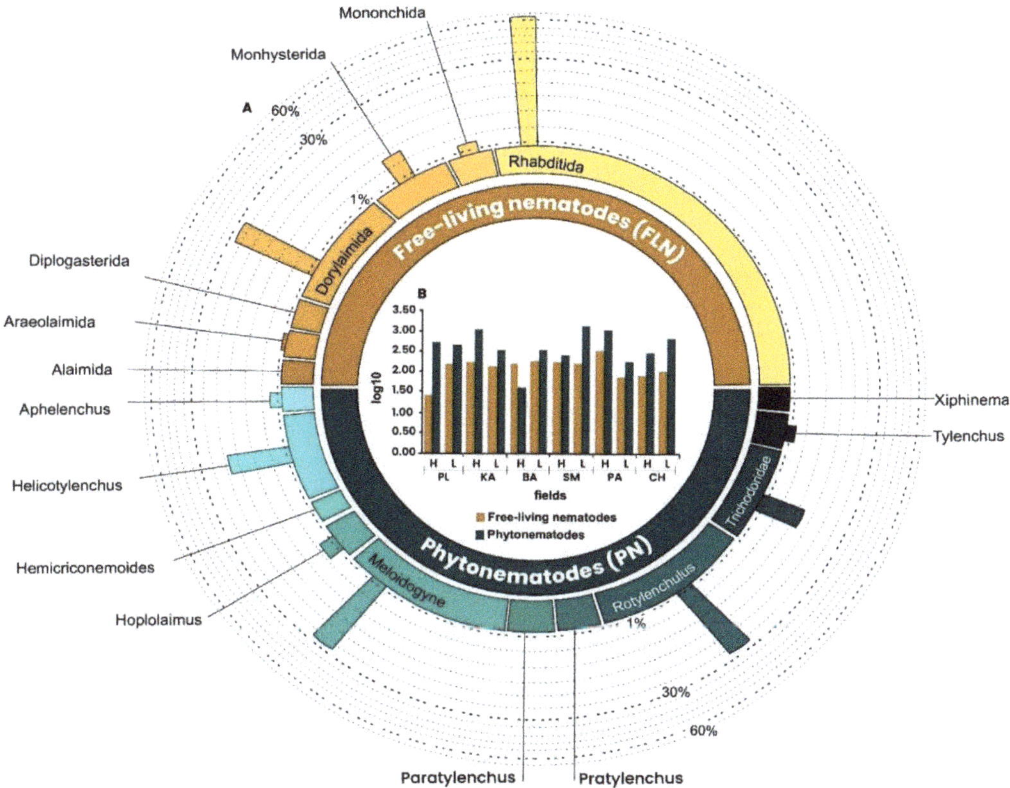

Figure 1. (**A**) Proportion of free-living nematodes (FLN) and Phytonematodes (PN) order/genre in the global banana sites. Vertical bars indicate the proportion of species in a group/total number of species in that group×100 (on a logarithmic scale); (**B**) Number of FLN and PN (log10) according to the level of banana productivity: high (H) and Low (L) in the six banana sites: PL, KA, BA, SM, PA, and CH.

Nematodes are the most abundant metazoan and they vary in their sensitivity to environmental disturbances. Free-living and plant-parasitic nematodes are effective ecological indicators, contributing to nutrient cycling and having important roles as primary, secondary, and tertiary consumers in food webs. Tillage, cropping patterns and nutrient management may have strong effects on nematodes, with changes in communities reflecting soil disturbance [30]. In the soils sampled in our study, the bacteria-consuming nematodes were the most representative, followed by the omnivores and lastly the predators of the order Mononchida. The low proportion of this last trophic group belonging to the predatory FLNs could reflect the presence of soil alteration due to intensification and poor soil management in the experimental points evaluated, as pointed out by Olivares et al. [9]; apparently these do not play a significant role as population regulators of the Phytonematodes present. According to Talavera et al. [31], the suppression of phytoparasitic nematodes by predatory nematodes is significant in overly complex soil food webs and agricultural management leads to a reduction in the suppressive capacity of the soil food web. Our results agree with those of Castilla-Díaz et al. [32] which show the genus c.f. *Dorylaiminae* (Omnivorous) as the most abundant nematode in Colombia. Other studies have found that omnivore and predator nematodes are less abundant in disturbed soils or intensive banana plantations [18].

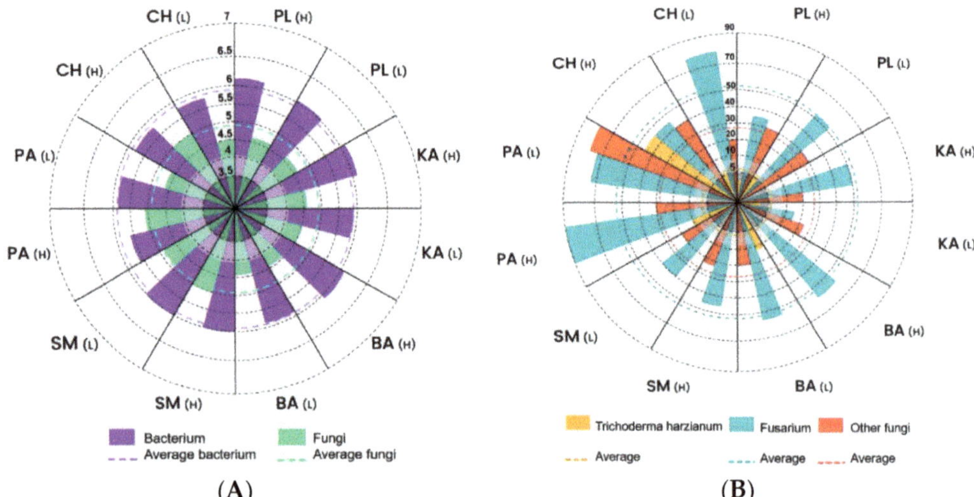

(**A**)   (**B**)

Figure 2. Logarithmic representation of: (**A**) bacteria and fungi according to banana productivity levels: high (H) and low (L) in the six evaluated banana sites; (**B**) *Trichoderma harzianum*, *Fusarium* sp. and other fungi in the six evaluated banana sites: PL, KA, BA, SM, PA, and CH.

Our results agree with those of Landi et al. [33] who found that representatives of the order Rhabditida (bacterivores) were the most abundant nematode under different land uses in Italy. Similar results were obtained by Arie-Vonk et al. [34] in the Netherlands with different management practices and soil types. This is because rhabditids, having short life cycles and high reproductive potential, respond more quickly to organic enrichment than other groups of nematodes [35].

Regarding the plant-parasitic nematodes, the most representative genera in this study were: *Rotylenchulus*, *Meloidogyne* and *Helicotylenchus*, respectively. These nematodes are very frequent in banana cultivation, especially in the central region of the country. The most important species are *R. reniformis*, *M. incognita*, *H. dihystera* and *H. multicinctus*. In Aragua State, a combination of *H. multicinctus* and *M. incognita* severely reduces banana yields [36,37].

The study by Davide [38] established that excess K in banana soils increased the penetration and damage of *Meloidogyne* spp., *Helicotylenchus* spp. and *Tylenchorhyndrus* spp. Low levels of potassium 610 mg / l reduced the impact of *R. similis* on banana roots [39]. Likewise, increases in nitrate content in the soil increased the percentage of parasitic nematodes in banana. On the other hand, in banana, a positive correlation was found between the presence of *Meloidogyne* spp. and *H. multicinctus* and the content of silt and sand, as well as a negative correlation with clay [40].

3.2. Identification of the Most Important Variables with PLS-DA

The results of the PLS-DA showed the first components of the PLS (accuracy accumulated: 0.61; R^2y: 0.50 and Q^2: 0.50). Figure 3A shows the scatter charts of the scores and the loading between component 1 and component 2. For the representation, an ellipse with 95% confidence level was constructed from Hotelling's T^2 statistic. In Figure 3A, in the plane formed by the 1st and 2nd components, PLS-DA could not distinguish the high and low productivity levels of bananas (Figure 3A). The first two main components (PC) explained 49.9% of the variables; however, no trends were detected in banana productivity differences. In our case, the optimal number of components is two; new components that are added collect information independent of that of the previous components, so it is to be expected that each time they cover less variability and introduce more noise in the graphs.

According to the criterion of selecting only those variables with | loading values| > 0.2, our results establish that only seven variables meet this condition in component 1, they are: FLN bacterivores (-0.57), FLN predators (-0.52), Ks (-0.42), FLN omnivorous (0.32), C total (0.27), NO_3 (-0.25) and MR (-0.21) (Figure 3B). Only corresponding loadings greater than the absolute value of 0.2 are the best option for model interpretation and revealing the most important variables regarding the explanation of the response [41].

On the other hand, two variable importance measures in PLS-DA were used in our study. The first, VIP is a weighted sum of squares of the PLS loadings considering the amount of explained Y-variation in each dimension. VIP scores are calculated for each component. In our study, as the first two components were used to calculate the feature importance, the average of the VIP scores is used (Table 2), with only four variables being the most important: FLN predators, FLN bacterivores, saturated hydraulic conductivity (Ks) and FLN omnivorous. Second, the other important measure is based on the weighted sum of PLS. The weights are a function of the reduction of the sums of squares across the number of PLS components; the average of the feature coefficient is used to indicate the overall coefficient-based importance. The highest coefficients correspond to the variables whose VIP was greater than 1. Table 3 shows the results of the Kruskal Wallis Test and Figure 4 the box graphs where the differences of the significant variables in the evaluated banana farms are evidenced.

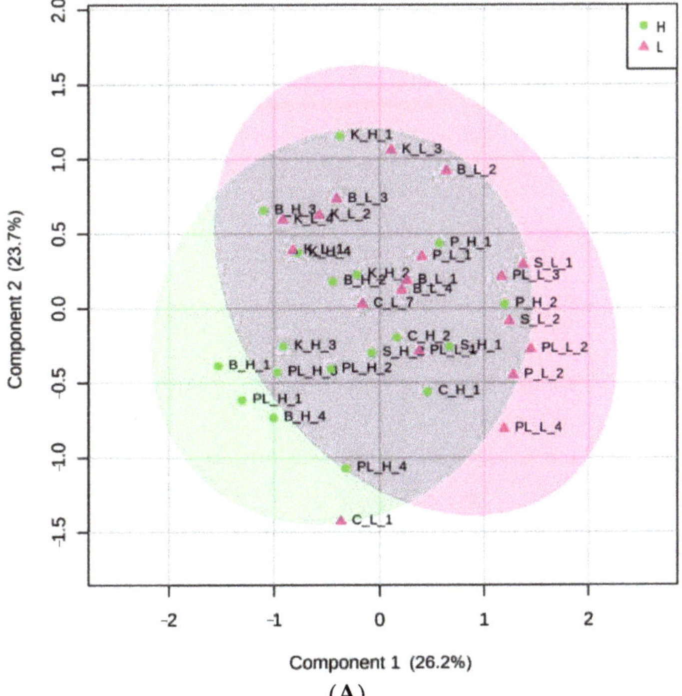

(A)

Figure 3. *Cont.*

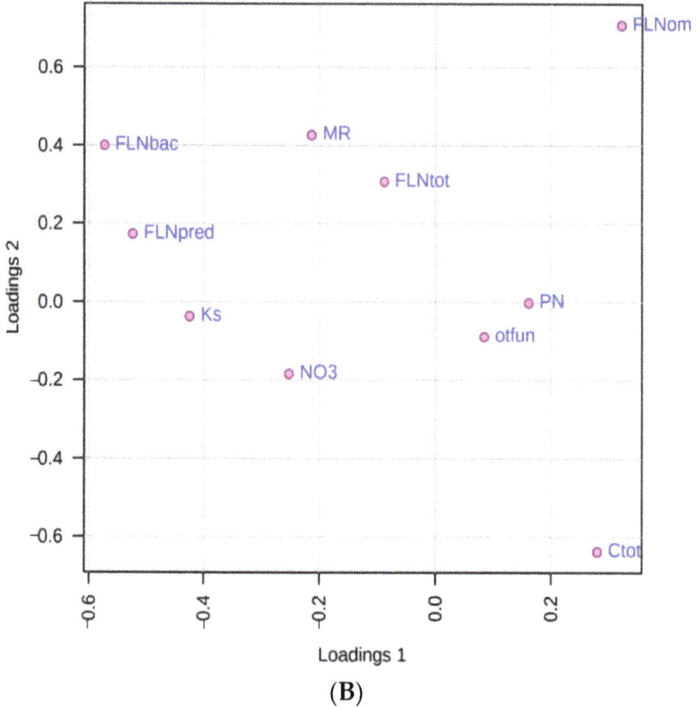

(B)

Figure 3. (**A**) Scores plot between component 1 and component 2 of the PLS-DA. Note: Data code composed of three digits, the first digit is the site (PL, KA, BA, SM, PA, and CH), followed by the level of productivity: (H) High and (L) low, and finally the repetition number (1–4). The explained variances are shown in brackets. Ellipse T2 de Hotelling (95%); (**B**) Loading plot between component 1 and component 2. (Accuracy accumulated: 0.61; R^2y: 0.450 and Q^2: 0.50). FLNtot: total free-living nematodes, FLNbac: free-living nematodes bacterivores, FLNpred: free-living nematodes predators, FLNom: free-living nematodes omnivores, Ks: saturated hydraulic conductivity, NO_3: nitrate, PN: Phytonematodes, Ctot: total organic carbon, MR: microbial respiration, otfun: other fungi.

Table 2. The variable importance in projection (VIP) in componente1, component 2, VIP average and coefficient score.

Variable	VIP 1	VIP 2	VIP Average	Coefficient Score
FLN predators	1.72	1.55	1.64	100.00
FLN bacterivore	1.48	1.40	1.44	77.58
Ks	1.41	1.27	1.34	76.43
FLN omnivores	1.29	1.22	1.26	75.68
NO_3	0.94	0.86	0.90	47.20
Phytonematodes	0.31	0.37	0.34	22.31
FLN total	0.27	0.25	0.26	15.16
Other fungi	0.20	0.20	0.20	14.95
Total organic carbon (Ctot)	0.16	0.94	0.55	11.91
Microbial respiration (MR)	0.08	0.76	0.42	0.00

FLN: free-living nematodes; Ks: saturated hydraulic conductivity; NO_3: nitrate; VIP: variable importance in projection.

Table 3. Important variables by Kruskal Wallis test result: chi-squared, *p*-value (<0.05) and false discovery rate (FDR).

Variable	Chi-Squared	*p*-Value	$-\log10(p)$	FDR
Microbial respiration (MR)	22.995	0.001	34.708	0.003
NO_3	19.706	0.001	28.481	0.007
FLN bacteriophage	17.479	0.003	24.347	0.010
Total organic carbon (Ctot)	17.169	0.004	23.778	0.010
FLN total	15.051	0.010	19.938	0.018
FLN omnivores	14.85	0.011	19.577	0.018

FLN: free-living nematodes; NO_3: nitrate; FDR: false discovery rate.

The result demonstrated how soil properties influence the production of a complex and variable system. The Ks, NO_3 and microbiological (biotic) properties such as microbial respiration, organic carbon and free-living nematodes were linked to biometric and productive responses in commercial Cavendish banana plantations. This relationship in production can also be related to the response of plants to diseases. However, the variability of the systems must be considered in the eventual implementation of practices to manage soil properties for both production and disease management.

The soil properties obtained through the model show that in some farms the differences between the high and low-productivity sites are very narrow, possibly due to the scarcely differentiated management carried out on the farms. Farm managers do not consider the association between productivity and soil properties, whether physical [11], chemical or biological. This suggests that fertilization is not based essentially on soil or foliar analysis and not on the use of soil conservation practices in the banana sites studied.

Soil management can be a critical factor in PL, BA, and KA banana farms, due to the relationship of the productivity index composed of the number of hands per bunch and the circumference of the mother plant, with the microbiological properties, which can be an indicator of the deterioration of banana production in the evaluated sites.

3.3. Debiased Sparse Partial Correlation Algorithm (DSPC)

The results of the deviant sparse partial correlation (DSPC) algorithm that is based on the de-parsified graph lasso modeling procedure [29] are presented. A key assumption underlying our modeling strategy is that the number of true connections between variables is much smaller than the available sample size, that is, the true network of partial correlations between variables is sparse. This assumption is strongly supported by both empirical evidence and theoretical calculations [42–44]. The algorithm DSPC reconstructed a graphical model and provided partial correlation coefficients and *p*-values for each pair of soil and microbial characteristics in the data set (Figure 5). Therefore, DSPC allowed discovering the connectivity between the number of variables and visualizing them as weighted networks where the nodes represent the microbial and soil variables, and the edges represent partial correlation coefficients or the associated *p* values.

Of all the variables involved, the total carbon has a degree of 4 and a betweenness of 7, which represents the importance of the relationships with the other variables of the system. In our case, the microbial activity was measured through the microbial biomass, which is the most active component of the soil; it comprises between 1 and 5% of total carbon and actively participates in the decomposition of organic matter in these banana soils. The variable FLN predators had a degree of 3 and betweenness of 4 in the correlation network, followed by NO_3. The network shows positive correlations between MR (r = 1.00; *p* = 0.014), FLN predators (r = 1.00; *p* = 0.032) and NO_3.

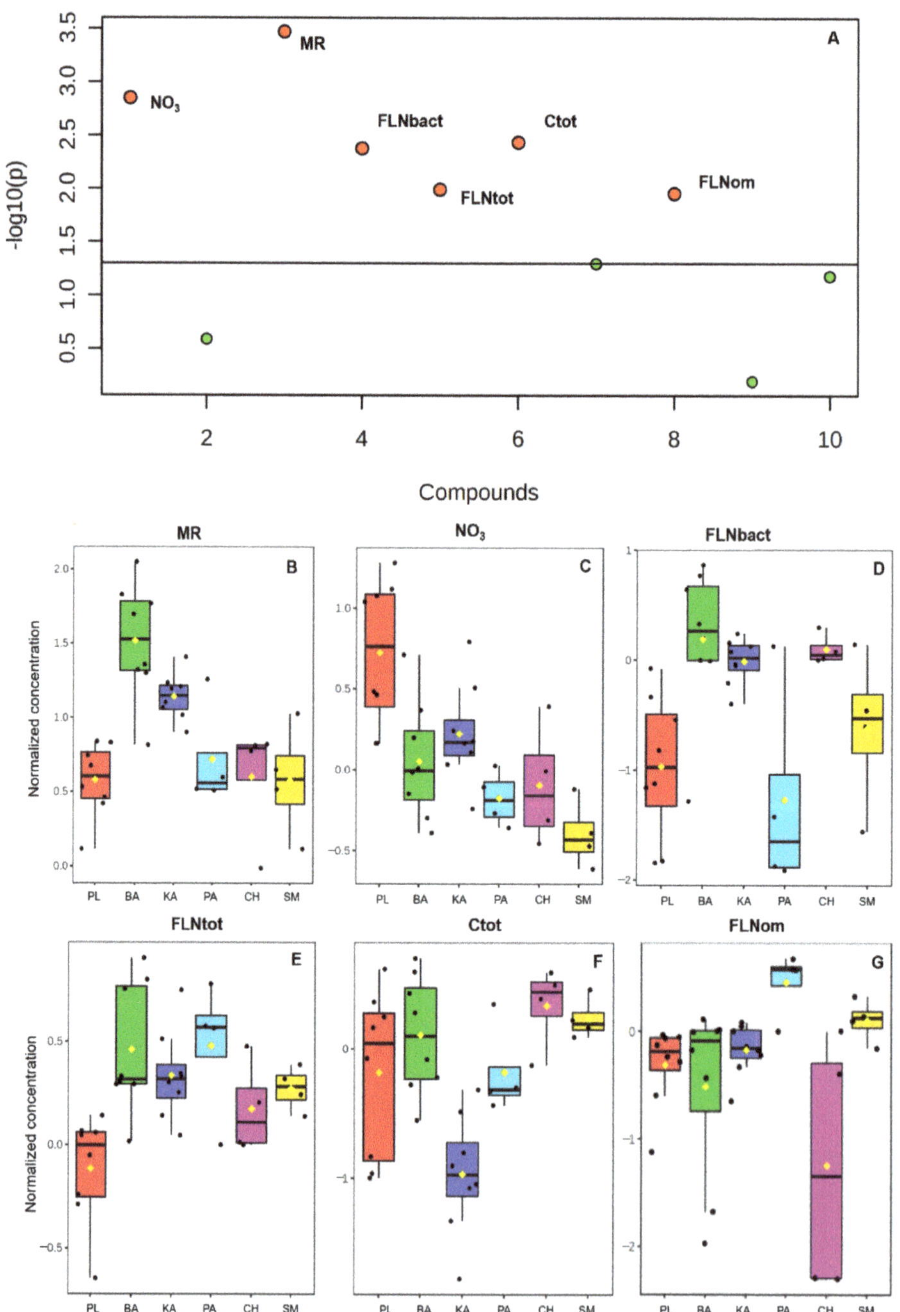

Figure 4. (**A**) Important features selected by Kruskal-Wallis's test plot with *p* value threshold 0.05; (**B–G**) Box plots of significant variables in the evaluated banana sites (PL, BA, KA, PA, CH, SM). MR: microbial respiration; NO_3: nitrate; FLNbac: free-living nematodes bacterivore; FLNtot: total free-living nematodes; Ctot: total organic carbon; FLNom: free-living nematodes omnivores.

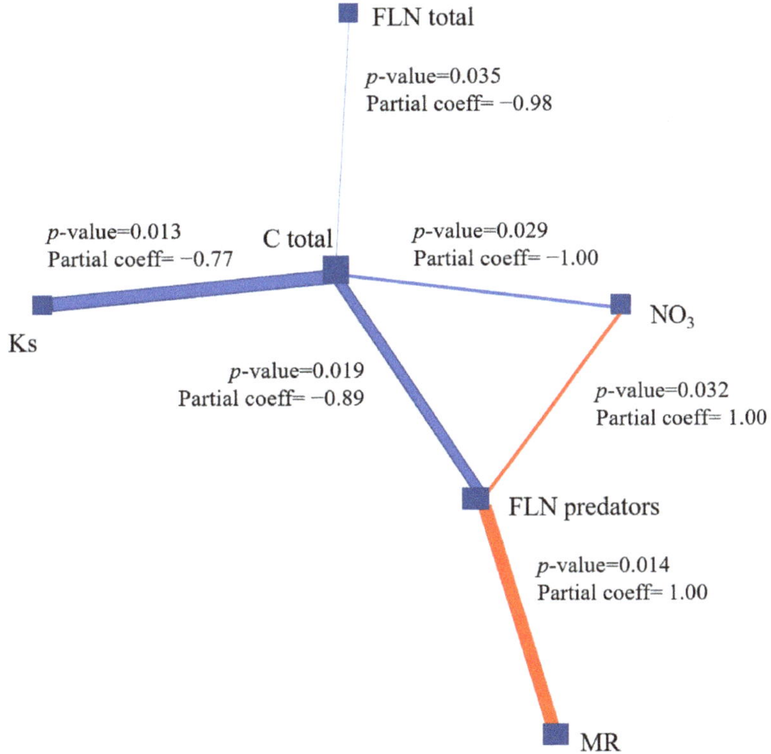

Figure 5. Partial correlation network of five evaluated variables. The size of the node indicates the direction of the change. The colored edges had a p value <0.05 and the false discovery rate (FDR) adjusted p value was <0.5. The red and blue borders show positive and negative correlations. FLNtot: total free-living nematodes; FLNpredators: free-living nematodes predators; Ks: saturated hydraulic conductivity; NO_3: nitrate; Ctot: total organic carbon; MR: microbial respiration.

The composition of nematode communities (Phytonematodes and predatory FLNs) is considered a good indicator of alterations such as the release of agrochemicals in the evaluated soils. These changes are correlated with indicators of the functioning of the ecosystem, such as the increase of NO_3 in the soil, the decrease in carbon, microbial biomass and the change in the structure of the trophic networks, according to Culman et al. [45]; this is reflected in the negative correlations with Phytonematodes (r = −0.99; p = 0.035) and predatory FLN (r = −0.89; p = 0.019) from our study (Figure 5). According to a study of the banana soils of Ecuador [46], the soil physical characteristics (sand, silt, and clay contents, associated with water retention and saturated hydraulic conductivity), 50% of the sampled areas (60 sites) showed a correlation of the weight of roots and with the number of total nematodes.

The study by Ferris et al. [14] revealed that nematodes in the family Tylenchidae, collectively referred to as root associates although some are fungivores, generally do not negatively impact plant growth and respond to increases in organic matter and root weight. Such nematodes contribute to the amplifiable prey pool. Besides providing resources for predator nematodes, they are also prey for micro invertebrate predators of nematodes which, in turn, increases predation pressure on target prey [14,17].

In this regard, [14,17] established that a functional guild of bacterivore and fungivore nematodes participate in organic matter decomposition pathways in the soil. Several species, often differing in ecological amplitude, may be present at the same time, each contributing to complementarity and continuity of their integral ecological function.

The results of the biological tests in the PL farm indicated a higher total microbial respiration for the high productivity level with respect to the low productivity site, showing that the soils with higher productivity have higher biological activity. In the case of BA and KA, the high-productivity soils presented a greater number of populations of bacteria and fungi, indicating a greater biological activity; this is related to the higher content of microbial carbon in these soils. However, total respiration was lower in high-productivity soils than in low-productivity soils.

In this research, high productivity plots (HH) showed low microbial respiration in all the sites evaluated, except in the PL farm, and in fact, soil organic carbon was not among the soil properties selected to be part of the predictive model of banana productivity based on soil properties. These results are consistent with the experiences of González-Pedraza et al. [8] that indicated the lower amount of microbial carbon and total organic carbon observed in Venezuelan soils with high-productivity plants.

In recent studies, no significant statistical differences were found between lots of different vigor or productivity for microbial carbon, basal soil respiration, and microbial and metabolic ratios [12]. The high microbial activity was closely related to the soil texture, and, in turn, this positively influenced the biometric parameters of the plants. The study by González-García et al. [13] and Rondón et al. [16] shows that the bulk density, content of fine particles, organic matter, and carbon and microbial coefficient, were favorable for the high productivity batches. In general, there were no statistical differences in fungal and bacterial colony-forming units (CFU) between vigor batches.

Higher water retention and higher content of organic matter promote better conditions for the development and activity of soil microorganisms. Results of the effect of these edaphic properties on microorganisms have been reported by Olivares et al. [9] and González-García et al. [13]. The results of Millán et al. [47] showed that the FLN genera found are sensitive to slight changes in soil pH, and show dependence on the porosity and soil moisture. In this regard, [48] showed that there is a significant and important correlation between the carbon of the microbial biomass with the macropores and the Ks ($r = 0.84$ and 0.59 respectively), that is, the carbon microbial biomass increased when there were good aeration conditions in the soil. In our study, the Ks values did not show significant differences, the highest mean being found in the SM farm with 5.18 ± 2.70 cm.h^{-1}.

The study by Olivares et al. [9] establishes that commercial banana plantations in the regions of Aragua and Trujillo (Venezuela) were characterized by the intensive use of agrochemicals, generating a considerable reduction in the productivity due to the change and deterioration of the physical, chemical and biological properties of the soil, with the content of Mg, penetration resistance, total microbial respiration, bulk density and omnivorous free-living nematodes being the most determining variables of the quality of banana soil in areas with different levels of productivity. Similarly, [11] indicated that in lacustrine and alluvial banana soils of Venezuela characterized by the change of land use from forest to plantation, the morphological properties of the soil such as biological activity, texture, dry consistency, reaction to HCl and structure type, allow identification of potentially suitable areas for high levels of productivity and long-term sustainability.

The results were compared with the findings of [16], which emphasizes that the high yields of commercial banana farms are associated with a high content of C in stable aggregates, as well as in the more labile fractions of macro-organic matter. These results highlight the importance of the use of less recalcitrant organic fertilizers as a strategy for the sustainable management of banana cultivation.

Likewise, this study represents an important contribution to the knowledge of the banana soils of Venezuela due to the current management of the banana systems in the country. The study can be improved through its systematic application in new locations. It is thus our intention and hopes that other research groups of the international scientific community join this task to produce improved versions.

The research also discusses a series of topics that reflect the thinking and logic of the primary choice of soil variables as the basis of the study. The main use of this research is intended for agricultural extension agents and service provider agents of the banana sector that assist producers of all types. It is also ideal for producers and technical personnel of large and medium-sized farms that design their own production alternatives and undertake problem-solving in Latin America and the Caribbean.

4. Conclusions

The methods used in this study to describe the productive capacity or potential of banana soils in Venezuela were based mainly on the study of physical and chemical properties and their relationships with special biological characteristics such as microbial respiration, total proportional free-living nematodes, bacterivores, omnivores, and total organic carbon, which were sufficient to explain the complex interactions of the soil and its rhizosphere.

Nematode populations in the roots and the content of elements in the soil are known to vary both temporally and spatially, which makes this interaction overly complex. The results suggest further studies in two lines: the effect of nutrition on the number of and damage caused by nematodes and their effect on the absorption of nutrients.

The high level of concordance observed in the network is encouraging, as it provides the basis for identifying new connections between soil properties that may represent still undiscovered or poorly studied regulatory interactions of microbial activity in banana soils. The discovery of such novel interactions may lead to a more complete picture of microbial activity in these Venezuelan banana soils.

Practices to increase and maintain soil quality and stimulate microbiological activity in banana soils in Aragua and Trujillo could have a positive effect on agricultural banana production, not only for low-productivity banana lots affected by disease but also for the sustainable use of banana lots of high productivity.

Author Contributions: Conceptualization, B.O.O.; methodology, B.O.O., J.C.R. and D.L.; software, B.O.O.; validation, D.L., J.C.R. and G.P.; formal analysis, B.O.O.; investigation, G.P. and B.O.O.; resources,. G.P., D.L.; data curation, J.C.R.; writing—original draft preparation, B.O.O., G.P. and D.L.; writing—review and editing, D.L.; visualization, B.O.O.; supervision, G.P. and D.L.; project administration, J.C.R.; funding acquisition, J.C.R. All authors have read and agreed to the published version of the manuscript.

Funding: This research was funded by the project "Technological innovations for the management and improvement of the quality and health of banana soils in Latin America and the Caribbean" financed by FONTAGRO and coordinated by Bioversity International 2007 (before INIBAP).

Institutional Review Board Statement: Not applicable.

Informed Consent Statement: Not applicable.

Data Availability Statement: Not applicable.

Acknowledgments: The financial support for international mobility of the Ibero-American Secretary General with Fundación Carolina and Action KA107 of Erasmus+ Program from Agrifood Campus of International Excellence (ceiA3) (2020).

Conflicts of Interest: The authors declare no conflict of interest. The funders had no role in the design of the study, in the collection. Analyses, or interpretation of data; in the writing of the manuscript; or in the decision to publish the results.

Appendix A

Table A1. Mean and standard deviation of bacteria, fungi, *Trichoderma harzianum*, *Fusarium* sp. and other fungi according to banana productivity levels: high (H) and low (L) in the six evaluated banana sites: PL, KA, BA, SM, PA, and CH.

Farm Code	Level	n	Lbact (Log)	Lfung (Log)	Trichoderma CFU·g^{-1}	Fusarium CFU·g^{-1}	Other Fungi CFU·g^{-1}
PL	H	4	6.06 ± 0.05	4.62 ± 0.05	2.50 ± 5.00	35.00 ± 31.09	30.01 ± 18.26
	L	4	6.13 ± 0.09	4.65 ± 0.06	5.00 ± 5.77	52.50 ± 22.17	30.00 ± 14.14
BA	H	4	6.15 ± 0.05	4.42 ± 0.15	0.00 ± 0.00	57.50 ± 30.96	0.00 ± 0.00
	L	4	5.81 ± 0.09	4.42 ± 0.15	12.50 ± 15.00	57.50 ± 35.94	10.00 ± 14.14
KA	H	4	6.17 ± 0.12	4.53 ± 0.15	0.00 ± 0.00	50.00 ± 14.14	20.00 ± 27.08
	L	4	5.92 ± 0.16	4.57 ± 0.06	2.50 ± 5.00	15.00 ± 12.91	22.50 ± 17.08
SM	H	2	6.02 ± 0.09	5.04 ± 0.62	0.00 ± 0.00	40.00 ± 56.57	20.00 ± 14.14
	L	2	6.02 ± 0.09	4.84 ± 0.08	0.00 ± 0.00	40.00 ± 0.00	20.00 ± 14.14
PZ	H	2	5.78 ± 0.11	5.04 ± 0.62	10.00 ± 14.14	90.00 ± 14.14	30.00 ± 14.14
	L	2	5.88 ± 0.04	4.60 ± 0.00	0.00 ± 0.00	70.00 ± 42.43	80.00 ± 0.10
CH	H	2	5.74 ± 0.06	5.04 ± 0.62	45.00 ± 21.21	45.00 ± 49.50	40.00 ± 0.00
	L	2	5.74 ± 0.06	4.50 ± 0.28	5.00 ± 7.07	80.00 ± 28.28	20.00 ± 0.00

Productivity levels: High (H) and Low (L). CFU = Colony-forming units.

Table A2. Mean and standard deviation of number of Free-living nematodes (FLN) and Phytonematodes (PN) (log10) according to the level of banana productivity: high (H) and Low (L) in the six banana sites: PL, KA, BA, SM, PA, and CH.

Farm Code	Level	n	log10FLN	log10PN
PL	H	4	1.41 ± 0.73	2.70 ± 1.86
PL	L	4	2.16 ± 1.40	2.64 ± 1.67
KA	H	4	2.21 ± 1.00	3.03 ± 2.30
KA	L	4	2.12 ± 1.30	2.52 ± 1.89
BA	H	4	2.19 ± 1.35	1.60 ± 1.09
BA	L	4	2.25 ± 1.58	2.54 ± 1.82
SM	H	2	2.20 ± 1.15	2.40 ± 2.03
SM	L	2	2.20 ± 1.93	3.10 ± 2.39
PA	H	2	2.48 ± 1.63	2.99 ± 1.93
PA	L	2	1.85 ± 0.85	2.23 ± 2.03
CH	H	2	1.90 ± 0.00	2.41 ± 2.2
CH	L	2	2.00 ± 1.15	2.77 ± 1.8

References

1. Martínez-Solórzano, G.E.; Rey-Brina, J.C. Bananos (*Musa* AAA): Importance, production, and trade in COVID-19 times. *Agron. Mesoam.* **2021**, *32*, 1034–1046. [CrossRef]
2. Pitti, J.; Olivares, B.O.; Montenegro, E.; Miller, L.; Ñango, Y. The role of agriculture in the Changuinola district: A case of applied economics in Panama. *Trop. Subtrop. Agroecosystems* **2021**, *25*, 017.
3. Montenegro, E.J.; Pitti-Rodríguez, J.E.; Olivares-Campos, B.O. Identification of the main subsistence crops of Teribe: A case study based on multivariate techniques. *Idesia* **2021**, *39*, 83–94. [CrossRef]
4. Olivares, B.; Rey, J.C.; Lobo, D.; Navas-Cortés, J.A.; Gómez, J.A.; Landa, B.B. Fusarium Wilt of Bananas: A review of agro-environmental factors in the Venezuelan production system affecting its development. *Agronomy* **2021**, *11*, 986. [CrossRef]
5. FAOSTAT. Food and Agriculture Organization of the United Nations (FAO). FAOSTAT Database. Available online: https://n9.cl/2axkm (accessed on 7 August 2021).
6. Hernández, Y.; Marín, M.; García, J. Response to the plantain (Musa AAB cv. Horn) yield as a function of the mineral nutrients and its phenologycal cycle. Part I. Growth and Production. *Rev. Fac. Agron. LUZ* **2007**, *24*, 607–626.
7. Delgado, E.; Trejos, J.; Villalobos, M.; Martinez, G.; Lobo, D.; Rey, J.C.; Rodriguez, G.; Rosales, F.E.; Pocasangre, L.E. Determination of a soil quality and health index for banana plantations in Venezuela. *Interciencia* **2010**, *35*, 927–933.
8. González-Pedraza, A.F.; Atencio, J.; Cubillán, K.; Almendrales, R.; Ramírez, L.; Barrios, O. Microbial activity in soils cultivated with plantain (Musa AAB plantain subgroup cv. Harton) with different vigor of plants. *Rev. Fac. Agron. LUZ* **2014**, *31*, 526–538.

9. Olivares, B.O.; Araya-Alman, M.; Acevedo-Opazo, C.; Rey, J.C.; Cañete-Salinas, P.; Kurina, F.G.; Balzarini, M.; Lobo, D.; Navas-Cortés, J.A.; Landa, B.B. Relationship Between Soil Properties and Banana Productivity in the Two Main Cultivation Areas in Venezuela. *J. Soil Sci. Plant Nutr.* **2020**, *20*, 2512–2524. [CrossRef]
10. Olivares, B.O.; Vega, A.; Calderón, M.A.R.; Rey, J.C.; Lobo, D.; Gómez, J.A.; Landa, B.B. Identification of Soil Properties Associated with the Incidence of Banana Wilt Using Supervised Methods. *Plants* **2022**, *11*, 2070. [CrossRef]
11. Olivares, B.O.; Calero, J.; Rey, J.C.; Lobo, D.; Landa, B.B.; Gómez, J.A. Correlation of banana productivity levels and soil morphological properties using regularized optimal scaling regression. *Catena* **2022**, *208*, 105718. [CrossRef]
12. González-García, H.; Pedraza, A.F.G.; Atencio, J.; Soto, A. Evaluación de calidad de suelos plataneros a través de la actividad microbiana en el sur del lago de Maracaibo, estado de Zulia, Venezuela. *Rev. Fac. Agron. LUZ* **2021**, *38*, 216–240. [CrossRef]
13. García, H.G.; González, A.F.; Pineda, M.; Escalante, H.; Yzquierdo, G.A.R.; Bracho, A.S. Microbiota edáfica en lotes de plátano con vigor contrastante y su relación con propiedades del suelo. *Bioagro* **2021**, *33*, 143–148. [CrossRef]
14. Ferris, H.; Pocasangre, L.E.; Serrano, E.; Muñoz, J.; Garcia, S.; Perichi, G.; Martinez, G. Diversity and complexity complement apparent competition: Nematode assemblages in banana plantations. *Acta Oecologica* **2012**, *40*, 11–18. [CrossRef]
15. Sial, T.A.; Khan, M.N.; Lan, Z.; Kumbhar, F.; Ying, Z.; Zhang, J.; Sun, D.; Li, X. Contrasting effects of banana peels waste and its biochar on greenhouse gas emissions and soil biochemical properties. *Process Saf. Environ. Prot.* **2019**, *122*, 366–377. [CrossRef]
16. Rondon, T.; Hernandez, R.M.; Guzman, M. Soil organic carbon, physical fractions of the macro-organic matter, and soil stability relationship in lacustrine soils under banana crop. *PLoS ONE* **2021**, *16*, e0254121. [CrossRef]
17. Ferris, H. Contribution of nematodes to the structure and function of the soil food web. *J. Nematol.* **2010**, *42*, 63–67.
18. Bautista, L.G.; Bolaños, M.M.; Massae Asakawa, N.; Villegas, B. Plantain Musa AAB simmonds phytonematodes response to soil integrated management strategies and nutrition. *Luna Azul.* **2015**, *40*, 69–84. [CrossRef]
19. Sun, J.; Zou, L.; Li, W.; Wang, Y.; Xia, Q.; Peng, M. Soil microbial and chemical properties influenced by continuous cropping of banana. *Sci. Agric.* **2018**, *75*, 420–425. [CrossRef]
20. Olivares, B.O.; Paredes, F.; Rey, J.C.; Lobo, D.; Galvis-Causil, S. The relationship between the normalized difference vegetation index, rainfall, and potential evapotranspiration in a banana plantation of Venezuela. *ST-JSSA* **2021**, *18*, 58–64. [CrossRef]
21. Olivares, B.O. Tropical rainfall conditions in rainfed agriculture in Carabobo, Venezuela. *GRANJA Rev. Cienc. Vida* **2018**, *27*, 86–102. [CrossRef]
22. Olivares, B.O.; Cortez, A.; Lobo, D.; Parra, R.; Rey, J.; Rodriguez, M. Evaluation of agricultural vulnerability to drought weather in different locations of Venezuela. *Rev. Fac. Agron. LUZ* **2017**, *34*, 103–129.
23. Rosales, F.E.; Pocasangre, L.E.; Trejos, J.; Serrano, E.; Peña, W. *Guía de Diagnóstico de la Calidad y Salud de Suelos Bananeros*; Bioversity International: Roma, Italy, 2008; pp. 48–80.
24. Chong, J.; Wishart, D.S.; Xia, J. Using MetaboAnalyst 4.0 for comprehensive and integrative metabolomics data analysis. *Curr. Protoc. Bioinform.* **2019**, *68*, e86. [CrossRef] [PubMed]
25. Gu, Z.; Gu, L.; Eils, R.; Schlesner, M.; Brors, B. Circlize implements and enhances circular visualization in R. *Bioinformatics* **2014**, *30*, 2811–2812. [CrossRef] [PubMed]
26. Mevik, B.; Wehrens, R. R Package: 'pls': Partial Least Squares and Principal Component regression. *J. Stat. Softw.* **2015**, *18*, 1–23.
27. Li, R.; Zeng, L.; Xie, S.; Chen, J.; Yu, Y.; Zhong, L. Targeted metabolomics study of serum bile acid profile in patients with end-stage renal disease undergoing hemodialysis. *PeerJ* **2019**, *7*, e7145. [CrossRef]
28. Szymańska, E.; Saccenti, E.; Smilde, A.K.; Westerhuis, J.A. Double-check: Validation of diagnostic statistics for PLS-DA models in metabolomics studies. *Metabolomics* **2012**, *8*, 3–16. [CrossRef]
29. Jankova, J. Confidence intervals for high-dimensional inverse covariance estimation. *Electron. J. Stat.* **2015**, *9*, 1205–1229. [CrossRef]
30. Lu, Q.; Liu, T.; Wang, N.; Dou, Z.; Wang, K.; Zuo, Y. A review of soil nematodes as biological indicators for the assessment of soil health. *Front. Agric. Sci. Eng.* **2020**, *7*, 275–281. [CrossRef]
31. Talavera, M.; Thoden, T.C.; Vela-Delgado, M.D.; Verdejo-Lucas, S.; Sánchez-Moreno, S. The impact of fluazaindolizine on free-living nematodes and the nematode community structure in a root-knot nematode infested vegetable production system. *Pest Manag. Sci.* **2021**, *77*, 5220–5227. [CrossRef]
32. Castilla-Díaz, E.; Millán-Romero, E.; Mercado-Ordoñez, J.; Millán-Páramo, C. Relation of soil parameters on the diversity and spatial distribution of free-living nematodes. *Tecnol. Marcha* **2017**, *30*, 24–34. [CrossRef]
33. Landi, S.; D'errico, G.I.A.D.A.; Papini, R.; Gargani, E.; Simoncini, S.; Amoriello, T.; Ciccoritti, R.; Carbone, K. Communities of plant parasitic and free-living nematodes in Italian hop crops. *Redia* **2019**, *102*, 141–148. [CrossRef]
34. Arie-Vonk, J.; Breure, A.M.; Mulder, C. Environmentally-driven dissimilarity of trait-based indices of nematodes under different agricultural management and soil types. *Agric. Ecosyst. Environ.* **2013**, *179*, 133–138. [CrossRef]
35. Holterman, M.H.; Korthals, G.W.; Doroszuk, A.; van Megen, H.H.; Bakker, J.; Bongers, T.; Helder, J.; van der Wurff, A. A strategy in searching for stress tolerance-correlated characteristics in nematodes while accounting for phylogenetic interdependence. *Nematology* **2011**, *13*, 261–275. [CrossRef]
36. Crozzoli, R. Nematodes of tropical fruit crops in Venezuela. In *Integrated Management of Fruit Crops and Forest Nematodes*; Ciancio, A., Mukerji, K.G., Eds.; Springer: Berlin/Heidelberg, Germany, 2009; pp. 63–83. [CrossRef]
37. Crozzoli, R. *La Nematología Agrícola en Venezuela*; Ediciones Facultad de Agronomía; UCV: Maracay, Venezuela, 2014; p. 500.

38. Davide, R.G. Influence of cultivar, age, soil texture, and pH on *Meloidogyne incognita* and *Radopholus similis* on banana. *Plant Dis.* **1980**, *64*, 571–573. [CrossRef]

39. Bwamiki, D.P.; Duxbury, J.M.; Esnard, J. Effect of banana nutritional status on host tolerant to the burrowing nematode *Radopholus similis*. In *42nd Annual Meeting of Society of Nematologists*; Cornell University, Ithaca: New York, NY, USA, 2003; pp. 328–329.

40. Cássia Ferreira, R.R.; Aparecida, A.; Mizobutsi, E.H.; Pereira, F.R.; Ribeiro, H.B.; Alexandre, P.A.A.; Ferraz, S. Influencia de fatores edáficos sobre a população de Meloidogyne javanica, Helicotylenchus multicinctus e Radopholus similis em bananeira. Proceedings of XVII ACORBAT Meeting, Joinville, Santa Catarina, Brazil; ACORBAT. Soprano, E., Tcacenco, F.A., Lichtemberg, L.A., Silva, M.C., Eds.; 2006; pp. 813–817.

41. Yang, B.; Zhang, C.; Cheng, S.; Li, G.; Griebel, J.; Neuhaus, J. Novel Metabolic Signatures of Prostate Cancer Revealed by 1H-NMR Metabolomics of Urine. *Diagnostics* **2021**, *11*, 149. [CrossRef] [PubMed]

42. Gardner, T.S.; Di Bernardo, D.; Lorenz, D.; Collins, J.J. Inferring genetic networks and identifying compound mode of action via expression profiling. *Science* **2003**, *301*, 102–105. [CrossRef]

43. Jeong, H.; Mason, S.P.; Barabási, A.L.; Oltvai, Z.N. Lethality and centrality in protein networks. *Nature* **2001**, *411*, 41–42. [CrossRef]

44. Basu, S.; Duren, W.; Evans, C.R.; Burant, C.F.; Michailidis, G.; Karnovsky, A. Sparse network modeling and metscape-based visualization methods for the analysis of large-scale metabolomics data. *Bioinformatics* **2017**, *33*, 1545–1553. [CrossRef]

45. Culman, S.W.; Young-Mathews, A.; Hollander, A.D.; Ferris, H.; Sánchez-Moreno, S.; O'Geen, A.T.; Jackson, L.E. Biodiversity is associated with indicators of soil ecosystem functions over a landscape gradient of agricultural intensification. *Landsc. Ecol.* **2010**, *25*, 1333–1348. [CrossRef]

46. Chávez-Velazco, C.; Araya-Vargas, M. Correlation between soil characteristics and banana root nematodes (Musa AAA) in Ecuador. *Agron. Mesoam.* **2009**, *20*, 361–369. [CrossRef]

47. Millán, E.; Castilla, D.; Millán, C. Comunidades de nematodos de vida libre del suelo y su correspondencia con la calidad. *Ing. Reg.* **2016**, *16*, 25–34. [CrossRef]

48. Jaurixje, M.; Torres, D.; Mendoza, B.; Henríquez, M.; Contreras, J. Physical, and chemical soil properties and their relationship with biological activity under different soil managements in Quibor, Lara State, Venezuela. *Bioagro* **2013**, *25*, 47–56.

 *sustainability*

Article

Using Time-Series Generative Adversarial Networks to Synthesize Sensing Data for Pest Incidence Forecasting on Sustainable Agriculture

Chen-Yu Tai, Wun-Jhe Wang and Yueh-Min Huang *

Department of Engineering Science, National Cheng Kung University, Tainan 70403, Taiwan
* Correspondence: huang@mail.ncku.edu.tw

Abstract: A sufficient amount of data is crucial for high-performance and accurate trend prediction. However, it is difficult and time-consuming to collect agricultural data over long periods of time; the consequence of such difficulty is datasets that are characterized by missing data. In this study we use a time-series generative adversarial network (TimeGAN) to synthesize multivariate agricultural sensing data and train RNN (Recurrent Neural Network), LSTM (Long Short-Term Memory), and GRU (Gated Recurrent Unit) neural network prediction models on the original and generated data to predict future pest populations. After our experiment, the data generated using TimeGAN and the original data have the smallest *EC* value in the GRU model, which is 9.86. The results show that the generative model effectively synthesizes multivariate agricultural sensing data and can be used to make up for the lack of actual data. The pest prediction model trained on synthetic data using time-series data generation yields results that are similar to that of the model trained on actual data. Accurate prediction of pest populations would represent a breakthrough in allowing for accurate and timely pest control.

Keywords: time series; data augmentation; deep learning; pest forecasting; generative adversarial network (GAN)

Citation: Tai, C.-Y.; Wang, W.-J.; Huang, Y.-M. Using Time-Series Generative Adversarial Networks to Synthesize Sensing Data for Pest Incidence Forecasting on Sustainable Agriculture. *Sustainability* **2023**, *15*, 7834. https://doi.org/10.3390/su15107834

Academic Editors: Barlin Orlando Olivares Campos, Edgloris E. Marys and Miguel Araya-Alman

Received: 17 April 2023
Revised: 6 May 2023
Accepted: 8 May 2023
Published: 10 May 2023

1. Introduction

The global demand for sustainable development is increasing, and agriculture is no exception. Sustainable agriculture is a production method that emphasizes sustainable development in economic, social, and environmental aspects [1,2]. It primarily focuses on protecting the environment, improving the productivity of crops and animals, enhancing farmers' income, and improving the quality of life in local communities. In order to improve agricultural productivity and reduce production costs, agricultural production is increasing thanks to the Internet of Things (IoT) and big data analytics [3]. For example, IoT devices receive sensor data [4], RGB images, or multispectrum images and then analyze these data for pest control [5], crop yield prediction [6], precise agricultural irrigation [7], and so on.

Pest prediction is vital in agriculture. A lack of effective pest control directly affects crop yields [8]. In general, it is difficult for farmers to gain a complete picture of crop growth due to the large area of farmland. In recent years, weather conditions have become more severe due to climate change, making pests more adaptable to environmental changes. Multiple sensors placed in remote fields would enable farmers to constantly monitor the environmental data in each area. By using environmental data and an understanding of pest species, experts could build a pest prediction system [9,10] for integrated pest management (IPM). Farmers could plan their control operations in advance given early warnings generated by the prediction system.

Effective analysis of meteorological and pest data is crucial for accurate pest prediction systems. Since both types of data are constantly monitored in the field, early warning is key to pest forecasting [11]. For sudden increases in pest populations, immediate control may not be possible. Technological advances have yielded more sophisticated methods for time-variant analysis for data prediction [12]. However, such prediction relies on a large amount of historical data. Historical data is rich in time dynamics. Greater amounts of sensing data facilitates more accurate pest prediction. Conversely, insufficient sensing data degrades the effectiveness of the prediction system.

Environmental sensing data is now obtained generally from IoT edge devices. The effectiveness of such edge devices deployed in farms can be degraded by climate, flora, fauna, and other natural factors, resulting in incorrect sensor values or even damage to the devices [13], causing problems in subsequent data analysis, such as data errors during preprocessing or data losses due to edge device damage; these conditions must be corrected to ensure continued data availability.

Data augmentation algorithms are one way to resolve data deficiencies. Since prediction models are affected by the amount of data, it is more important to address the fundamental problem of data deficiency than to use complex prediction models [14]. Data augmentation increases the variability of data by generating data samples. For instance, augmentation methods for image recognition use rotation, translation, and scaling to produce multiple images from the same image [15]. In contrast, a prediction system for linear problems uses time series data, for which data augmentation is dissimilar to that for images [16]. Here, we investigate time series data augmentation to address the problem of insufficient data for pest prediction.

Samples with missing data can be divided into those that are missing for a long period of time and those that are missing for only a short instant. The amount of missing agricultural sensing data can significantly affect the accuracy of the amplification [17]. When a small sample of data is missing, common scaling methods generate simulations with similar results. However, when more data samples are missing and the data are multivariate, such scaling does not preserve both the temporal features and the characteristics of multivariate data. Therefore, we take into account both temporal and multivariate data features, and generate data that retains both features.

Time-series data is a regression problem in which sequence values change according to time dynamics. Clearly, data generation requires that we preserve the temporal characteristics of the original data and increase the number of samples. However, depending on the number of data samples to be augmented and the number of data variables, basic augmentation cannot be used to generate long-term or multivariate sensing data. In this case, a time-series generative adversarial network (TimeGAN) is an effective model for generating long-term data [18]. The temporal features of the data are learned by the autoregressive (AR) model, and a large amount of data is synthesized by the generative adversarial network (GAN). In this study, we use TimeGAN for data augmentation to synthesize multivariate sensing data.

The target pest in this study is *Scirtothrips dorsalis* Hood (*S. dorsalis*), which in Taiwan primarily affects mangoes. We seek to use agricultural sensing data as the main analysis. Firstly, we use multivariate agricultural sensing data to better understand the environmental conditions, and further use a deep generative model to synthesize more time-series data. We then verify the effect of the generated data using time prediction models such as recurrent neural networks (RNN), long short-term memory (LSTM) models, and gated recurrent unit (GRU) models [19]. Our purpose is to predict future trends and alert farmers of them in order to control pest populations in a timely manner. We analyze the effectiveness of time series data augmentation using data visualization and evaluation metrics for regression.

In summary, we have put forth the following two hypotheses and designed corresponding experimental procedures to verify them:

(1) To examine whether TimeGAN can generate synthetic data with the same temporal characteristics as the original data.
(2) To investigate whether the pest prediction model trained using the synthetic data yields comparable results to the model trained using the original data.

The remaining sections of this paper are organized as follows: Section 2 surveys IoT smart agriculture applications for pest prediction and time series data augmentation, Section 3 describes the system architecture and our experimental methods, and Section 4 presents a comparison and discussion of the experimental results, including the processing of environmental data, a correlation analysis of meteorological factors, the construction of the generation model, and an evaluation of the prediction models. In Section 5, we conclude and suggest future research directions for pest prediction.

2. Related Literature

Below, we describe the current agricultural sensing data to implement time series tasks, after which we explore practical techniques for pest prediction based on image recognition and time series data. We then discuss time series data augmentation, focusing particularly on data augmentation methods based on deep generative models.

2.1. Agricultural Sensing Data

Smart agriculture uses sensors to collect agriculture-related data and transfer them to a cloud database. The stored data are analyzed to identify potential problems and to advance agricultural technology. Big data time analysis plays a crucial role in this because agricultural sensors often produce huge amounts of time-series data [20–22]. As the method used for data processing can affect downstream results, the challenge is how to analyze past data, adjust the dataset appropriately, and come up with the best processing strategy. Ren et al. [23] use data preprocessing methods to clean up specific features, transform structures and formats, and reduce complexity. By using multiple data preprocessing steps, we retain important features and improve efficiency when constructing deep learning models.

Analysis methods in agriculture vary in terms of the sensing data format and the needs of the task. Time-series tasks can be divided into classification, anomaly detection, and prediction. In time-series classification, sensor data identify crop species [24] and monitor growth status [25]; values from the spectral sensors are converted into a vegetation index which reveals information about crop growth and environmental health. In time-series abnormality detection, sensor data can reveal abnormalities in the environment or in crop yields [26]. In time-series prediction, past dynamic series provide potential trends by which to estimate future crop yields [27] and pest populations [28]. Selecting important features in multivariate agricultural data and improving the rationality of predictions facilitates subsequent crop and pest management.

2.2. Pest Prediction

The yield of high-value crops in the agricultural sector is of concern. As pest infestation during the production season can significantly reduce crop yields [29], pest infestation is an important and immediate problem. Delayed control or low doses of pesticides can cause economic crop losses [30]. According to Heeb et al. [31], up to 40% of crop production in the world is currently lost due to unmanaged pest damage. If experts could provide farmers with advance warnings of increases in pest populations, farmers would have enough time to apply the recommended amount of pesticides, and could thus effectively control the pest situations in their fields and prevent the endless spread of pest populations. Below, we survey research on pest prediction.

2.2.1. Image Recognition-Based Approaches

In recent years, the dilemma of traditional pest prediction has changed due to the advanced development of artificial intelligence (AI) and two technological innovations: intelligent image recognition technology [32] and time-series data prediction [33]. In image recognition technology, features of visible or multispectral images are analyzed to select and create pest-specific features by resolving color features and texture features that are different from those of the crop [34]. Deep learning models learn the characteristics of the pest, mark the location of the pest in the image using object detection, classify the pest species, and report the probability that it is in fact this pest. Given the ease of image collection, deep research is now available both for close-up images from cameras or cell phones [35,36], as well as for overhead drone imagery [37,38].

2.2.2. Time-Series Data-Based Approaches

In time-series data prediction, it is common to combine multiple types of sensing devices to monitor and record real-time environmental and pest conditions in the field [39,40]. Prediction given time-series data is similar to weather prediction in that the prediction model learns past environmental and pest development trends from a large amount of historical data [41,42]. Since potential feature changes are sequential, temporal dynamics can be used to estimate the number of pests that may occur in the future. Figure 1 shows the number of studies on pest prediction from time-series data from 2012 to 2021. We retrieved the keywords *pest forecasting*, *pest prediction*, *time-series* from the Web of Science (WOS) database and recorded the number of papers published in each year.

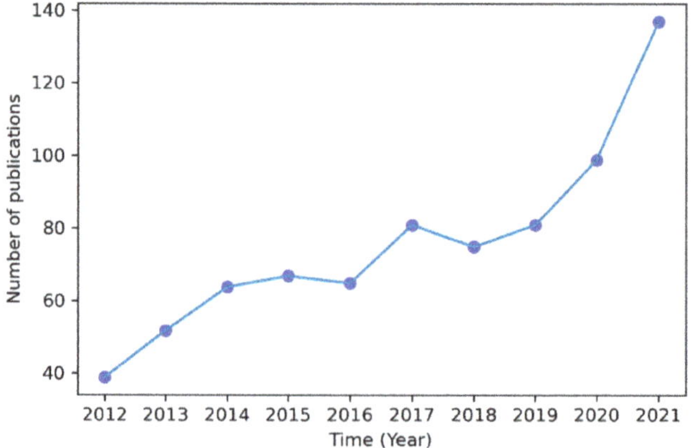

Figure 1. Publications on time-series data for pest prediction.

As shown in Figure 1, the publication of pest prediction-related papers increased gradually and steadily between 2012 and 2018. Pest prediction research then exploded beginning in 2019, both in terms of the stability of the values collected by IoT devices [43] and the more in-depth research on prediction models [44], which have contributed to the development of pest prediction. However, bottlenecks in pest prediction remain, including the use of a small number of variable features, the failure to analyze the validity of weather factors on prediction, the failure to investigate the effects of different model architectures on prediction, and the scarcity of dataset samples [45]. Therefore, we review the literature to further analyze and discuss pest prediction studies from recent years.

Techniques applied for data prediction include statistical methods, machine learning, and deep learning. Pest prediction using statistical methods is often achieved using the autoregressive integrated moving average (ARIMA) model, which does not take into account variations in relevant random variables. Narava et al. [46] use an ARIMA model to predict the population dynamics of *Helicoverpa armigera* and show that ARIMA is easier to implement for pest prediction.

Machine learning-based pest prediction is often implemented using the k nearest neighbors algorithm, which searches for the k closest samples in the feature space. Gómez et al. [47] build and compare the performance of six machine learning algorithms to predict desert locusts based on soil moisture and demonstrate that the model prediction performance is limited by the space and time of the data.

Deep model structures are often used to predict pests in deep learning. Tan et al. [48] use deep autoregressive (DeepAR) models in deep learning to predict *Chilo suppressalis*, a major threat to rice production, and show that deep learning predicts pest dynamics more accurately by integrating meteorological data, pest data, and temporal features. Therefore, we used a deep learning model for pest prediction due to our need to focus on the temporal dynamics of the sensing data and due to the ease of implementing pest prediction.

2.3. Time-Series Data Augmentation

Recent time-series prediction models include a large number of additional neural structures and use deeper architectures. Since the model is trained using a deeper structure, the dataset size affects the accuracy of the model prediction, resulting in overfitting for small datasets [49]. Note that the use of real-time datasets is limited by factors such as the completeness, ownership, and access of the dataset. Therefore, various methods are used to augment the original dataset and expand the amount of training data to increase its comprehensiveness. Wen et al. [50] has classified augmentation methods for time series data into two categories: basic approaches and advanced approaches. The basic approaches encompass techniques such as time domain, frequency domain, and time-frequency domain. Meanwhile, the advanced approaches consist of decomposition methods, statistical generative models, and learning methods. The learning methods can further be categorized into embedding space, deep generative models, and automated data augmentation. In the next section, we introduce basic time data augmentation methods and describe data augmentation algorithms based on the depth generation model.

2.3.1. Basic Data Augmentation

The most important feature of timeseries data are their time and frequency characteristics. Data values are fused with time and frequency to increase the amplitude of the signal to produce a time-series dataset. Therefore, data augmentation algorithms include time- and frequency-domain variants, for which data are transformed to the appropriate domain for subsequent task requirements. In time-series prediction, the effectiveness of time- and frequency-domain transformations has been demonstrated in the literature [51]. Common time-domain methods include window slicing, window warping, flipping, and noise injection; frequency-domain conversion algorithms include amplitude and phase perturbations (APP) and Fourier transforms (FT).

Um et al. [52] state that a single augmentation method has limited effectiveness, and that it is crucial to select and combine the right combination of transformations for the task to improve the performance via subsequent model analysis. Thus, it is important to select and combine methods for different tasks to develop a combined strategy for the task. Standard practices for data augmentation in the time domain include flipping, down-sampling, and adding slope. In the frequency domain, methods such as amplitude-adjusted Fourier transform (AAFT), iterated amplitude-adjusted Fourier transform (IAAFT), amplitude-phase permutation (APP), and short-time Fourier transform (STFT) are commonly used.

2.3.2. Deep Generative Model Approaches

Time-series data augmentation should not be limited to the generation of diverse data; for the dataset to be used with confidence, the features of the generated data should resemble the distribution of the actual data features. With the rise of deep learning, deep generative models (DGMs) have been created that combine the advantages of generative models and deep neural networks. The most advanced techniques in DGM are variational auto-encoders (VAEs), normalizing flow (NF), and generative adversarial networks (GANs) [53]. Many different architectures are GAN derivatives, including recurrent GAN (RGAN) and recurrent conditional GAN (RCGAN) [54]. Given the recent trend of using GAN approaches to synthesize simulated training sets, we select the time-series generative adversarial network (TimeGAN) [18] as the basis for the data augmentation model based on the characteristics of the sensing dataset.

3. Research Method

Here we describe the system architecture and the experimental implementation process. We present the specific design of the experimental framework, after which we describe the implementation method from the selection of the depth generation model and the data augmentation and model training to the final prediction and analysis by time-prediction models. Then, we list common evaluation metrics and explain the usage of validation metrics to adjust the training parameters.

3.1. System Design

The system records environmental data at fixed time intervals using sensor equipment at the experimental site and monitors pest conditions at various locations in the field as agricultural sensor data. After preprocessing this time-series data, the time units of the data are converted to weeks to merge the two datasets. The multidimensional time series is then augmented and analyzed for predictive purposes. Here we estimate the timing of pest emergence and recommend subsequent plans for pest control. This requires four steps: (1) agricultural sensing data collection and production, (2) time series data preprocessing and fusion, (3) multivariate time series data augmentation, and (4) pest prediction from synthetic data. The system architecture of this study is shown in Figure 2.

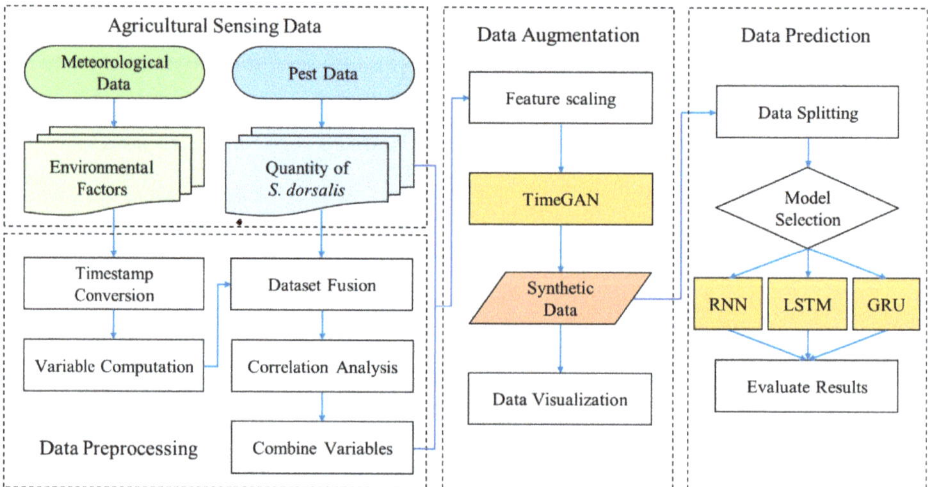

Figure 2. Proposed pest prediction system.

3.2. Time-Series Generative Adversarial Network

This study uses TimeGAN, a deep generative model, to build a deep learning model for agricultural sensing data [18]. TimeGAN was first used to improve the temporal correlation of raw time-series data that are not taken into account when GAN generates data. TimeGAN employs a generator-discriminator architecture for sequential generative adversarial training while fusing the embedding and recovery functions to learn encoding features of the data and generating representations in a large number of training epochs. The embedding and recovery functions construct a hidden space in which the adversarial networks are trained so that the feature space and the hidden space echo each other, generating features of the currently generated data with representations of the original sequence, and finally learning the temporal features of the time series.

3.2.1. TimeGAN Training Data

The training data comprised 59 instances of raw data with 22 characteristics, 21 including ground temperature, dew point temperature, relative humidity, and other weather data, along with one pest count. The weather data came from the Agricultural Weather Observation Network Monitoring System of the Central Weather Bureau [55], and the pest data came from the Bureau of Agriculture, Kaohsiung City Government [56]. More details about the dataset are provided in Section 4.1.

Since the system uses the current week's sensing data to predict the next week's pest population (Section 3.3, TimeGAN's training data are produced as two-week samples and contain all agricultural sensing features, of which there are 22 sensing features after preprocessing. Therefore, TimeGAN generates a realistic random vector by setting the window size of the real data to 2, which means the generated data simulates two weeks of real data, and sets the feature length to 22, which means that the generated data simulates changes in the real 22 sensing data features.

Sliding the entire dataset completely with 59 samples of the original data using a moving window yields 57 two-week matrices for subsequent model training, where each matrix contains two weeks of sensing data, and the sensing data is preprocessed to obtain 22 features. As described above, real-time series input data trained by TimeGAN is produced to obtain a dataset of dimensions (57, (2, 22)) with 57 samples, each with 2 columns (weeks) and 22 variables (agricultural sensing features). Figure 3 shows a schematic of the TimeGAN training data, the features of which are normalized to [0, 1] to facilitate subsequent model training.

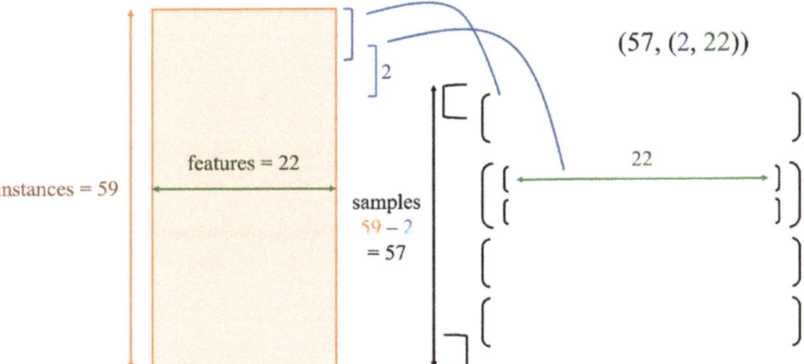

Figure 3. Schematic diagram of TimeGAN training data.

3.2.2. TimeGAN Training

The TimeGAN training process is shown in Figure 4. Augmented model training is composed of three stages. In stage 1, the embedding and recovery stage, actual data are used to achieve the best reconstruction of original data and obtain historical data features. In stage 2, the supervised stage, samples are generated to learn temporal features from real samples in matrix space. In stage 3, the joint stage, all of the networks (embedding, generator, recovery, and discriminator) are trained simultaneously to optimize all loss functions based on backpropagation.

In Figure 4, the TimeGAN model in this study uses a different loss function for optimization in each stage. The embedding stage trains using reconstruction loss ($\mathcal{L}_R$), the supervised stage optimizes using supervised loss ($\mathcal{L}_S$) and determines the time correlation between the training samples and the generated samples, and the joint stage optimizes all loss functions ($\mathcal{L}_R$, $\mathcal{L}_S$) and adjusts the generation network ($\mathcal{L}_U$) using unsupervised loss feedback; thus, this last stage is the most time-consuming.

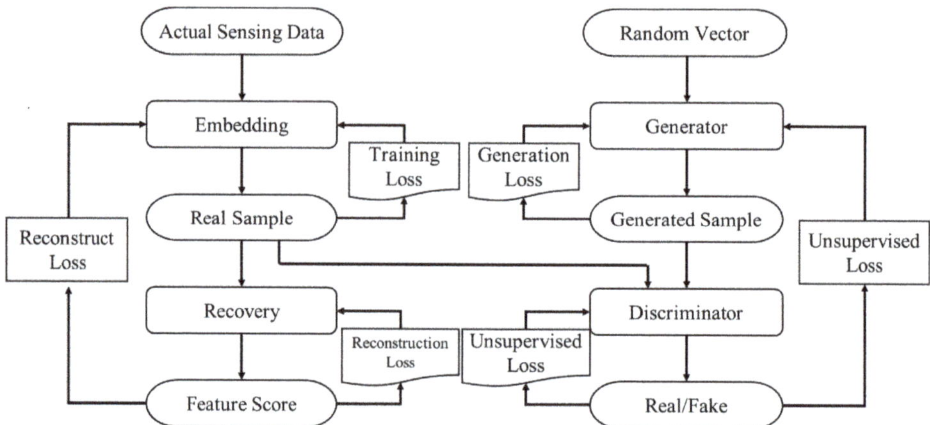

Figure 4. Training the TimeGAN model to generate data.

Here, the reconstruction loss ($\mathcal{L}_R$) is calculated as the difference between the actual data (X_t) and the restored training sequence ($\widetilde{X}_t$), as shown in (1). The original data are downscaled and restored to the original data. Loss reconstruction means that the real data are similar to the restored data after downscaling.

$$\mathcal{L}_R = \mathbb{E}_{S,X(x,y)\sim\mathcal{P}data(x,y)}\left[\sum_t \left|\left|X_t - \widetilde{X}_t\right|\right|_2\right] \tag{1}$$

In this study, the supervision loss ($\mathcal{L}_S$) is calculated as the difference between the temporal features of the training sequence (h_t) and the temporal features of the generated sequence ($\widetilde{h}_t$), as shown in (2). The original data is downscaled to retain the temporal features of the data, and the generator adjusts to and learns the temporal features. Low supervised loss means that the training sequence has temporal features that are similar to the generated sequence.

$$\mathcal{L}_S = \mathbb{E}_{S,\ X(x,y)\sim\mathcal{P}data(x,y)}\left[\sum_t \left|\left|h_t - \widetilde{h}_t\right|\right|_2\right] \tag{2}$$

Here, unsupervised loss ($\mathcal{L}_U$) indicates the correct classification of the generated sequences as shown in (3). Temporal features retained after the original data is downscaled are used as a basis by which to identify whether the generated data created by the generator fit the true temporal features.

$$\mathcal{L}_U = \mathbb{E}_{X(x,y)\sim\mathcal{P}data(x,y)}\left[\sum_t \log y_t\right] + \mathbb{E}_{S,X(x,y)\sim\mathcal{P}data(x,y)}\left[\sum_t \log(1 - \hat{y}_t)\right] \tag{3}$$

As such, the objective function of this study is to minimize the very large generator and the identifier (G^*, D^*) by maximizing $\mathcal{L}_U$ (for the identifier) and minimizing $\mathcal{L}_R$ and $\mathcal{L}_S$ (for the generator), as shown in (4). Since it is more important to preserve the temporal characteristics of the original data, in the final stage we seek the best generator and identifier solutions simultaneously.

$$G^*, D^* = \min_{\theta_e,\theta_g,\theta_r}\left(\mathcal{L}_R + (\lambda + \eta)\mathcal{L}_S + \max_{\theta_d}\mathcal{L}_U\right) \tag{4}$$

3.2.3. TimeGAN Architecture

The multivariate agricultural sensing data augmentation model constructed in this study is implemented by the ydata-synthetic Python package [57]. The TimeGAN model is composed of an embedding network, a recovery network, a generator, and a discriminator, which are connected through a serial connection. Below, we describe the network architecture and settings of the four components.

Here, the generator input is a random sequence of size (2, 22) as described in Section 3.2.1, and the generator and discriminator are both networks with three layers of GRU neurons stacked on top of each other. The number of GRU neurons cascaded in each layer is hidden_dim, the hidden dimension of the network. Since we seek to expand the data in two-week increments, we set hidden_dim to 2, which causes the TimeGAN model to generate and identify two-week sample data. In addition, the generator input is a random vector of size (2, 22), and the generator output and the discriminator input are also sequences of size (2, 22), but the discriminator output is a discriminative score of size 1, used to determine whether the current generated sequence consists of true or fake data.

In this study, the embedded and recovered networks are trained with real data of size (2, 22), as discussed in Section 3.2.1. As the size of the generated target data must be the same as the real data, the input size of the network is the same. The embedding and recovery network is a three-layer LSTM stacked autoencoder because the dimension of the original data is too large to be used directly as the discriminator's judgment criterion; hence, the embedding and recovery network is used to downscale the data, and the training sequence is fitted through multiple generations to provide the discriminator for the generated sequence. The hidden_dim is set to 2 to match the data size of the generator and the discriminator. In addition, the input of the embedded network consists of real data of size (2, 22) and the output of the embedded network and the input of the recovered network are training sequences of size (2, 22), but the output of the recovered network has a size of 22 features scored for the coding and decoding effect of the autoencoder.

3.2.4. TimeGAN Training Validation

The purpose of generating new agricultural sensing data based on a train-on-synthetic, test-on-real (TSTR) [58] methodology and augmented data is to provide input to the subsequent time-series prediction model. The pest predictions are used to evaluate whether the new data generated are superior to the original data in the area of agricultural prediction tasks where data are not available. In this phase, therefore, the real model is trained independently from the synthetic model, where the real model represents training with real data and validation with real data, and the synthetic model represents validation with real data but training with augmented data. The model is evaluated to measure the effectiveness of generated data against real data.

3.2.5. Visualization of Synthetic Data

Data visualization makes it possible to present the generated multivariate agricultural sensing data as a picture of their model training effect and then analyze them for consistency between the generated and real data. Generally, basic visualization is used to compare the curves of the generated data with the real data. However, as the experimental agricultural sensing data is multivariate, the data dimensionality is reduced using algorithms such as principal components analysis (PCA) and t-distributed stochastic neighbor embedding (t-SNE) [59] to observe the distribution of the dataset generated by TimeGAN and that of the original agricultural sensing data.

3.3. Predictive Models

Here, we generate agricultural sensing data and then use a simple model to verify the difference between generated and real data. The initial validated evaluation metrics are real and simulated data, which include all weather variables but not pest prediction; thus, models for pest prediction must be designed. The two-stage model is used to evaluate the indicators and the combination of different independent and dependent variables. Results demonstrate the effect of multivariate time-series data augmentation, and experiments reveal which factors are significant in predicting *S. dorsalis* population trends.

Prediction of future pest populations requires time-series data as inputs to the model. The input data are a combination of independent variables based on weather factors investigated in the correlation analysis, and the output data are weekly pest counts. After confirming the input and output data, the data are normalized, and the experimental dataset is partitioned into a training dataset for input during model training and a test dataset to evaluate the performance of a trained model. We use three time-series prediction models (RNN, LSTM, and GRU), and set their parameters, architectures, and optimizers accordingly. Finally, we use the evaluation metrics to compare the prediction errors of the three models and calculate the prediction errors of the original and generated data. The time-series model prediction flow is shown in Figure 5.

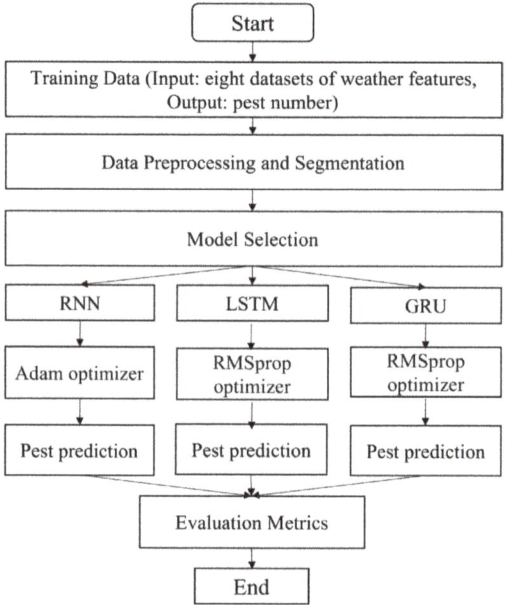

Figure 5. Pest prediction based on weather factors.

3.4. Evaluation Metrics

The experimental dataset combines observed environmental meteorological data and field surveys of pest infestation by investigators. Assuming that there is too few data, we use the TimeGAN model to expand the time-series data with more variables, and compare the three prediction models (RNN, LSTM, GRU) before and after this expansion. We measure the model effectiveness using time-series prediction error metrics. Since future pest populations are predicted as a continuous value, the predicted value should be close to the real data. Therefore, we use four common regression metrics: mean absolute error (*MAE*), mean absolute percentage error (*MAPE*), root mean squared error (*RMSE*), and the coefficient of determination (R^2) [60]. In addition, since the original data and the generated data are used as training data, the error change (*EC*) is used to compare the predicted error change before and after data augmentation. The above index equations are as follows:

$$MAE = \frac{1}{n} \sum_{t=1}^{n} |y_i - \hat{y}_i| \tag{5}$$

$$MAPE = \frac{100\%}{n} \sum_{t=1}^{n} \left| \frac{\hat{y}_i - y_i}{y_i} \right| \tag{6}$$

$$RMSE = \sqrt{\frac{1}{n}\sum_{t=1}^{n}(y_i - \hat{y}_i)^2} \tag{7}$$

$$R^2 = 1 - \sum_i (y_i - f_i)^2 \big/ \sum_i (y_i - \hat{y})^2 \tag{8}$$

$$EC = MAE_{w/\,aug} - MAE_{w/o\,aug} \tag{9}$$

As *MAE* is the sum of the absolute values of the difference between the actual and predicted values, it reflects the actual difference between the predicted and sample values, as shown in (5); *MAPE* explains the difference between the predicted and actual values intuitively in terms of a percentage, as shown in (6); *RMSE* is the sum of the squares of the difference between the actual and predicted values, and is used to evaluate the error between observed and actual values. Since *RMSE* is calculated by squaring, it amplifies the error in addition to maintaining the absolute value estimation, as shown in (7); R^2 represents the strength of the correlation between the dependent variable and the independent variable, and is used to explain the degree of fit between the predicted and actual values in the model, as shown in (8); *EC* is the difference between the *MAE* of the augmented data and the *MAE* of the original data. *MAE* results are compared to analyze the performance of the training model between the augmented data and the original data, as shown in (9). The smaller the value of the first three error indicators (*MAE*, *MAPE*, and *RMSE*), the lower the error, and the larger the fourth validation indicator (R^2), the better. The smaller the *EC* value, the better the fit of the expansion to the original data. Therefore, for regression, the same evaluation metric can be used for different network architectures. The evaluation metrics correspond directly to the model performance and accuracy.

4. Experimental Procedure and Discussion

4.1. Research Site

The site for this experiment was an orchard in Liugui District, Kaohsiung City, Taiwan, as shown in Figure 6. According to the annual Kaohsiung City Agricultural Statistics report [61], the total production of mangoes in Kaohsiung City is about 13,300 tons, of which the production in Liugou District is about 3100 tons. Since this area is the highest yielding area in Kaohsiung, we chose Liugui as the most representative site for pest threats. In addition, since mango is the most abundant fruit in the experimental area, the main research issue is how to maintain the quality of the mangoes and reduce the threat of pests and diseases. In summary, we seek to investigate the historical data of this area and estimate future pest trends via smart agriculture to provide farmers with real-time spraying guidance to control pest populations.

4.1.1. Pest Data

The pest surveyed in this study was *Scirtothrips dorsalis* Hood (*S. dorsalis*), also known as the yellow tea thrip. The experimental pest dataset was obtained from public data monitored from 2019 to 2021 by the Kaohsiung City Bureau of Agriculture [56]. Since the current *S. dorsalis* control focuses mainly on issuing early warnings when the number of insects rises to remind farmers to prepare for pest control in advance, the annual monitoring period ends when the mangoes are in season, so no monitoring was conducted from September to December. Therefore, the purpose of this study is to estimate the timing of pest emergence so farmers know when to spray insecticides.

The pest dataset is a collection of data collected from farmland locations in response to experimental needs. Eight mango orchards larger than one hectare were recorded, and for each mango orchard we selected, the most representative locations were affected by only the small yellow thistle and no other pests. The number of crops in the vicinity of each site was approximately the same. Ten sticky boards were deployed at each site for trapping and counting, and the boards were replaced at regular intervals every week. The data were recorded in terms of the number of pests, and the average of each sticky board was taken, so the pest data are expressed as the average number of pests at the site for that week. The sum of the eight sites represents the pest level corresponding to the weather information for that week.

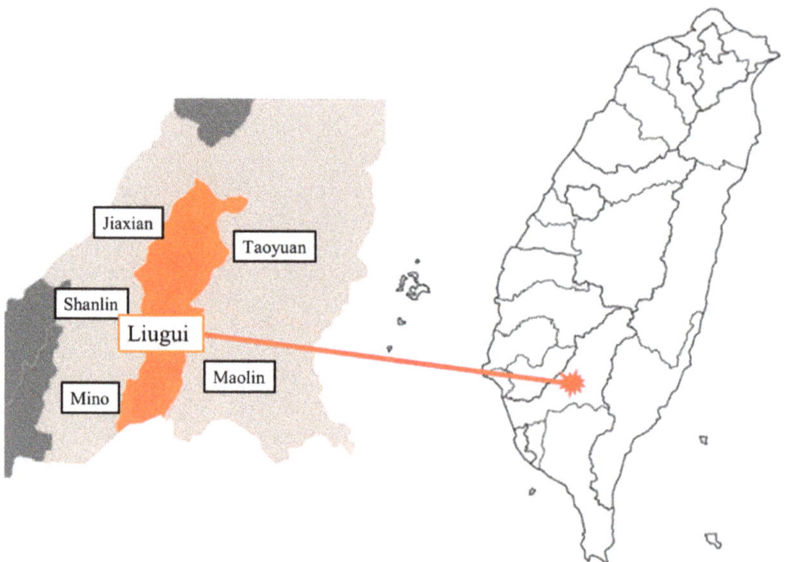

Figure 6. Experimental site: Liugui District, Kaohsiung City.

4.1.2. Meteorological Data

As pest populations are affected by environmental variables such as *temperature, humidity*, and *light*, we used meteorological data to construct a multivariate agricultural dataset. Since the number of pests and the weather dataset must be monitored at the same time and location, for the meteorological data we also used the environmental data for Liugui District, Kaohsiung City. The experimental meteorological dataset was obtained from the Agricultural Weather Watch monitoring system of the Central Weather Bureau [55] of Taiwan. On the CWB's weather monitoring website, there is one station for Liugui District (code E2P980), so the data recorded for this station were used as the meteorological data, as they are most representative of the weather in the orchards in Liugui District.

We filtered the meteorological data from 2020 to 2021. The weather data were sampled according to time attributes, and the raw data included 16 weather factors, among which are *temperature, precipitation, relative humidity, atmospheric pressure, all-sky solar radiation*, and *wind speed*, as shown in Table 1. The weather data were monitored at hourly intervals, and there were 17,000 data instances. We rescaled the time units of the environmental features to match the temporal attributes of the meteorological data and the pest data, and fused the data to generate the agricultural sensing data for this experiment.

4.2. Data Preprocessing

For data preprocessing, we used data cleaning, missing value processing, resampling, and data fusion. In addition, we conducted a correlation analysis to better understand the relationship between variables of agricultural sensing data.

Table 1. Meteorological data of weather variables and units.

Weather Variable	Unit	Weather Variable	Unit
Temperature (*T*)	°C	Rainfall (*RF*)	mm
Ground temperature 0 cm (*GT0*)	°C	Relative humidity (*RH*)	%
Ground temperature 5 cm (*GT5*)	°C	Vapor pressure (*VP*)	hPa
Ground temperature 10 cm (*GT10*)	°C	Sea level pressure (*SLP*)	hPa
Ground temperature 20 cm (*GT20*)	°C	Solar irradiance (*LUX*)	MJ/m^2
Ground temperature 50 cm (*GT50*)	°C	Insolation duration (*LUXTIME*)	h
Ground temperature 100 cm (*GT100*)	°C	Mean wind speed (*WS*)	m/s
Dew point temperature (*DPT*)	°C	Maximum gust wind speed (*MWS*)	m/s

4.2.1. Data Preprocessing

As is typical, the environmental monitoring data were characterized by missing values and outliers, which were preprocessed using data purification and missing value processing. Since no outliers were found in the weather data samples, data purification did not change the original dataset. We observed missing values in each feature, so we removed the fields in which the missing values were located. After preprocessing and data correction, we converted all feature data from hourly to weekly dimensions. We resampled using four functions: average, maximum, minimum, and cumulative. After resampling, the amount of data for each feature was reduced from 17,000 to 105 samples. The temporal attributes of the meteorological data matched that of the pest data, and the complexity of the meteorological dataset was reduced and potential trends were identified.

4.2.2. Correlation Analysis

As shown in Table 2, we used Pearson's *r* to calculate the correlation between meteorological variables and pest populations based on agricultural sensing data from the site and pest data surveyed by agricultural experts. *Temperature, light,* and *wind speed* were positively correlated with pest populations, whereas *humidity, pressure,* and *rainfall* were negatively correlated. These correlations constitute weather indicators of pest incidence for farmers. In addition, the positive correlations of *temperature, light,* and *wind speed* were combined with eight input data sets to compare the performance of the predictive models.

Table 2. Pearson's *r* coefficients between meteorological variables and pest population.

Variable	*r*	Variable	*r*	Variable	*r*
T_AVG	0.426 **	GT20_AVG	0.589 ***	VP_MIN	−0.179
T_MAX	0.421 **	GT50_AVG	0.551 ***	SLP_AVG	−0.269
T_MIN	0.319 *	GT100_AVG	0.469 ***	RF_SUM	−0.107
DPT_AVG	0.281 *	RH_AVG	−0.274	LUX_SUM	0.322 *
GT0_AVG	0.571 ***	RH_MIN	−0.014	LUXTIME_SUM	0.400 **
GT5_AVG	0.575 ***	VP_AVG	−0.266	WS_AVG	0.474 ***
GT10_AVG	0.583 ***	VP_MAX	−0.280 *	WS_MAX	0.122

Note. *: $p < 0.05$, ** $p < 0.01$, ***: $p < 0.001$.

4.3. Data Augmentation Analysis

We investigated the data augmentation from three aspects: the parameter settings of the TimeGAN model, a visual presentation of the generated data, and an evaluation of the training effectiveness of the generated and real data.

4.3.1. TimeGAN Parameters

First, we set the parameters of the TimeGAN model. As described in Section 3.2.1, we set the size of the training data to (57, (2, 22)). We used trial and error to set *batch_size, num_layer, hidden_dim, noise_dim,* and *learning_rate.*

After testing and adjusting the TimeGAN parameters, the hyperparameter settings were as follows: we used a GRU generator and authenticator architecture and an LSTM embedding and recovery network architecture; *num_layer* = 3, *hidden_dim* = 2, *noise_dim* = 32, *batch_size* = 32, *learning_rate* = 0.0005, and the number of generations (*train_epoch*) was set to 50,000.

4.3.2. Visualization of Synthetic Data

After training the TimeGAN model, we used it to generate a new serialized object (pickle file, .pkl) for each iteration. For example, a data sample of 64 samples was generated by reading the pickle file, where the size of the new dataset was (64, (2, 22)): 64 samples with 2 weeks and 22 data features (21 weather variables and 1 pest count).

The generated data were evaluated by visualizing the data distribution. First, we directly compared the generated data with the original data, where all the features were selected at the same time and a step size of 2 was used to perform the graphing. The generated TimeGAN data is represented by the basic visualization line graph shown in Figure 7, in which the solid blue line represents the real data, and the dashed orange line represents the generated data. In this case, given the multivariate nature of the trained agricultural sensing data, it was not possible to visualize all of the features and present a dataset with a step size of 1.

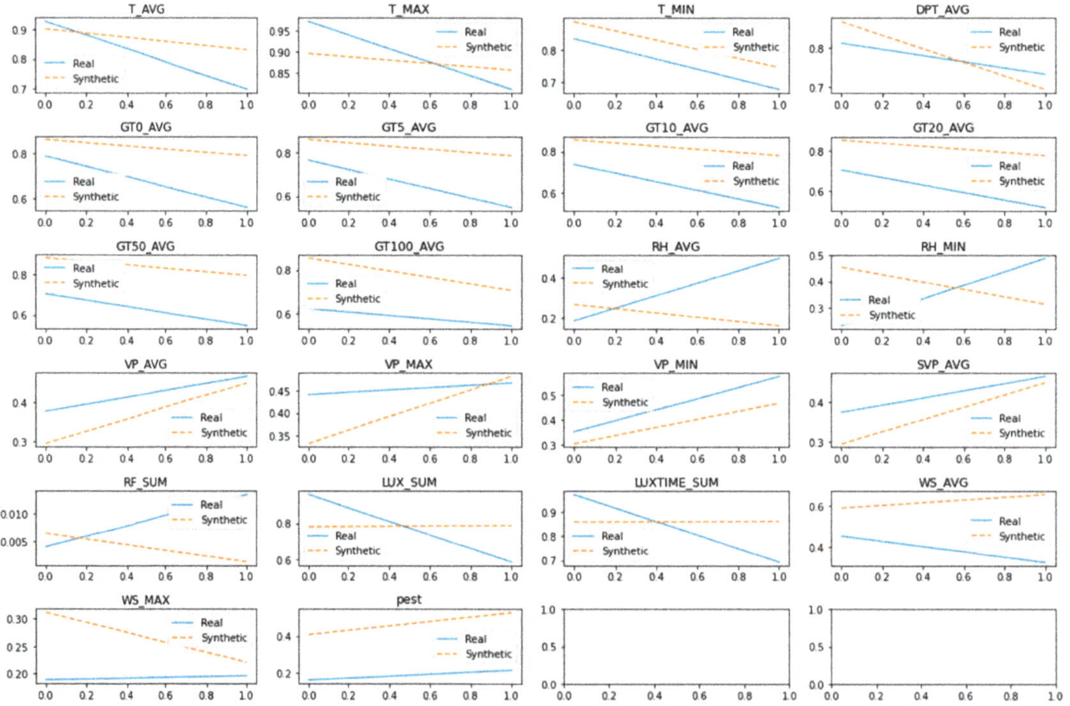

Figure 7. Basic visualization of generated data (*train_epoch* = 50,000).

The second visualization approach used the PCA and t-SNE algorithms to reduce the dimensionality of the dataset from 22 to 2. Figure 8 shows the distribution of the TimeGAN data: the left panel shows PCA results, and the right panel shows the t-SNE results. Since PCA uses linear downscaling, many features are discarded, resulting in underfitting (the distribution is not well-fitted); conversely, as t-SNE is nonlinear, the similarity between sample points in high and low dimensions is defined by chance. The black and red points in t-SNE almost overlap.

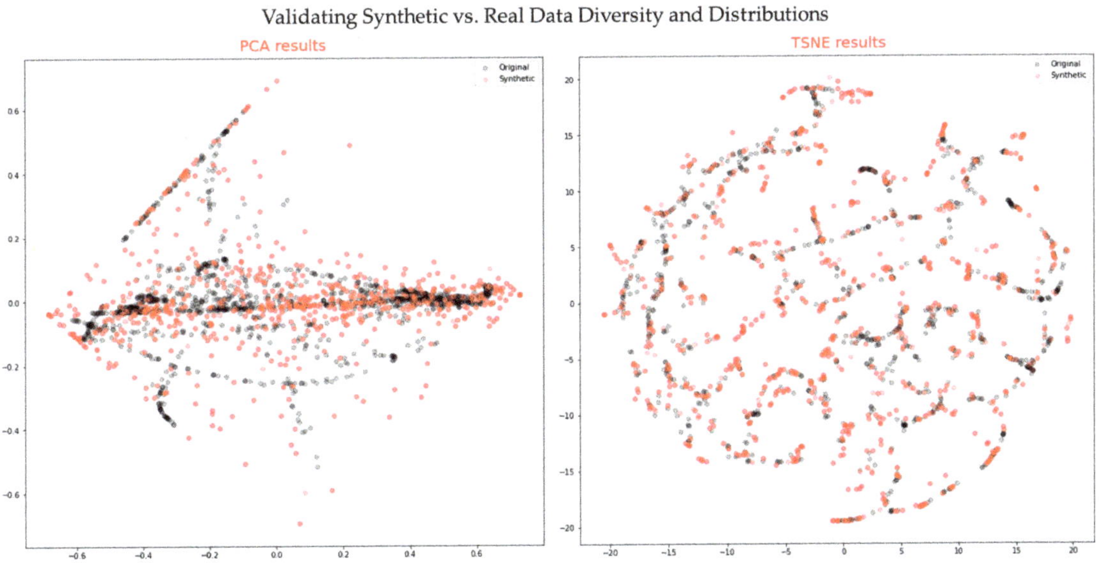

Figure 8. PCA and t-SNE visualizations of generated data (*train_epoch* = 50,000).

4.3.3. Evaluation of Synthetic Data

After training the TimeGAN model, we evaluated the generated data. We thus trained two models, one on real data and one on generated data. For this phase, the test set used real data; the evaluation metrics confirm the difference between the generated data and the model trained with real data. Table 3 shows the results of experiments using trial and error to find a good dataset cut ratio and model parameter settings based on the TSTR premise. After comparing the effectiveness of various dataset cutting ratios, we set the cut ratio between the training set and the test set to 8:2. We built a single-layer GRU to evaluate the model, where the number of neurons (*num_cell*) was 16, *batch_size* was 64, we used the Adam optimizer, and we used early stopping to prevent overfitting. The R^2, *MAE*, and *RMSE* results in Table 3 reveal a small difference between *MAE* and *RMSE*, and show that the R^2 of the model with generated data is close to that of the model with real data.

Table 3. Performance of TSTR-based models.

Type of Data	R^2	*MAE*	*RMSE*
Actual data	0.335843	0.142426	0.204147
Synthetic data	0.379916	0.141800	0.196959

4.4. Data Forecast Analysis

We divided the pest prediction experiment into four parts. First, we optimized the pest prediction model parameters. Second, based on the results of variable correlation analysis, we calculated the evaluation indicators by the prediction model under eight types of independent variables, and selected suitable independent variable data for subsequent multivariate data augmentation experiments. Third, based on the resultant prediction model and independent variable data, we calculated evaluation indicators using three prediction models, and observed the degree of fit between the augmented and original data in the model prediction. Fourth, we presented the overall pest prediction system.

4.4.1. Optimization of Predictive Models

The optimal model parameters use GRU as the base model and a one-to-one framework to design a predictive model that uses all weather variables of the current week to predict the next week's pest population. Recall that the training data accounts for 80% of the total data in this study. GRU model parameters including *batch_size*, *num_cell*, *num_hidden* layer, and dropout were adjusted sequentially by scaling the *RMSE* of the unreturned data to the same evaluation metric to optimize pest prediction, thus optimizing the parameters and structure of the model. Figure 9 shows the *RMSE* curves for optimizing the parameters of the GRU pest prediction model.

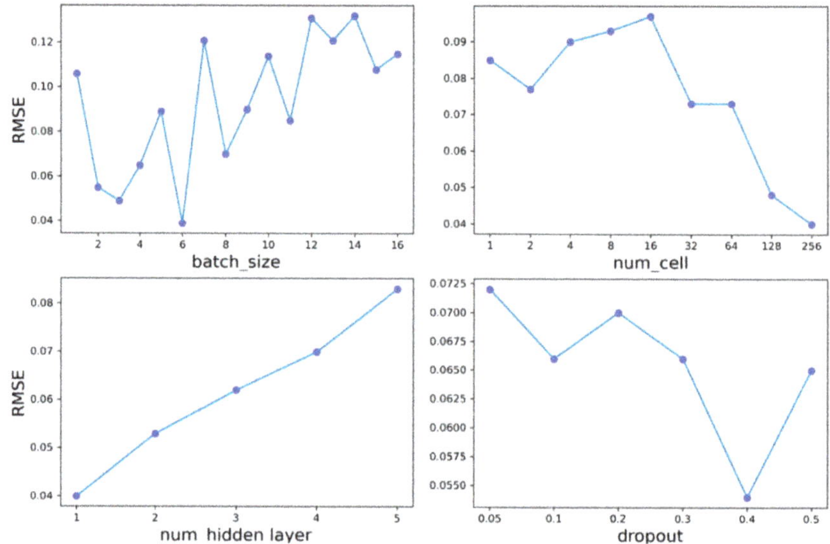

Figure 9. Optimization of model parameter curves.

In Figure 9, the optimal *batch_size* for prediction model training is 6, *num_cell* of GRU is 256, GRU cells are not stacked, the dropout of cells is 0.4, the training model optimizer was RMSprop, and the number of epochs is controlled by early stopping to avoid overfitting. The parameters of the pest prediction model were optimally combined with those of the GRU model, and the RNN and LSTM models used the same parameters and architecture. In addition, we used the Adam optimizer for RNN.

4.4.2. Selection of Input Variables

We selected three sets of positively correlated variables (*temperature*, *light*, and *wind speed*) and combined them to classify the pest prediction independent variables into eight types of data. We used three models (RNN, LSTM, and GRU) to evaluate and predict pest populations. We selected the best-performing independent variables for subsequent experiments to augment the independent variable data. Figure 10 shows the results of pest prediction using the three models. The black dashed line represents the real data, and the green solid line, red solid line, and blue solid line are the predicted values of the RNN model, the LSTM model, and the GRU model, respectively. The y-axis unit is the unnormalized pest counts. Table 4 shows the evaluation results of the eight types of independent variables using the three models.

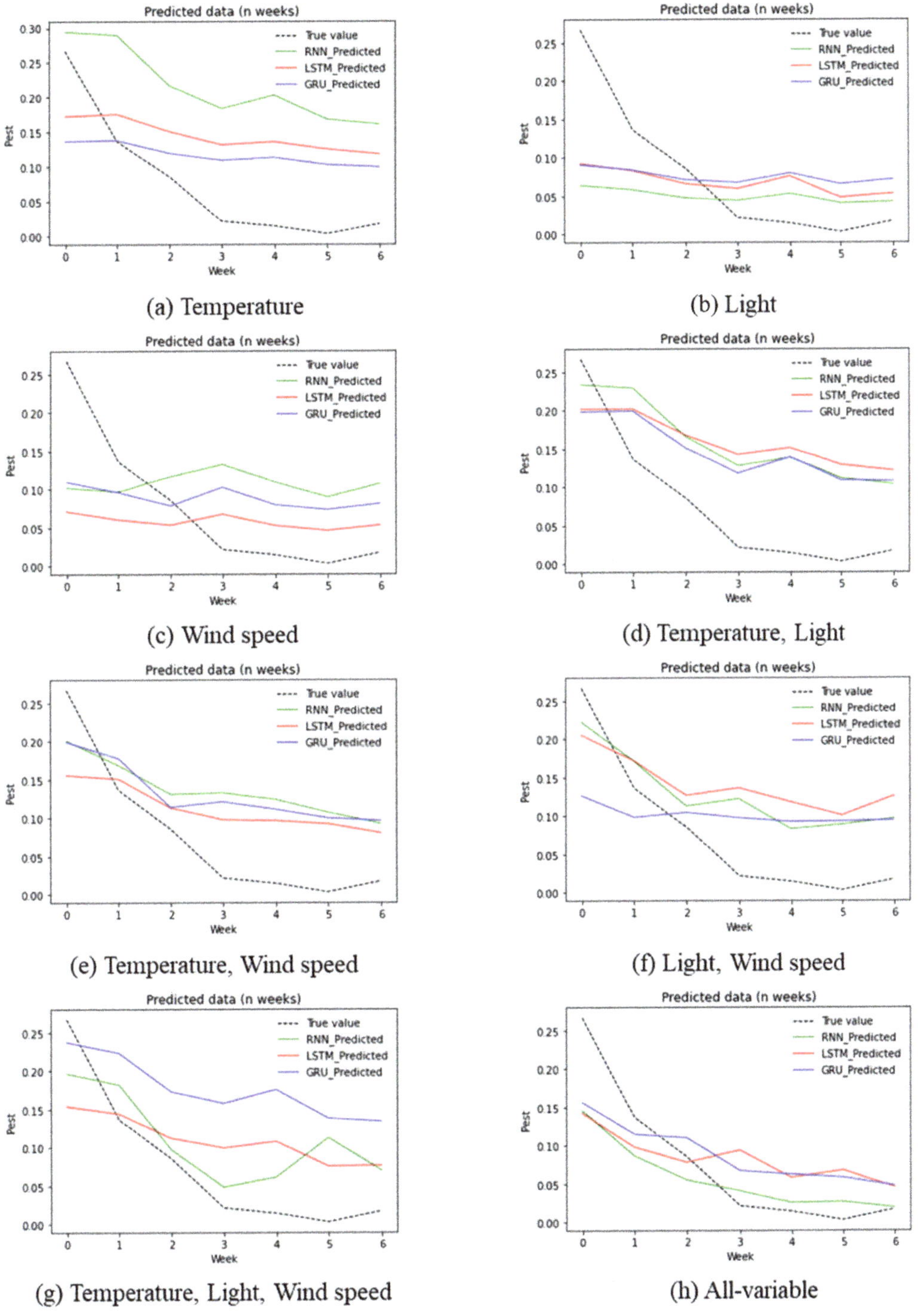

Figure 10. Weather feature prediction results.

Table 4. Results for weather feature datasets.

Input Factor	Model	*MAE*	*MAPE*	*RMSE*
	RNN	153.12	11.89%	162.07
Temperature	LSTM	103.34	8.42%	107.98
	GRU	84.80	6.84%	95.82
	RNN	70.12	2.66%	95.52
Light	LSTM	67.85	3.34%	86.03
	GRU	74.78	4.39%	91.00
	RNN	97.78	6.48%	107.96
Wind Speed	LSTM	74.08	3.15%	95.23
	GRU	77.00	5.00%	90.13
	RNN	100.03	7.76%	104.43
Temperature, Light	LSTM	110.81	8.90%	114.84
	GRU	97.38	7.58%	100.25
	RNN	86.23	7.25%	92.17
Temperature, Wind Speed	LSTM	73.77	5.96%	81.63
	GRU	81.02	6.74%	86.11
	RNN	70.32	5.96%	75.87
Light, Wind Speed	LSTM	89.35	7.20%	95.69
	GRU	82.32	6.11%	91.29
	RNN	58.04	6.09%	66.36
Temperature, Light, Wind Speed	LSTM	71.71	5.30%	80.87
	GRU	118.94	9.76%	127.04
	RNN	41.47	1.51%	58.27
All Variables	LSTM	60.52	4.17%	71.87
	GRU	53.85	3.65%	62.07

In Figure 10a–c, we observe a single combination of independent variables (*temperature, light,* and *wind speed*) where the predicted values curve smoothly and differ more from the actual values; in Figure 10d–g, the combined meteorological variables are predicted with a larger magnitude and gradually approximate the real data curve, but still produce large errors at some moments; in Figure 10h, when training with the agricultural sensing data that covers all variables, the predicted trend curve is clearly closer to the real line than the other seven results. We thus infer that the complete data is more comprehensive except for the variables, and the combination of positive correlation variables does not reduce the prediction error. Therefore, for the data augmentation experiment we use the complete multivariate agricultural sensor data as input values and compare the training set with the unaugmented and augmented improvement.

Note in Table 4 that for models without augmentation (RNN, LSTM, and GRU), the *MAPE* values are significantly smaller than those for the *temperature* features when *light* and *wind speed* are the independent variables, but the predicted curves in Figure 10b,c do not approximate the real data, indicating that a single variable does not contain enough information to fit real-world situations. For [*temperature, light*], [*temperature, wind speed*], [*light, wind speed*], and [*temperature, light, wind speed*], the *MAPE* values are slightly higher. Likewise, due to the additional data features, the results in Figure 10d–g more closely approximate the true values. When the independent variable is [*all variables*], the *MAPE* results for RNN, LSTM, and GRU are 1.51%, 4.17%, and 3.65%, respectively, which are the lowest values in all cases, and the *MAE* and *RMSE* values are the lowest among all combinations. Thus, using all variables yields the most complete characteristics for model learning; note that both predictions are close to the real situation in Figure 10h. Consequently, we generate data using all variables and compare the prediction errors before and after augmentation in RNN, LSTM, and GRU.

4.4.3. Comparison of Predictive Metrics

We independently tested the prediction effects of different independent variables and selected all variables as the original augmented input data; thus, we used the all of the multivariate agricultural sensing data, and used RNN, LSTM, and GRU to compare the six results in terms of the prediction error of the two data sets (MAE) and the error difference (EC). Table 5 shows the performance of the time-series prediction model using different data.

Table 5. Prediction error for different training data.

Model	Training Data			Test Data		
	$MAE_{w/o\ aug}$	$MAE_{w/\ aug}$	EC	$MAE_{w/o\ aug}$	$MAE_{w/\ aug}$	EC
RNN	13.05	91.91	78.86	79.86	120.39	40.53
LSTM	19.62	76.67	57.05	73.29	107.34	34.05
GRU	17.43	85.34	67.91	77.67	87.53	9.86

In Table 5, when the original and augmented data are used for training, the MAE values of the RNN model are 13.05 and 91.91 and the EC value is 78.86. The LSTM model trained with augmented data has the lowest prediction error, and its MAE value of 76.67 is the lowest of the three models. The difference between the prediction errors calculated from the real and simulated data using the LSTM model is also the smallest, and the EC value of 57.05 is the best of the three models.

In addition, as shown in Table 5, the MAE values of the RNN model are 79.86 and 120.39 and the EC value is 40.53 when the original and augmented data is used for the test data. The MAE values of the LSTM model are 73.29 and 107.34 and the EC value is 34.05. The MAE values of the GRU model are 77.67 and 87.53 and the EC value is 9.86. The prediction error of the GRU model trained with the augmented data is the lowest, and the MAE value of 87.53 is the lowest of the three models. The difference between the prediction error of real and simulated data using the GRU model is also the smallest, and the EC value of 9.86 is the lowest of the three models.

4.4.4. Overall Pest Prediction System

The final objective of this study was to present a multivariate pest prediction system for agricultural sensing data. After optimizing the prediction model parameters, we selected the independent variables for pest prediction, used these independent variables to generate data, and then used the generated data together with the original data as the training set for the prediction models. After the initial optimization of the pest prediction model parameters, we constructed six pest prediction models after the model overlay training: the original RNN model, the original LSTM model, the original GRU model, an RNN model trained with generated data, an LSTM model trained with generated data, and a GRU model trained with generated data. Figure 11 shows the prediction results on the training set using the six pretrained models. Figure 12 shows the prediction results of the pretrained models on the test set. The x-axis unit is weeks, and the y-axis unit is the number of pests that were restored without data normalization.

In Figures 11 and 12, the black dashed line represents the real data, the green solid line represents the original RNN model, the green dashed line represents the RNN model trained with the generated data, the red solid line represents the original LSTM model, the red dashed line represents the LSTM model trained with the generated data, the blue solid line represents the original GRU model, and the blue dashed line represents the GRU model trained with the generated data.

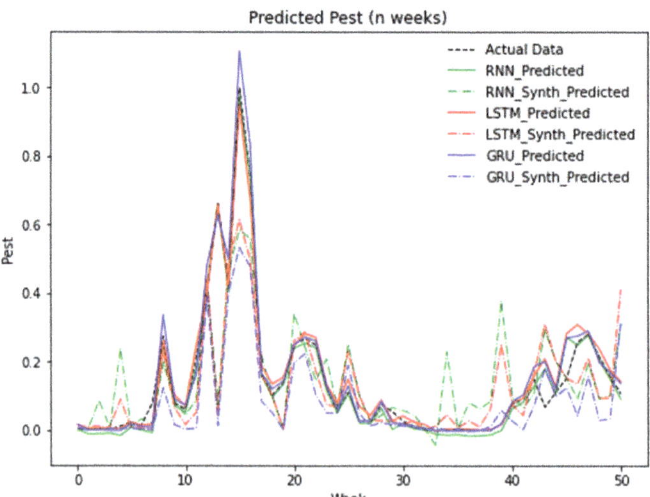

Figure 11. Pest prediction on training set.

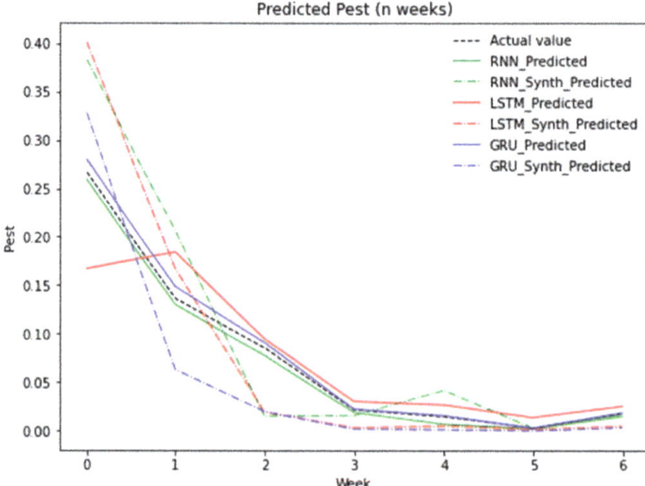

Figure 12. Pest prediction on test set.

5. Conclusions and Prospects

In Section 4.2.1, we use downsampling to convert the raw data into more weather characteristics and then use the prediction model to learn more complete weather trends. In addition, to highlight which combination of weather factors is most beneficial for the task, we analyze eight combinations of independent variables on pest trends by evaluating the results and visualization predictions in Section 4.4.2: when we use all of the variables as the independent variables for the prediction model, the *MAE*, *MAPE*, and *RMSE* are best compared to the other combinations, and the prediction curves most closely approximate the real data. In Section 4.4.1, we further optimize the model parameter configuration and architecture of the predictive model in terms of the same evaluation metrics. Finally, in Sections 4.4.2–4.4.4, we compare and collate the differences in model effectiveness between the optimized models by considering the effects of multiple models simultaneously in different pest prediction experiments.

Data augmentation has not yet been applied to agricultural sensing data. Nevertheless, as the agricultural sensing data is in the form of a time series, they are suited for data augmentation in the form of data generation of sensing data. We generate the original time-series data and use this data for pest trend prediction. The experimental results in Section 4.3.3 show that the generated data can be used to simulate real situations in the absence of sufficient real data. Moreover, in Sections 4.4.3 and 4.4.4, we conduct experiments to evaluate the effect of the generated data on pest prediction. Finally, in Section 4.3.2, we visualize the augmented sensing data and graphically represent the differences between the data via dimension reduction. The t-SNE algorithm shows that the markers of the generated data are generally the same as those for the real data.

Currently, most of the applications of Generative Adversarial Networks (GANs) in agriculture are focused on image generation. To improve the usability of generated data, many variations of GANs have emerged, such as conditional GANs and CycleGANs. While these applications have the potential to enhance various aspects of agriculture, there is a growing emphasis on sustainable agriculture practices that prioritize the long-term health of the environment and ecosystems. TimeGAN, which is mainly used for generating time-series data, is highly suitable for sustainable agriculture applications. It can help generate data that inform sustainable agricultural practices, such as optimizing crop yields, improving resource efficiency, and reducing the environmental impact of agricultural activities. To the best of our knowledge, there has not yet been any research on generating time-series data in this field, but the potential for TimeGAN to contribute to sustainable agriculture practices is promising.

In summary, we address the fundamental problem of insufficient data for pest prediction through time-series data augmentation. We use three common temporal prediction models to evaluate the differences between a model trained on large amounts of generated data and a model trained on raw data. The results show that complex LSTM and GRU models produce prediction results that are similar to those produced by models trained on the raw data. For future work, we suggest using a more complete dataset to predict pest occurrences. In addition, the scaling method proposed in this study compares only the prediction difference between the original data and the generated data, and does not consider the scaling principle of other time-series data. Future studies could investigate the generation effects of different augmentation methods and the improvement in pest prediction results due to augmentation methods such as interpolation, which is commonly used to augment linear problems.

Author Contributions: Conceptualization, Y.-M.H. and C.-Y.T.; methodology, C.-Y.T.; software, C.-Y.T.; validation, C.-Y.T.; investigation, W.-J.W.; data curation, W.-J.W.; writing—original draft preparation, C.-Y.T.; writing—review and editing, Y.-M.H.; visualization, C.-Y.T.; supervision, Y.-M.H.; project administration, Y.-M.H. All authors have read and agreed to the published version of the manuscript.

Funding: This research received no external funding.

Institutional Review Board Statement: Not applicable.

Informed Consent Statement: Not applicable.

Data Availability Statement: Data available in a publicly accessible repository that does not issue DOIs. Publicly available datasets were analyzed in this study. This data can be found here: [https://agr.cwb.gov.tw], [https://data.kcg.gov.tw/organization/3ff96cb3-b16d-4612-98cf-88e63aa6a012?tags=%E5%B0%8F%E9%BB%83%E8%96%8A%E9%A6%AC]. All accessed on 15 June 2022.

Conflicts of Interest: The authors declare no conflict of interest.

References

1. Reganold, J.P.; Papendick, R.I.; Parr, J.F. Sustainable agriculture. *Sci. Am.* **1990**, *262*, 112–121. [CrossRef]
2. Harwood, R.R. A history of sustainable agriculture. In *Sustainable Agricultural Systems*; CRC Press: Boca Raton, FL, USA, 2020; pp. 3–19.

3. Qazi, S.; Khawaja, B.A.; Farooq, Q.U. IoT-equipped and AI-enabled next generation smart agriculture: A critical review, current challenges and future trends. *IEEE Access* **2022**, *10*, 21219–21235. [CrossRef]

4. Jamil, F.; Ibrahim, M.; Ullah, I.; Kim, S.; Kahng, H.K.; Kim, D.-H. Optimal smart contract for autonomous greenhouse environment based on IoT blockchain network in agriculture. *Comput. Electron. Agric.* **2021**, *192*, 106573. [CrossRef]

5. Tian, E.; Li, Z.; Huang, W.; Ma, H. Distributed and Parallel simulation methods for pest control and crop monitoring with IoT assistance. *Acta Agric. Scand. Sect. B Soil Plant Sci.* **2021**, *71*, 884–898. [CrossRef]

6. van Klompenburg, T.; Kassahun, A.; Catal, C. Crop yield prediction using machine learning: A systematic literature review. *Comput. Electron. Agric.* **2020**, *177*, 105709. [CrossRef]

7. García, L.; Parra, L.; Jimenez, J.M.; Lloret, J.; Lorenz, P. IoT-Based Smart Irrigation Systems: An Overview on the Recent Trends on Sensors and IoT Systems for Irrigation in Precision Agriculture. *Sensors* **2020**, *20*, 1042. [CrossRef] [PubMed]

8. Boudwin, R.; Magarey, R.; Jess, L. Integrated Pest Management Data for Regulation, Research, and Education: Crop Profiles and Pest Management Strategic Plans. *J. Integr. Pest Manag.* **2022**, *13*, 13. [CrossRef]

9. Ibrahim, E.A.; Salifu, D.; Mwalili, S.; Dubois, T.; Collins, R.; Tonnang, H.E. An expert system for insect pest population dynamics prediction. *Comput. Electron. Agric.* **2022**, *198*, 107124. [CrossRef]

10. Bradhurst, R.; Spring, D.; Stanaway, M.; Milner, J.; Kompas, T. A generalised and scalable framework for modelling incursions, surveillance and control of plant and environmental pests. *Environ. Model. Softw.* **2021**, *139*, 105004. [CrossRef]

11. Aharoni, R.; Klymiuk, V.; Sarusi, B.; Young, S.; Fahima, T.; Fishbain, B.; Kendler, S. Spectral light-reflection data dimensionality reduction for timely detection of yellow rust. *Precis. Agric.* **2020**, *22*, 267–286. [CrossRef]

12. Sasikala, V.; Venkatesan, G. Time Variant Multi Feature Census Analysis for Efficient Prediction of Migration Risks in Agriculture. *J. Comput. Theor. Nanosci.* **2020**, *17*, 5323–5333. [CrossRef]

13. Morais, R.; Mendes, J.; Silva, R.; Silva, N.; Sousa, J.J.; Peres, E. A Versatile, Low-Power and Low-Cost IoT Device for Field Data Gathering in Precision Agriculture Practices. *Agriculture* **2021**, *11*, 619. [CrossRef]

14. Ayyoubzadeh, S.M.; Zahedi, H.; Ahmadi, M.; Kalhori, S.R.N. Predicting COVID-19 Incidence Through Analysis of Google Trends Data in Iran: Data Mining and Deep Learning Pilot Study. *JMIR Public Health Surveill.* **2020**, *6*, e18828. [CrossRef] [PubMed]

15. Su, D.; Kong, H.; Qiao, Y.; Sukkarieh, S. Data augmentation for deep learning based semantic segmentation and crop-weed classification in agricultural robotics. *Comput. Electron. Agric.* **2021**, *190*, 106418. [CrossRef]

16. Singhal, A.; Cheriyamparambil, A.; Jha, S.K. Spatial extrapolation of statistically downscaled weather data over the Northwest Himalayas at major glacier sites. *Environ. Model. Softw.* **2022**, *149*, 105317. [CrossRef]

17. Zhang, Y.; Thorburn, P.J. Handling missing data in near real-time environmental monitoring: A system and a review of selected methods. *Futur. Gener. Comput. Syst.* **2021**, *128*, 63–72. [CrossRef]

18. Yoon, J.; Jarrett, D.; Van der Schaar, M. Time-series generative adversarial networks. *Adv. Neural Inf. Process. Syst.* **2019**, *32*, 5508–5518.

19. Dupuis, A.; Dadouchi, C.; Agard, B. Predicting crop rotations using process mining techniques and Markov principals. *Comput. Electron. Agric.* **2022**, *194*, 106686. [CrossRef]

20. Bhat, S.A.; Huang, N.-F. Big Data and AI Revolution in Precision Agriculture: Survey and Challenges. *IEEE Access* **2021**, *9*, 110209–110222. [CrossRef]

21. Husin, N.A.; Khairunniza-Bejo, S.; Abdullah, A.F.; Kassim, M.S.M.; Ahmad, D. Multi-temporal analysis of terrestrial laser scanning data to detect basal stem rot in oil palm trees. *Precis. Agric.* **2021**, *23*, 101–126. [CrossRef]

22. Du, X.; Elbakidze, L.; Lu, L.; Taylor, R.G. Climate Smart Pest Management. *Sustainability* **2022**, *14*, 9832. [CrossRef]

23. Ren, C.; Kim, D.-K.; Jeong, D. A survey of deep learning in agriculture: Techniques and their applications. *J. Inf. Process. Syst.* **2020**, *16*, 1015–1033.

24. Xu, L.; Zhang, H.; Wang, C.; Zhang, B.; Liu, M. Crop Classification Based on Temporal Information Using Sentinel-1 SAR Time-Series Data. *Remote Sens.* **2018**, *11*, 53. [CrossRef]

25. Htitiou, A.; Boudhar, A.; Lebrini, Y.; Lionboui, H.; Chehbouni, A.; Benabdelouahab, T. Classification and status monitoring of agricultural crops in central Morocco: A synergistic combination of OBIA approach and fused Landsat-Sentinel-2 data. *J. Appl. Remote Sens.* **2021**, *15*, 014504. [CrossRef]

26. Rembold, F.; Meroni, M.; Urbano, F.; Csak, G.; Kerdiles, H.; Perez-Hoyos, A.; Lemoine, G.; Leo, O.; Negre, T. ASAP: A new global early warning system to detect anomaly hot spots of agricultural production for food security analysis. *Agric. Syst.* **2018**, *168*, 247–257. [CrossRef] [PubMed]

27. Filippi, P.; Jones, E.J.; Wimalathunge, N.S.; Somarathna, P.D.S.N.; Pozza, L.E.; Ugbaje, S.U.; Jephcott, T.G.; Paterson, S.E.; Whelan, B.M. An approach to forecast grain crop yield using multi-layered, multi-farm data sets and machine learning. *Precis. Agric.* **2019**, *20*, 1015–1029. [CrossRef]

28. Dong, Y.; Xu, F.; Liu, L.; Du, X.; Ren, B.; Guo, A.; Geng, Y.; Ruan, C.; Ye, H.; Huang, W.; et al. Automatic System for Crop Pest and Disease Dynamic Monitoring and Early Forecasting. *IEEE J. Sel. Top. Appl. Earth Obs. Remote Sens.* **2020**, *13*, 4410–4418. [CrossRef]

29. van Mourik, S.; van der Tol, R.; Linker, R.; Reyes-Lastiri, D.; Kootstra, G.; Koerkamp, P.G.; van Henten, E.J. Introductory overview: Systems and control methods for operational management support in agricultural production systems. *Environ. Model. Softw.* **2021**, *139*, 105031. [CrossRef]

30. Sharma, R. Artificial Intelligence in Agriculture: A Review. In Proceedings of the 2021 5th International Conference on Intelligent Computing and Control Systems (ICICCS), Madurai, India, 6–8 May 2021; pp. 937–942.

31. Heeb, L.; Jenner, E.; Cock, M.J.W. Climate-smart pest management: Building resilience of farms and landscapes to changing pest threats. *J. Pest Sci.* **2019**, *92*, 951–969. [CrossRef]

32. Meshram, V.; Patil, K.; Meshram, V.; Hanchate, D.; Ramkteke, S. Machine learning in agriculture domain: A state-of-art survey. *Artif. Intell. Life Sci.* **2021**, *1*, 100010. [CrossRef]

33. Issad, H.A.; Aoudjit, R.; Rodrigues, J. A comprehensive review of Data Mining techniques in smart agriculture. *Eng. Agric. Environ. Food* **2019**, *12*, 511–525. [CrossRef]

34. Yang, J.; Ma, S.; Li, Y.; Zhang, Z. Efficient Data-Driven Crop Pest Identification Based on Edge Distance-Entropy for Sustainable Agriculture. *Sustainability* **2022**, *14*, 7825. [CrossRef]

35. Chen, J.-W.; Lin, W.-J.; Cheng, H.-J.; Hung, C.-L.; Lin, C.-Y.; Chen, S.-P. A Smartphone-Based Application for Scale Pest Detection Using Multiple-Object Detection Methods. *Electronics* **2021**, *10*, 372. [CrossRef]

36. Liu, L.; Wang, R.; Xie, C.; Yang, P.; Wang, F.; Sudirman, S.; Liu, W. PestNet: An End-to-End Deep Learning Approach for Large-Scale Multi-Class Pest Detection and Classification. *IEEE Access* **2019**, *7*, 45301–45312. [CrossRef]

37. Roosjen, P.P.; Kellenberger, B.; Kooistra, L.; Green, D.R.; Fahrentrapp, J. Deep learning for automated detection of *Drosophila suzukii*: Potential for UAV -based monitoring. *Pest Manag. Sci.* **2020**, *76*, 2994–3002. [CrossRef] [PubMed]

38. Yu, R.; Luo, Y.; Zhou, Q.; Zhang, X.; Wu, D.; Ren, L. A machine learning algorithm to detect pine wilt disease using UAV-based hyperspectral imagery and LiDAR data at the tree level. *Int. J. Appl. Earth Obs. Geoinf.* **2021**, *101*, 102363. [CrossRef]

39. Mekonnen, Y.; Namuduri, S.; Burton, L.; Sarwat, A.; Bhansali, S. Review—Machine Learning Techniques in Wireless Sensor Network Based Precision Agriculture. *J. Electrochem. Soc.* **2019**, *167*, 037522. [CrossRef]

40. Materne, N.; Inoue, M. IoT Monitoring System for Early Detection of Agricultural Pests and Diseases. In Proceedings of the 2018 12th South East Asian Technical University Consortium (SEATUC), Yogyakarta, Indonesia, 12–13 March 2018; pp. 1–5. [CrossRef]

41. Bourhis, Y.; Bell, J.R.; Bosch, F.V.D.; Milne, A.E. Artificial neural networks for monitoring network optimisation—A practical example using a national insect survey. *Environ. Model. Softw.* **2020**, *135*, 104925. [CrossRef]

42. Chen, P.; Xiao, Q.; Zhang, J.; Xie, C.; Wang, B. Occurrence prediction of cotton pests and diseases by bidirectional long short-term memory networks with climate and atmosphere circulation. *Comput. Electron. Agric.* **2020**, *176*, 105612. [CrossRef]

43. Saleem, M.H.; Potgieter, J.; Arif, K.M. Automation in agriculture by machine and deep learning techniques: A review of recent developments. *Precis. Agric.* **2021**, *22*, 2053–2091. [CrossRef]

44. Jiang, J.-A.; Syue, C.-H.; Wang, C.-H.; Liao, M.-S.; Shieh, J.-S.; Wang, J.-C. Precisely forecasting population dynamics of agricultural pests based on an interval type-2 fuzzy logic system: Case study for oriental fruit flies and the tobacco cutworms. *Precis. Agric.* **2022**, *23*, 1302–1332. [CrossRef]

45. Zhang, J.; Huang, Y.; Pu, R.; Gonzalez-Moreno, P.; Yuan, L.; Wu, K.; Huang, W. Monitoring plant diseases and pests through remote sensing technology: A review. *Comput. Electron. Agric.* **2019**, *165*, 104943. [CrossRef]

46. Narava, R.; Sai Ram Kumar, D.V.; Jaba, J.; Anil Kumar, P.; Ranga Rao, G.V.; Srinivasa Rao, V.; Mishra, S.P.; Kukanur, V. Development of Temporal Model for Forecasting of *Helicoverpa armigera* (Noctuidae: Lepidopetra) Using Arima and Artificial Neural Networks. *J. Insect Sci.* **2022**, *22*, 2. [CrossRef]

47. Gómez, D.; Salvador, P.; Sanz, J.; Rodrigo, J.F.; Gil, J.; Casanova, J.L. Prediction of desert locust breeding areas using machine learning methods and SMOS (MIR_SMNRT2) Near Real Time product. *J. Arid. Environ.* **2021**, *194*, 104599. [CrossRef]

48. Tan, S.; Liang, Y.; Zheng, R.; Yuan, H.; Zhang, Z.; Long, C. Dynamic Prediction of *Chilo suppressalis* Occurrence in Rice Based on Deep Learning. *Processes* **2021**, *9*, 2166. [CrossRef]

49. Delgado, J.M.D.; Oyedele, L. Deep learning with small datasets: Using autoencoders to address limited datasets in construction management. *Appl. Soft Comput.* **2021**, *112*, 107836. [CrossRef]

50. Wen, Q.; Sun, L.; Yang, F.; Song, X.; Gao, J.; Wang, X.; Xu, H. Time Series Data Augmentation for Deep Learning: A Survey. *arXiv* **2020**, arXiv:2002.12478.

51. Barunik, J.; Krehlik, T.; Vacha, L. Modeling and forecasting exchange rate volatility in time-frequency domain. *Eur. J. Oper. Res.* **2016**, *251*, 329–340. [CrossRef]

52. Um, T.T.; Pfister, F.M.J.; Pichler, D.; Endo, S.; Lang, M.; Hirche, S.; Fietzek, U.; Kulić, D. Data augmentation of wearable sensor data for parkinson's disease monitoring using convolutional neural networks. In Proceedings of the 19th ACM International Conference on Multimodal Interaction, Glasgow, UK, 13–17 November 2017; pp. 216–220.

53. Dumas, J.; Wehenkel, A.; Lanaspeze, D.; Cornélusse, B.; Sutera, A. A deep generative model for probabilistic energy forecasting in power systems: Normalizing flows. *Appl. Energy* **2022**, *305*, 117871. [CrossRef]

54. Asimopoulos, D.C.; Nitsiou, M.; Lazaridis, L.; Fragulis, G.F. Generative Adversarial Networks: A systematic review and applications. *SHS Web Conf.* **2022**, *139*, 03012. [CrossRef]

55. Agricultural Meteorological Observation Network Monitoring System. Available online: https://agr.cwb.gov.tw/ (accessed on 15 June 2022).

56. Government. Active Investigation and Monitoring of Epidemic Diseases and Insect Pests in Kaohsiung City—Information Table of the Number of Scirtothrips dorsalis. Available online: https://data.kcg.gov.tw/organization/3ff96cb3-b16d-4612-98cf-88e6 3aa6a012?tags=%E5%B0%8F%E9%BB%83%E8%96%8A%E9%A6%AC (accessed on 15 June 2022).

57. YData. YData-Synthetic. Available online: https://github.com/ydataai/ydata-synthetic (accessed on 15 June 2022).

58. Jordon, J.; Yoon, J.; Van Der Schaar, M. PATE-GAN: Generating Synthetic Data with Differential Privacy Guarantees. In International Conference on Learning Representations. 2018. Available online: https://openreview.net/forum?id=S1zk9iRqF7 (accessed on 15 June 2022).
59. Anowar, F.; Sadaoui, S.; Selim, B. Conceptual and empirical comparison of dimensionality reduction algorithms (pca, kpca, lda, mds, svd, lle, isomap, le, ica, t-sne). *Comput. Sci. Rev.* **2021**, *40*, 100378. [CrossRef]
60. Chicco, D.; Warrens, M.J.; Jurman, G. The coefficient of determination R-squared is more informative than SMAPE, MAE, MAPE, MSE and RMSE in regression analysis evaluation. *PeerJ Comput. Sci.* **2021**, *7*, e623. [CrossRef] [PubMed]
61. Government. Kaohsiung City Agricultural Statistics Annual Report in 110 Years of the Republic of China. Available online: https://agri.kcg.gov.tw/FileDownLoad/FileUpload/20220502105956424644.pdf (accessed on 15 June 2022).

 *sustainability*

Article

Study on the Resistance of 'Cabernet Sauvignon' Grapevine with Different Rootstocks to *Colomerus vitis*

Wenchao Shi [1,2], Wang He [3], Zhijun Zhang [1,2], Junli Sun [1,2], Chunmei Zhu [1,2], Zhiyu Liu [1,2], Yeqing Xu [1,2] and Baolong Zhao [1,2,*]

1. Department of Horticulture, Agricultural College, Shihezi University, Shihezi 832000, China
2. Key Laboratory of Special Fruits and Vegetables Cultivation Physiology and Germplasm Resources Utilization of Xinjiang Production and Construction Crops, Shihezi 832000, China
3. Agriculture Technology Service Center of Qianjiang District, Chongqing 409700, China
* Correspondence: zhaobaolong@shzu.edu.cn; Tel.: +86-180-0993-5729

Abstract: In recent years, *Colomerus vitis* has caused serious economic losses due to reduced grape production in Xinjiang (northwest China). Several rootstock varieties have been reported to improve the resistance of Cabernet Sauvignon to *Colomerus vitis*. This study explored the influence of Cabernet Sauvignon with different rootstocks on the resistance to *Colomerus vitis*. In particular, Cabernet Sauvignon/Cabernet Sauvignon (CS/CS) was selected as the control, and Cabernet Sauvignon grafted with five resistant rootstocks (3309C, 1103P, 140R, SO4, and 5C) was employed as the treatment. The infestation rate and injury index of *Colomerus vitis* to grapevines was investigated, and insect-resistant types of grapevines with different rootstocks were determined. The resveratrol (Res) content, the gene expression of resveratrol synthase (*RS*), and the activities of peroxidase (POD), polyphenol oxidase (PPO), catalase (CAT), and superoxide dismutase (SOD) in the leaves of each rootstock grapevine were measured. The activity of the four enzymes and the content of Res were negatively correlated with the injury index. The results revealed the ability of the rootstock to improve the resistance of grapevines to *Colomerus vitis* by increasing the enzyme activity or Res content. In particular, 140R, SO4, and 5C rootstocks can be employed as rootstocks of the 'Cabernet Sauvignon' grapevine with resistance to *Colomerus vitis*. The contents of Res and the four resistance enzymes studied here can be used as indexes to evaluate the insect resistance of rootstock–scion combinations.

Keywords: rootstock; grapevines; *Colomerus vitis*; oxidase; resveratrol; mite

Citation: Shi, W.; He, W.; Zhang, Z.; Sun, J.; Zhu, C.; Liu, Z.; Xu, Y.; Zhao, B. Study on the Resistance of 'Cabernet Sauvignon' Grapevine with Different Rootstocks to *Colomerus vitis*. *Sustainability* **2022**, *14*, 15193. https://doi.org/10.3390/su142215193

Academic Editors: Barlin Orlando Olivares Campos, Edgloris E. Marys and Miguel Araya-Alman

Received: 13 October 2022
Accepted: 14 November 2022
Published: 16 November 2022

1. Introduction

The widespread use of resistant rootstocks of grapevines (*Vitis* spp.) began in the 19th century during the fight against the grape phylloxera [1]. This, consequently, prompted the beginning of research on grapevine resistant rootstocks [2]. Previous work has reported that rootstocks can improve the resistance of grapevine scions to Bois noir (BN) disease [3] and Pierce's disease (PD) [4,5]. As a high-quality grape production base, Xinjiang (northwest China) has unique environmental and geographical advantages. In addition, the grape industry has developed rapidly in recent years. 'Cabernet Sauvignon' is a high-quality red grape of the wine-making variety, which has become the main wine-making grape variety in Xinjiang. However, the increase in grapevine planting area has caused a rise in diseases and insect pests, which negatively impacts grape production, as well as the quality and yield of grapevines. Using rootstocks can effectively improve the stress and disease resistance of grapevine scions [3]. Current research on rootstocks focuses on the resistance to root aphids and nematodes, while advancements have also been made in screening rootstocks with salt, alkali, drought, and cold resistance [6,7]. However, reports on the resistance of grafted grapes with different rootstocks to gall mites are limited.

Colomerus vitis (Pagenstecher) (Acari: Eriophyidae) (Grape erineum mite, GEM) eats grapevine leaves, leading to leaf deformities and grape production losses [8]. GEM-infested grapevine leaves initially show white spots on the back. This is followed by the appearance of a bubble bulge. The leaves then twist, become brittle, and experience the formation of a layer of white fluff plaque at the back. Similar to rusts, brown, felt-like patches form in the later period, which is commonly known as 'Mao Zhanbing'. When GEM infestation is serious, the grapevines cannot regenerate and grow new shoots, which inhibits the light and ability of leaves and reduces the quality and yield of berries [9]. GEM infestation in grapevine leaves facilitates the spread of grape berry necrosis virus [10] or grapevine Pinot gris virus [11]. Furthermore, GEM, as a common gall mite in wine grape plantations in Xinjiang, has proved to be serious in numerous years, causing great production losses. Therefore, the control of grape gall mites in Xinjiang should not be underestimated.

When plants are under stress and infected by pathogens, their disease resistance systems start to respond. The accumulation of reactive oxygen species (ROS) is closely related to the initial defense response of plants. For example, H_2O_2, as a signal of plant-induced defense responses, can activate the disease-related protein (PR) gene and induce phytoalexin or other defense responses [12]. In order to protect cells and reduce the pressure of reactive oxygen species, the activities of SOD, CAT, POD, and other enzymes increase [13–16]. Moreover, resveratrol (Res) is a phytoalexin produced by plants under stress and is widely studied due to its key role in grapevine disease resistance. In the response of grapevines to downy mildew, the accumulation of Res and H_2O_2 is related to programmed cell death (PCD), and the faster the response speed, the stronger the resistance of the plant to downy mildew [17]. Grapevines are a key source of natural Res, a stilbene compound that usually exists as a trans-structure in nature and is typically synthesized by phenylalanine metabolism in plants [18]. Res is also one of the main substances related to the health care function of wine. The content of Res in grapes gradually increases with ripening after it enters the ripening stage, with significant differences among grapevine varieties [19]. At present, breeding experts improve the disease resistance of grapevines through intraspecific or interspecific breeding, including increasing the Res content of plants [20,21]. Related studies have found that the content of Res in scions can also be increased by grafting [22,23]. However, research shows that resveratrol itself does not have a high antimicrobial activity, and its polymers, such as Pterostilbene, are the main substances that can resist microbial infection. Res is usually accumulated at a high concentration in response to induction or pathogen infection [24]. Stilbene compounds may be involved in cell wall strengthening through peroxidase-mediated cross-linking with cell wall components [25]. After the leaves of grapevines are infected by fungal diseases, Res is rapidly synthesized in large quantities at the infected site and is then synthesized by the rootstock to be transported upward to the infected site through the phloem. Moreover, large amounts of Res produced by infected grapevine varieties will be quickly glycosylated to form polydatin (Pd) with less toxicity, while large amounts of Res produced by infected resistant varieties are rapidly oxidized into more toxic viniferin [26,27]. Therefore, the Res content can be used as a reference index to measure the disease or insect resistances of grafted grapevines.

This study explores the effects of different rootstocks on the resistance of 'Cabernet Sauvignon' vines to GEM. In particular, 'Cabernet Sauvignon' grapevines with different rootstocks were used as the experimental material, and GEM was inoculated artificially in the field in order to investigate the infestation rate and infestation index of new shoot leaves infested with GEM. The activity of related resistance enzymes, Res content, and RS gene expression in the leaves of different 'Cabernet Sauvignon' grapevine rootstocks with different GEM infestation grades were measured. The correlation between the infestation index and enzyme activity and the Res content of scion grapevine leaves at the initial stage of infestation (infestation grade 1) was analyzed. The resistance of 'Cabernet Sauvignon' grapevines with different rootstocks to GEM was then evaluated. This work provides a reference to explore the mechanism of GEM resistance in 'Cabernet Sauvignon' grapevines with different rootstocks and acts as a theoretical basis for the selecting of grapevine rootstocks with GEM resistance in Xinjiang.

2. Materials and Methods

2.1. Field Experiments

The test site is located in the experimental station of the Agricultural College of Shihezi University, Xinjiang, China. Six-year-old grapevine combinations with different rootstocks of 'Cabernet Sauvignon' were used as the experimental material. 'Cabernet Sauvignon' grafted with 'Cabernet Sauvignon' (CS/CS) was used as the control (CK). The planting row spacing was set as 1 m × 3 m, with hedgerow cultivation. Rootstocks and scions were all from the Zhengzhou Fruit Research Institute, Chinese Academy of Agr-cultural Sciences. The selected rootstocks were 3309C (*V. riparia* × *V. rupestris*), 1103P (*V. berlandieri* × *V. rupestris*), SO4 (*V. berlandieri* × *V. riparia*), 140R (*V. berlandieri* × *V. rupestris*), and 5C (*V. berlandieri* × *V. riparia*). Six grapevine plants with similar growth were selected from each rootstock–scion combination in the experimental station as the experimental material. Each rootstock–scion combination was isolated from other plants in the experimental station using a 300-mesh (84.6 µm aperture) nylon gauze. In 2019, abamectin was used to control mites across the entire experimental station (including the experimental plants) to ensure that there were no GEM and predatory natural enemies on the grapevines in the experimental station. In April 2020, the experimental plants isolated by nylon netting were unearthed separately and placed on shelves. On 3 May, a chemical mixture (abamectin and mancozeb) was used to ensure that there were no pests or injuries on the experimental plants. Following this, the field trimming, soil, fertilizer, and water management of the test materials were standardized and unified. On 30 May, the main shoot was cored, while the auxiliary shoot was not cored.

On 15 June, at 20:30 (Beijing time), the branches of GEM-infected leaves were collected in the 'Cabernet Sauvignon' Vineyard, Shihezi City (13 km away from the experimental station) and kept fresh at 25 °C in a sampling box. Whole leaves exhibiting felted injury spots were used as the GEM inoculation material for inoculation. In the evening, the GEM inoculation material and young leaves of the secondary shoots of the experimental plants were fixed back-to-back with paper clips at the experimental station. The total area of the GEM-infected leaves of each grapevine was approximately 10 cm^2. On 30 June, the grapevine plants exhibited symptoms of GEM infestation. Follow-up management ensured that the nylon net was in a closed state, with the exception of the tester entrance and exit. The resistance of 'Cabernet Sauvignon' grapevines with different rootstocks to gall mites was then investigated in the field.

In 2020, 36 grapevine plants of a similar growth stage were selected as the test material. Identical rootstocks were combined into pairs and considered as a single treatment. All treatments were repeated three times. At the peak period of the GEM damage (18 August), five accessory shoots (new shoots) were selected from each grapevine, and five functional leaves larger than 1/3 of the accessory shoot and above (Upper morphology) were selected from each new shoot. A total of 50 leaves were selected to investigate the GEM damage and related indexes. As no leaves with a GEM infestation level of 4 were identified in the statistical analysis, the activities of SOD, CAT, PPO, and POD, the contents of Res and Pd, and the relative expression of the *RS* gene in the leaves were measured using functional leaves with infestation levels of 0, 1, 2, and 3. Leaves of the same rootstock and scion combinations and the same infestation grade were collected and mixed for sampling. Each test index was repeated three times, and the significant differences of related indexes of different rootstock and scion combinations under the same injury grade were analyzed. The index measured at the earliest stage of worm development following the appearance of the GEM infestation symptoms (i.e., at the level 1 infestation stage) was selected for the correlation analysis with the injury index.

2.2. Infestation Evaluation

The damage degree of GEM to grapevine leaves can be divided into different injury grades based on the following criteria: level 0, no bubble bulge; level 1, bubble bulge area accounts for less than 1/4 of leaf area; level 2, bubble bulge area accounts for 1/4–1/2 of whole leaf area; level 3, bubble bulge area accounts for 1/2–3/4 of whole leaf area; and level 4, bubble bulge area accounts for more than 3/4 of whole leaf area. Leaf area surveying was performed with a single-lens reflex (SLR) camera, and the affected area was calculated with Photoshop CC2017 (Adobe Inc., San Jose, CA, USA) as follows:

$$\text{Incidence (\%)} = \text{Incidence/Investigation} \times 100\% \tag{1}$$

$$\text{Injury index} = \sum \text{(number of cases at all levels} \times \text{value at all levels)/(total number of investigations} \times \text{highest grade value)} \times 100 \tag{2}$$

The types of grapevine resistance to gall mites were divided into five levels: (i) immunity (IM), with an injury index of 0; (ii) high resistance (HR), with an injury index of 0.1–5.0; (iii) injury resistance (R), with an injury index of 5.1–25.0; (iv) injury (I), with an injury index of 25.1–50.0; and (v) high injury (HI), with an injury index greater than 50.0.

2.3. Biochemistry Methods

The enzyme activities of SOD, CAT, PPO, and POD in scion leaves with different injury grades were determined according to [28]. Three technical repetitions were performed. The Res and Pd contents were quantified via HPLC (high-performance liquid chromatography) using Shimadzu High Performance Liquid Chromatography (Kyoto, Japan) (LC-2010AHT) of Japan. The equipment included: an ultraviolet detector, a HPLC 2D workstation, the Kunshan ultrasonic instrument, a freezing centrifuge, the SEG ProteCol-PC18 reverse chromatographic column, etc. The chromatographic bottle used for the high-efficiency liquid phase was Agilent Technology Co., Ltd. (Santa Clara, CA, USA). The specific process employed is described as follows: take a 1 g sample and grind it into powder in liquid nitrogen; transfer to a centrifuge tube with methanol to a constant volume of 10 mL; swirl for 2 min; extract with ultrasonic for 30 min; centrifuge it at 12,000 rpm for 10 min; repeat twice; filter the supernatant with a 0.45 μm filter membrane (organic phase) in the chromatographic bottle (Agilent); and place it in a 4 °C refrigerator for later use. The mobile phase was selected as acetonitrile/0.2% phosphoric acid = 45/55, the detection wavelength as 306 nm, and the flow rate as 0.8 mL/min. Column temperature, sample volume, and elution condition were set as: 25 °C, 10 μL, and low elution, respectively.

The related primers were designed according to the literature on the synthesis pathway of resveratrol from a grapevine (Table 1). The primers of the *RS* gene and internal reference gene (*actin1* [29]) were designed with Primer Premier 6.0. The designed primers were synthesized by Shanghai Shenggong Co., Ltd. (Shanghai, China).

Table 1. Primer Sequences.

Genes	Forward Primer	Reverse Primer
actin1	CTTGCATCCCTCAGCACCTT	TCCTGTGGACAATGGATGGA
RS	GCTATGCAGGTGGAACTGTCCTTC	CTCAGAGCACACCACAAGAACTCG

An RNA extraction kit (Meiji Biotechnology Co., Ltd., Guangzhou, China) was used for the total RNA extraction. The quality of the RNA was identified by a micro-ultraviolet spectrophotometer and agarose gel electrophoresis. An RT-5-UHUUK reverse transcription kit (Shanghai Yisheng biotechnology Co., Ltd., Shanghai, China) was employed to synthesize cDNA strands by reverse transcription, and was stored in the refrigerator at −20 °C.

qRT-PCR reactions were performed with the Green Realtime PCR Master Mix kit (Toyobo, Japan). Each sample was repeated three times. The total volume of the amplification system was 20 μL, with a cDNA 2 μL template, SYBR mixture of 10 μL, ddH2O of 7.2 μL, and forward and reverse primers of 0.4 μL each. The amplification procedure is described as follows: pre-denaturation at 95 °C for 5 min; denaturation at 95 °C for 5 s; annealing at 60 °C for 10 s; extension at 72 °C for 15 s; and a duration of 40 cycles. The relative expression of genes was calculated using the $2^{-\Delta\Delta Ct}$ method.

2.4. Data Analysis

The experimental data were analyzed and visualized using Photoshop CC2017, Excel 2010 (Microsoft Corp., Redmond, WA, USA), SPSS 19.0 (IBM), and GraphPad Prism 9 (GraphPad Software, Inc., San Diego, CA, USA). Among them, SPSS 19.0 was used for Tukey's HSD test, and GraphPad Prism 9 was used for the normal distribution test and Peel correlation analysis.

3. Results

3.1. GEM Field Identification Results of 'Cabernet Sauvignon' with Different Rootstocks

We investigated the damage of GEM to 'Cabernet Sauvignon' grapevines with different rootstocks in the field (Figure 1). GEM was generally observed to damage the new shoots and leaves of grapevines and occurred in the new shoots and leaves of 'Cabernet Sauvignon' under all rootstock–scion combinations. The GEM damage rates of leaves in all rootstock combinations were compared (Table 2). The CS/3309C, CS/1103P, and CK combinations exhibited the highest incidence. Compared with the CK, there were no significant differences in the GEM resistance of 3309C and 1103P rootstocks to the 'Cabernet Sauvignon' grapevine, and the incidence of GEM infestation in the three combinations was much higher than that of other rootstock combinations. The incidence of the CS/5C combination was the lowest, reaching just 9%. The injury index reflects the severity of the injury. Comparing the injury index of leaves in each rootstock–scion combination identifies the CK as the most seriously damaged, and the resistance type of the variety is injury (I). The SO4, 5C, and 140R rootstocks were able to significantly improve the GEM resistance of the 'Cabernet Sauvignon' grapevine, and the corresponding morbidity and injury indexes were much lower than those of the CK. The SO4 and 140R varieties were determined to have injury resistance (R). The CS/5C combination had the lowest infestation degree and injury index (4.89), and the strongest resistance to GEM, with a high resistance (HR) type.

Figure 1. GEM infection on leaves of different rootstocks of 'Cabernet Sauvignon' (18 August 2020).

Table 2. GEM field identification results of different rootstocks of 'Cabernet Sauvignon' grapevine leaves. The letters above infestation rate and injury index denote significant differences ($p < 0.05$) attested using Tukey's HSD test.

Processing Combination	Infestation Rate	Injury Index	Injury Resistance Degree
CS/CS	68% c	39.78 d	I
CS/3309C	65% c	33.11 cd	I
CS/1103P	66% c	34.22 d	I
CS/SO4	44% b	23.01 bc	R
CS/140R	29% b	14.00 ab	R
CS/5C	9% a	4.89 a	HR

3.2. Effects of Different Rootstocks on Activities of SOD, PPO, CAT, and POD in Leaves of 'Cabernet Sauvignon' under Different Levels of GEM Injury

3.2.1. Analysis of SOD Activity in Leaves of 'Cabernet Sauvignon' with Different Rootstocks and GEM Injury Levels

CS/140R, CS/5C, CS/SO4, and CS/1103P can significantly increase the SOD activity of scion leaves compared with CS/CS (Figure 2). The SOD activity of the CS/5C and CS/140R combinations with different rootstock and scion combinations was higher in three injury grades. Compared with CS/CS, the combinations CS/1103P and CS/3309C were unable to significantly improve SOD activity. At the injury level of 1, the SOD activity of each rootstock–scion combination increased rapidly, with CS/CS (49.03 U/(g·min)) and CS/1103P (4.81 U/(g·min)) exhibiting the highest increase and activity (27%). With the exception of CS/CS, the activities of SOD in other rootstock and scion combinations increased again at the injury level of 2. Following this, the SOD activity of each rootstock–scion combination decreased in the third stage of injury.

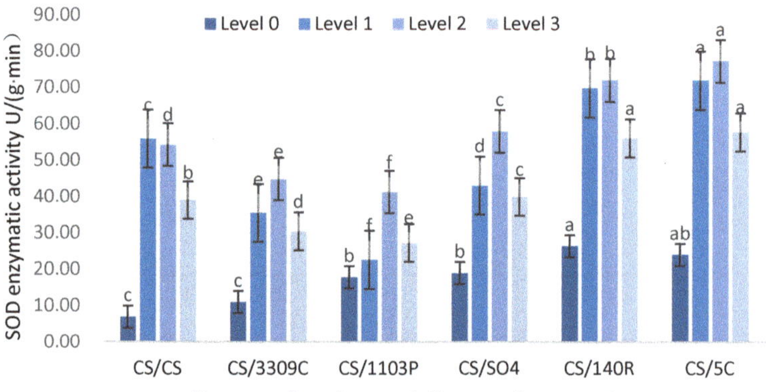

Figure 2. Analysis of SOD activity in leaves of different rootstocks of a 'Cabernet Sauvignon' grapevine in different injury grades of GEM. Lowercase letters above the bars denote significant differences ($p < 0.05$) attested using Tukey's HSD test. Error bars represent SE (n = 3).

3.2.2. Analysis of PPO Enzyme Activity in Leaves of 'Cabernet Sauvignon' with Different Rootstocks and GEM Injury Grades

Significant differences were observed in the PPO activity of 'Cabernet Sauvignon' grapevine leaves with different rootstocks (Figure 3). The PPO activity of the same rootstock–scion combination under different injury grades also exhibited differences. Each rootstock–scion combination exhibited significantly higher PPO activity than that of CS/CS in different injury grades, and the CS/5C activity was the highest among the three injury grades, followed by CS/140R. The PPO activity of leaves of the same rootstock–scion combination increased rapidly after being infested with GEM, and subsequently decreased with the increasing injury grade. In the healthy period, the PPO enzyme activities of CS/5C and CS/140R were higher in six combinations (the enzyme activities were all over 100 U/(g·min)), and in particular, they were 128.61% and 86% higher than those of CS/CS, respectively. At the GEM injury level 1, the PPO enzyme of CS/5C was significantly higher than other rootstock–scion combinations of the same level, reaching 209.85 U/(g·min). Compared with the healthy period, the PPO activity of CS/5C and CS/140R in the six rootstock combinations at level 1 did not exceed 100% (43% and 62%, respectively), yet their enzyme activities were higher in all rootstock combinations (at 209.85 U/(g·min) and 193.17 U/(g·min), respectively). CS/SO4 exhibited the highest increase (109%), with an enzyme activity that ranked third among the six combinations in the same grade. At the injury level of 2, CS/140R, CS/5C, and CS/3309C presented the lowest decrease in enzyme activity, while those with the highest activity were CS/5C, CS/140R, and CS/309C. At the grade 3 injury level, the enzyme activities of CS/CS, CS/140R, and CS/5C decreased the least compared with those of grade 2, while the top three combinations for PPO enzyme activity in descending order were CS/5C (166.44 U/(g·min)), CS/140R (152.44 U/(g·min)), and CS/3309C (120.48 U/(g·min)).

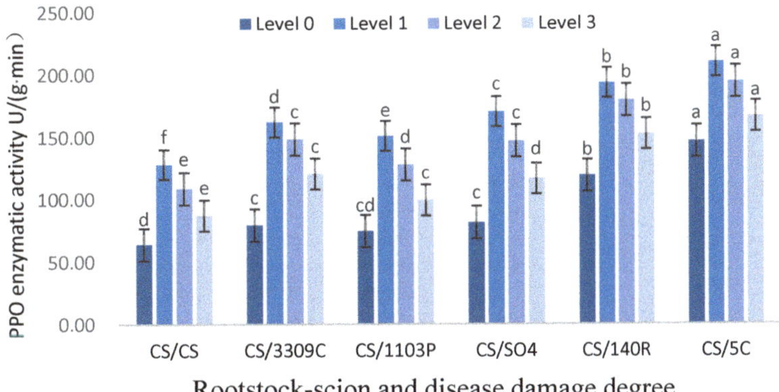

Figure 3. Analysis of PPO enzyme activity of different rootstocks of 'Cabernet Sauvignon' grapevine leaves in different injury grades of GEM. Lowercase letters above the bars denote significant differences ($p < 0.05$) attested using Tukey's HSD test. Error bars represent SE (n = 3).

3.2.3. Analysis of CAT Enzyme Activity in Leaves of 'Cabernet Sauvignon' with Different Rootstocks and GEM Injury Grades

Prior to the GEM injury, CS/SO4 could significantly improve CAT enzyme activity compared with CS/CS, while other rootstocks did not have any obvious effects on the CAT enzyme activity of the scion (Figure 4). However, following the infestation, the CAT activity increased greatly, with varying degrees of differences among the rootstock–scion combinations. Compared with the control, different rootstocks improved the CAT enzyme activity of the scion at varying degrees. After the infestation, the CAT activity of each rootstock–scion combination increased rapidly and subsequently decreased with the increasing injury grade. The CS/5C enzyme activity was the highest among all rootstock and scion combinations across the three injury grades. Compared with the healthy period, the CAT activity of each rootstock and scion combination increased by: (i) 81.19 U/(g·min) (22.14 times) for CS/CS; (ii) 140.97 U/(g·min) (32.75 times) for CS/3309C; (iii) 139.82 U/(g·min) (30.01 times) for cs/1103p; (iv) 245.83 U/(g·min) (47.65 times) for CS/5C; (v) 206.82 U/(g·min) (41.95 times) for CS/140R; and (vi) 245.83 U/(g·min) for CS/5C, which was the highest CAT activity.

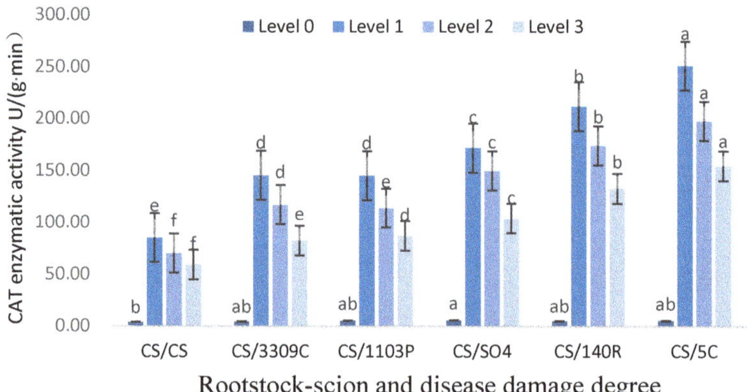

Figure 4. Analysis of CAT enzyme activity of different rootstocks of 'Cabernet Sauvignon' grapevine leaves in different injury grades of GEM. Lowercase letters above the bars denote significant differences ($p < 0.05$) attested using Tukey's HSD test. Error bars represent SE (n = 3).

3.2.4. Analysis of POD Enzyme Activity in Leaves of 'Cabernet Sauvignon' with Different Rootstocks and GEM Injury Grades

CS/SO4, CS/5C, and CS/140R were observed to affect the POD enzyme activity of scion (CS) leaves to different degrees compared with the control (Figure 5). In addition, the POD enzyme activity of CS/5C was significantly higher than that of other rootstock combinations at all levels of GEM injury, with a maximum enzyme activity during the healthy period (105.09 U/(g·min)) between groups and within groups. The POD activity in each combination decreased with the increasing injury grade. The POD enzyme activities of CS/CS, CS/3309C, and CS/1103P initially increased at the injury level of 1, and subsequently decreased as injury level increased. CS/SO4 and CS/140R did not exert any obvious changes in the POD activity at the injury level of 1 compared with the healthy period. The POD activity for these two combinations also exhibited a downward trend with the increasing injury level.

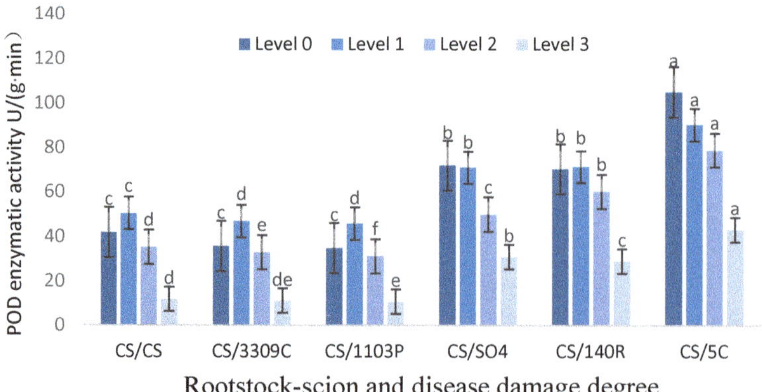

Figure 5. Analysis of POD enzyme activity of different rootstocks of 'Cabernet Sauvignon' grapevine leaves in different injury grades of GEM. Lowercase letters above the bars denote significant differences ($p < 0.05$) attested using Tukey's HSD test. Error bars represent SE (n = 3).

3.3. Effects of Different Rootstocks on the Relative Expression of the Res and RS Genes in Leaves of 'Cabernet Sauvignon' Grapevine under Different GEM Injury Levels

3.3.1. Changes of Res Content in GEM Leaves of 'Cabernet Sauvignon' with Different Rootstocks and Injury Grades

The rootstocks increased the contents of Res and Pd in 'Cabernet Sauvignon' leaves by varying degrees (Figure 6). At the grade 1 infection level, the Res content in the leaves of each rootstock–scion combination increased rapidly, with increasing rates of 486% (CS/CS), 476% (CS/3309C), 422% (CS/1103P), 321% (CS/5C), 256% (CS/SO4), and 195% (CS/140R). CS/140R exhibited the highest Res content, with a value that was significantly different from other rootstock combinations, followed by CS/5C. At the injury grade of 2, the Res content of CS/SO4 increased greatly for the second time, reaching 12.83 µg/g, which is 3.60 µg/g higher than that of the first grade (by 39%), and significantly different from other rootstock and scion combinations in the same injury grade. The Res content of other combinations decreased or remained stabled. The Res content of CS/3309C increased slightly at the injury grade of 3, exceeding the grade 2 value by 0.65 µg/g to reach 10.31 µg/g (increase of 7%). This was significantly different from other rootstock combinations at the same grade.

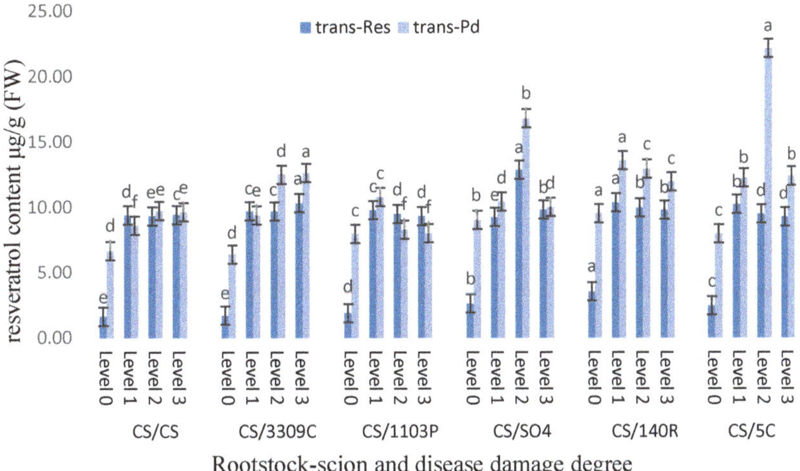

Figure 6. Changes of Res and Pd contents in GEM leaves of different rootstocks of 'Cabernet Sauvignon'. Lowercase letters above the bars denote significant differences ($p < 0.05$) attested using Tukey's HSD test. Error bars represent SE (n = 3).

The Pd content exhibited marked changes in different rootstock–scion combinations and injury grades. At the injury level of 1, the Pd content in leaves of 'Cabernet Sauvignon' grapevine with different rootstocks increased greatly. At grade 2, the changes in Pd content for different rootstock and scion combinations began to show variations. The CS/5C content was the highest (22.12 µg/g).

3.3.2. Expression of the RS Gene under Different GEM Injury Grades in the 'Cabernet Sauvignon' Leaves of Different Rootstocks

Resveratrol synthase, as the last enzyme in the pathway of Res synthesis, has a strong relationship with the content of Res. Thus, we measured the relative expression of the *RS* gene in different samples (Figure 7). During the healthy period, the rootstocks could increase the relative expression of the *RS* gene in scion leaves to different degrees, with CS/140R exerting the most significant impact. The relative expression of the *RS* gene in each rootstock combination increased rapidly at the injury level of 1, and CS/140R exhibited the highest (and significantly different) expression level among all rootstock combinations, followed by CS/5C. As the injury grade increased, the relative expression of the *RS* gene in different rootstock–scion combinations varied. The relative expression of the *RS* gene in the same rootstock combination is not completely consistent with the change in Res content (cf. Figure 6).

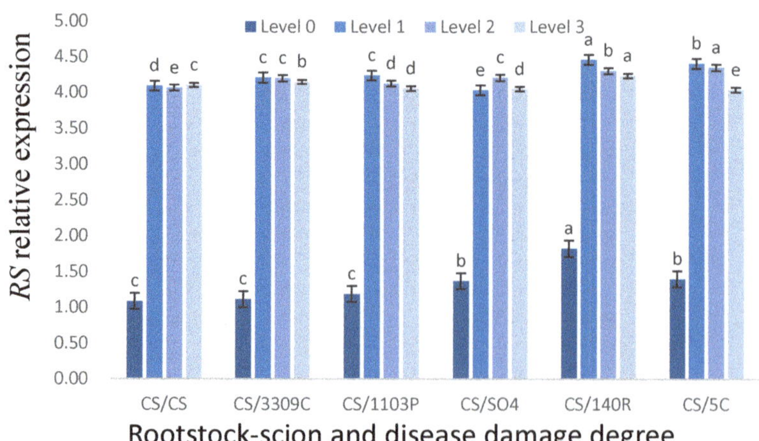

Figure 7. Relative expression of the *RS* gene of GEM in different injury grades of different rootstocks of 'Cabernet Sauvignon' grapevine leaves. Lowercase letters above the bars denote significant differences ($p < 0.05$) attested using Tukey's HSD test. Error bars represent SE (n = 3).

3.4. Correlation Analysis between Res Content, Resistance Enzyme Activity, and Injury Index of 'Cabernet Sauvignon' Grapevine Leaves with Different Rootstocks

Correlation analysis of the Res content, injury index and SOD, POD, PPO, and CAT activities of each rootstock–scion combination sample under injury grade 1 was performed (Figure 8). The distribution of the data was tested, and all the data combinations exhibited a skewness <2 and kurtosis <8, thus conforming to the normal distribution. The content of Res, Pd, and SOD in the leaves of each rootstock–scion combination was moderately negatively correlated with the injury index, with correlation values of -0.66, -0.83, and -0.68, respectively. There was a high negative correlation between POD activity and injury index (r = -0.93). The PPO and CAT activities were significantly negatively correlated with the injury indexes (-0.96 and -0.96, respectively), reaching significant levels. Thus, combinations with high CAT and PPO activity in scion grapevine leaves can effectively resist GEM infection.

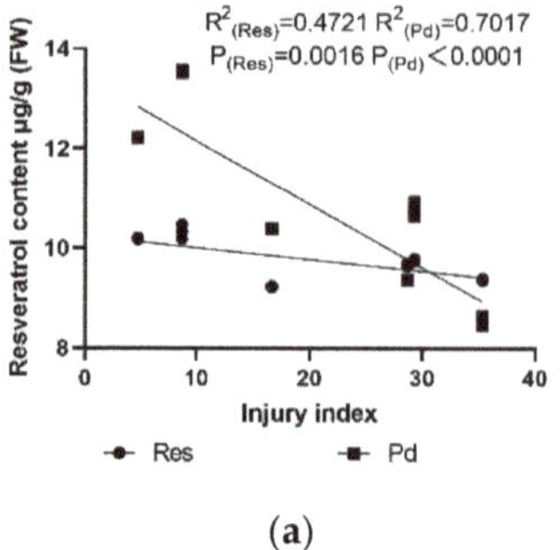

(a)

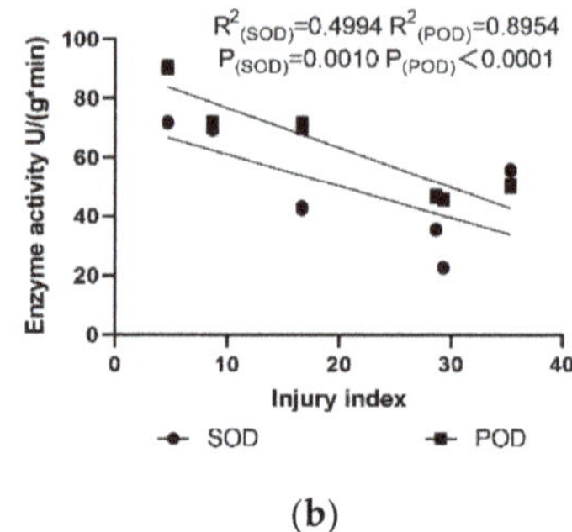

(b)

Figure 8. *Cont.*

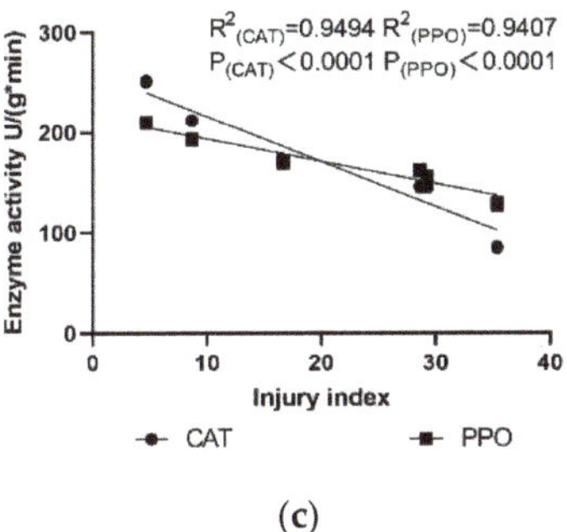

(**c**)

Figure 8. Correlation Analysis. (**a**) Resveratrol, polydatin and injury index. (**b**) SOD, POD, and injury index. (**c**) CAT, PPO, and injury index. Note: $0.5 < |r| \leq 0.8$, moderate correlation; $0.8 \leq |r| < 0.95$, highly correlated; $|r| \geq 0.95$, significant correlation.

4. Discussion

The grape is an important economic berry. In the process of studying the injury and insect resistance of grapevine rootstocks, it is indispensable to identify the injury and insect resistance of rootstocks. Field identification can accurately reflect the natural injury and insect resistance of grapevines. In this experiment, 'Cabernet Sauvignon' grafted with different rootstocks was investigated under the condition of artificial inoculation of GEM in the field. The incidence and injury index of the CS/5C, CS/140R, and CS/SO4 combinations caused by GEM infection were lower than those of other combinations. 5C, 140R, and SO4 rootstocks can significantly improve the GEM resistance of 'Cabernet Sauvignon' grapevines. Through the determination of the activity of four enzymes and the content of resveratrol in different GEM injury grades, it was confirmed that grafted rootstocks could indirectly affect the resistance of scions to GEM through secondary metabolites and enzyme activities.

Previous research shows that the activities of SOD, CAT, POD, and other enzymes are up-regulated after plants are subjected to abiotic or biotic stress [30–32]. The current study also found that the activities of four protective enzymes increased rapidly after being injured by GEM, but there were significant differences in enzyme activities at different stages of injury. The process adopted by sap-sucking pests when injuring plant leaves with probe mouthparts is similar to that of the invasion of plants by pathogens [33]. Therefore, there may be some similarities between the grapevine defense response activated by GEM invading grapevine leaves and the induced results of fungal injuries. After mites feed, leaves are damaged and reactive oxygen species (ROS) begin to accumulate. In order to prevent lipid peroxidation, SOD, CAT, and POD begin to participate in the reaction and finally convert O^{2-} into H_2O and O_2 [34,35]. In addition, POD can thicken the cell wall by participating in the cork and lignification of plant cells [36]. As the first barrier against pathogens, the change in the cell wall polymer can strengthen the structure of the cell wall [37]. This process can make it more difficult for gall mites to pierce cells. In addition, grapevine varieties with a thick waxy layer and high waxy content also have high resistance to GEM [38]. After infestation, plants will also increase the content of phenolic compounds in the feeding substances of gall mites, thus reducing the palatability of leaves and preventing gall mites from eating [39]. PPO can oxidize polyphenols in plants into quinones, which can inhibit the growth of pathogenic bacteria. Moreover, quinones and phenolic compounds have inhibitory, antifeedant, and toxic effects on gall mites [40]. In the current study, the PPO activity of CS/5C and CS/140R maintained a high level at all stages, which may be related to the low injury index of the two combinations. We found the CAT activity of each rootstock–scion combination to be low in the healthy period, and only CS/SO4 was different from CS/CS, while the other three enzyme activities were higher in combinations with strong resistance to GEM. The CS/5C enzyme activities were the highest in the tested rootstock–scion combinations or had no significant difference with the highest combination. The strong correlation between CAT and the injury index may be a result of H_2O_2 accumulation. The activities of CAT, SOD, POD, etc., have been reported to increase significantly or more rapidly when varieties insensitive to GEM are infected by GEM [28]. This agrees with our results. In addition, we also observed the activities of POD, CAT, and PPO gradually decrease with the development of GEM infection.

Previous studies have revealed the impact of rootstocks on the Res content of 'Cabernet Sauvignon' [41,42]. In this study, we also determined significant differences in the Res content of healthy leaves among several rootstock–scion combinations. The content of Res in leaves of all combinations increased rapidly after being injured by GEM. The combination with the top three Res contents is considered to have a high resistance to GEM. Therefore, we speculated that the content of resveratrol in plants is related to GEM resistance. This was confirmed by the correlation analysis. Res, as a type of plant protection element with a high content in grapevine, plays an important role in the injury resistance of grapevines, and acts as a signal molecule in plant injury resistance. Res is accumulated in high concentrations in response to pathogen infection, and its derivatives are related to POD through peroxidase-mediated cell wall strengthening [17,24]. Studies have also shown that calmodulin synergistically regulates cell division and proliferation in plant tissues, for example, the root gall of *Arabidopsis thaliana* caused by the parasitic root nematode *M. incognita infection*. The root gall of a grapevine injured by grape phylloxera is related to this pathway [43,44]. Therefore, the blistering of leaves caused by gall mites may be related to calmodulin, and whether Res is involved in this process as a signal molecule requires further research. The infestation of pathogens can induce the change in antimicrobial active substances in plants, such as *Erwinia carotovora*, which increases the mustard oil content in *Arabidopsis thaliana* [45]. Fungal infection can induce the synthesis of Res in grapevines [46]. In this experiment, GEM infestation also induced a rapid increase in the Res content of leaves. The contents of Res in the CS/140R, CS/5C, and CS/SO4 combinations at a low infestation rate were all high. This indicates the influence of Res on the resistance of grapevines to GEM. Res was observed to be negatively correlated with the injury index. At present, there is no direct research evidence on the toxic effect of Res GEM, yet a series of injury-related responses mediated by Res may play a role in the resistance to GEM. The derivatives of Res, such as Pterostilbene, have a higher toxicity than Res, and may also play a toxic role in GEM. When a scion is infected with powdery mildew, the Res in the infected region is rapidly synthesized and accumulated. Res in rootstocks can be polarly transported through the phloem to the injured site of the scion [20]. Therefore, we speculate that the change in Res content in CS/3309C scion leaves is inconsistent with the change in the relative expression of the RS gene. This may be due to the upward transportation of Res in rootstock, which enhances the Res content in the scion. The relative expression of the RS gene is different among rootstock and scion combinations at the same injury level, which may also be affected by rootstocks. CS is a susceptible variety, and the Pd content in CS/SO4 and CS/5C increased rapidly when the injury grade was 2. This may be attributed to the glycosylation of Res provided by rootstocks into Pd by the scions. This implies that the source of Res in the scion leaves of each rootstock combination is not only the synthesis of the leaves themselves, but also the contribution of rootstocks. According to the changes of the relative *RS* expression in scion leaves with different grades of GEM injury in each rootstock combination, the rootstocks are able to regulate the expression of the RS gene through several pathways during GEM development.

5. Conclusions and Prospects

In this experiment, the GEM resistance of 'Cabernet Sauvignon' grafted with different rootstocks was investigated. The activities of SOD, CAT, POD, and PPO, and the content of resveratrol in grapevine leaves of different GEM injury grades were determined. The results reveal the ability of GEM to induce changes in resistance-related factors in grapevine (Cabernet Sauvignon) leaves, such as the Res and Pd contents, the relative expression of the

RS gene, and the increase in SOD, CAT, POD, and PPO enzyme activities. In the current study, the activities of the four enzymes, and the Res and Pd contents were negatively correlated with the injury index of the GEM infection. This can be used as the basis for judging the rootstock resistance to GEM. This work clarifies the injury resistance mechanism of the rootstock–scion interaction, and also provides a theoretical basis for selecting GEM-resistant rootstocks of grapevines in Xinjiang.

At present, although the varieties of grapevine resistant rootstocks are gradually increasing, the breeding and excavation of the composite resistance of rootstocks are not at the required level. This leads to limitations and selectivity in the related applications. In recent years, the incidence of grapevine injuries and insect pests in various regions has become more and more complicated, which has brought severe challenges to the sustainable development of the industry. As an organic whole, the resistance of grafted plants to injuries and pests is a result of rootstocks and scions. This offers a powerful approach to reduce the use of fertilizer and medicine, improve plant quality, and employ the resistance of rootstock varieties to injuries and pests by selecting the best rootstock–scion combination with multiple resistances for key grapevine varieties. In the follow-up study, we will combine scanning electron microscopy, transmission electron microscopy, and metabolomics to conduct in-depth research.

Author Contributions: W.S.: methodology, investigation, formal analysis, writing—original draft preparation, and visualization; W.H.: methodology and writing—review and editing; Z.Z.: investigation, data curation, and software; C.Z.: writing—review and editing and visualization; Z.L.: data curation and visualization; Y.X.: writing—review and editing; J.S.: writing—review and editing, project administration, and supervision; B.Z.: conceptualization, methodology, writing—review and editing, supervision, and funding acquisition. All authors have read and agreed to the published version of the manuscript.

Funding: This work was supported by the Natural Science Foundation of China, grant numbers 31560542, 32060647 and 32060648.

Institutional Review Board Statement: This study does not involve human or animal research.

Informed Consent Statement: Research that does not involve humans.

Data Availability Statement: Data generated or analyzed during this study are included in this article.

Acknowledgments: Thanks to Zhengzhou Fruit Research Institute for providing grapevine rootstocks and scions. Thanks to EditorBar (www.editorbar.com) for providing English editing services during the preparation of this manuscript.

Conflicts of Interest: The authors declare no conflict of interest.

References

1. Riaz, S.; Pap, D.; Uretsky, J.; Laucou, V.; Boursiquot, J.-M.; Kocsis, L.; Andrew Walker, M. Genetic diversity and parentage analysis of grape rootstocks. *Theor. Appl. Genet.* **2019**, *132*, 1847–1860. [CrossRef] [PubMed]
2. Klimek, K.; Postawa, K.; Kapłan, M.; Kułażyński, M. Evaluation of the Influence of Rootstock Type on the Yield Parameters of Vines Using a Mathematical Model in Nontraditional Wine-Growing Conditions. *Appl. Sci.* **2022**, *12*, 7293. [CrossRef]
3. Moussa, A.; Quaglino, F.; Faccincani, M.; Serina, F.; Torcoli, S.; Miotti, N.; Passera, A.; Casati, P.; Mori, N. Grafting of recovered shoots reduces bois noir disease incidence in vineyard. *Crop Prot.* **2022**, *161*, 106058. [CrossRef]
4. Wallis Christopher, M.; Wallingford Anna, K.; Jianchi, C. Grapevine rootstock effects on scion sap phenolic levels, resistance to *Xylella fastidiosa* infection, and progression of Pierce's disease. *Front. Plant Sci.* **2013**, *4*, 502. [CrossRef]
5. Dandekar, A.M.; Jacobson, A.; Ibáñez, A.M.; Gouran, H.; Dolan, D.L.; Agüero, C.B.; Uratsu, S.L.; Just, R.; Zaini, P.A. Trans-graft protection against Pierce's disease mediated by transgenic grapevine rootstocks. *Front. Plant Sci.* **2019**, *10*, 84. [CrossRef] [PubMed]
6. Provost, C.; Campbell, A.; Dumont, F. Rootstocks impact yield, fruit composition, nutrient deficiencies, and winter survival of hybrid cultivars in eastern Canada. *Horticulturae* **2021**, *7*, 237. [CrossRef]
7. Zongmin, F.; Junli, S.; Baolong, Z.; Huaifeng, L.; Kun, Y.; Zhijun, Z.; Jingjing, L. Evaluationofcoldresistanceofone-yearshoots from'CabernetSauvignon'grape vine grafted on different rootstocks. *J. Fruit Sci.* **2020**, *37*, 215–225. [CrossRef]
8. Carew, M.E.; Goodisman, M.A.; Hoffmann, A.A. Species status and population genetic structure of grapevine eriophyoid mites. *Entomol. Exp. Appl.* **2004**, *111*, 87–96. [CrossRef]

9.   Cooper, M.; Hobbs, M.; Strode, B.; Varela, L. Grape erineum mite: Postharvest sulfur use reduces subsequent leaf blistering. *Calif. Agric.* **2020**, *74*, 94–100. [CrossRef]

10. Kunugi, Y.; Asari, S.; Terai, Y.; Shinkai, A. Studies on the grapevine berry inner necrosis virus disease, 2: Transmission of grapevine berry inner necrosis virus by the grape erineum mite, *Colomerus vitis* in Yamanashi [Japan]. *Bull. Yamanashi Fruit Tree Exp. Stn.* **2000**, *10*, 57–63.

11. Malagnini, V.; de Lillo, E.; Saldarelli, P.; Beber, R.; Duso, C.; Raiola, A.; Zanotelli, L.; Valenzano, D.; Giampetruzzi, A.; Morelli, M. Transmission of grapevine *Pinot gris virus* by *Colomerus vitis* (Acari: *Eriophyidae*) to grapevine. *Arch. Virol.* **2016**, *161*, 2595–2599. [CrossRef] [PubMed]

12. Feng, L.; Sun, J.; Jiang, Y.; Duan, X. Role of Reactive Oxygen Species against Pathogens in Relation to Postharvest Disease of Papaya Fruit. *Horticulturae* **2022**, *8*, 205. [CrossRef]

13. El-Gendi, H.; Al-Askar, A.A.; Király, L.; Samy, M.A.; Moawad, H.; Abdelkhalek, A. Foliar Applications of Bacillus subtilis HA1 Culture Filtrate Enhance Tomato Growth and Induce Systemic Resistance against Tobacco mosaic virus Infection. *Horticulturae* **2022**, *8*, 301. [CrossRef]

14. Liu, C.; Chen, L.; Zhao, R.; Li, R.; Zhang, S.; Yu, W.; Sheng, J.; Shen, L. Melatonin induces disease resistance to *Botrytis cinerea* in tomato fruit by activating jasmonic acid signaling pathway. *J. Agric. Food Chem.* **2019**, *67*, 6116–6124. [CrossRef]

15. Zhang, Q.; Yong, D.; Zhang, Y.; Shi, X.; Li, B.; Li, G.; Liang, W.; Wang, C. Streptomyces rochei A-1 induces resistance and defense-related responses against *Botryosphaeria dothidea* in apple fruit during storage. *Postharvest Biol. Technol.* **2016**, *115*, 30–37. [CrossRef]

16. Zhang, Y.; Huber, D.J.; Hu, M.; Jiang, G.; Gao, Z.; Xu, X.; Jiang, Y.; Zhang, Z. Delay of postharvest browning in litchi fruit by melatonin via the enhancing of antioxidative processes and oxidation repair. *J. Agric. Food Chem.* **2018**, *66*, 7475–7484. [CrossRef]

17. Wingerter, C.; Eisenmann, B.; Weber, P.; Dry, I.; Bogs, J. Grapevine Rpv3-, Rpv10-and Rpv12-mediated defense responses against Plasmopara viticola and the impact of their deployment on fungicide use in viticulture. *BMC Plant Biol.* **2021**, *21*, 470. [CrossRef]

18. Yamaguchi, T.; Kurosaki, F.; Suh, D.-Y.; Sankawa, U.; Nishioka, M.; Akiyama, T.; Shibuya, M.; Ebizuka, Y. Cross-reaction of chalcone synthase and stilbene synthase overexpressed in *Escherichia coli*. *FEBS Lett.* **1999**, *460*, 457–461. [CrossRef]

19. Gatto, P.; Vrhovsek, U.; Muth, J.; Segala, C.; Romualdi, C.; Fontana, P.; Pruefer, D.; Stefanini, M.; Moser, C.; Mattivi, F. Ripening and genotype control stilbene accumulation in healthy grapes. *J. Agric. Food Chem.* **2008**, *56*, 11773–11785. [CrossRef]

20. Li, R.; Xie, X.; Ma, F.; Wang, D.; Wang, L.; Zhang, J.; Xu, Y.; Wang, X.; Zhang, C.; Wang, Y. Resveratrol accumulation and its involvement in stilbene synthetic pathway of Chinese wild grapes during berry development using quantitative proteome analysis. *Sci. Rep.* **2017**, *7*, 9295. [CrossRef]

21. Wang, Z.; Wang, Y.; Cao, X.; Wu, D.; Hui, M.; Han, X.; Yao, F.; Li, Y.; Li, H.; Wang, H. Screening and Validation of SSR Molecular Markers for Identification of Downy Mildew Resistance in Intraspecific Hybrid F_1 Progeny (*V. vinifera*). *Horticulturae* **2022**, *8*, 706. [CrossRef]

22. Mengqi, L.; Fuli, M.; Fengying, W.; Changyue, J.; Yuejin, W. Expression of stilbene synthase VqSTS6 from wild Chinese Vitis quinquangularis in grapevine enhances resveratrol production and powdery mildew resistance. *Planta* **2019**, *250*, 1997–2007. [CrossRef]

23. Zhijun, Z.; Huaifeng, L.; Junli, S.; Baolong, Z.; Lizhong, P.; Wang, H.; Jingjing, L. Analysis of Differential Metabolites in Cabernet Sauvignon Skins from Different Rootstock-Scion Combinations by Non-Targeted Metabolomics. *Food Sci.* **2020**, *41*, 22–30. [CrossRef]

24. Adrian, M.; Jeandet, P.; Veneau, J.; Weston, L.A.; Bessis, R. Biological activity of resveratrol, a stilbenic compound from grapevines, against *Botrytis cinerea*, the causal agent for gray mold. *J. Chem. Ecol.* **1997**, *23*, 1689–1702. [CrossRef]

25. Calderon, A.; Pedreno, M.; Barcelo, A.R.; Munoz, R. A spectrophotometric assay for quantitative analysis of the oxidation of 4-hydroxystilbene by peroxidase-H2O2 systems. *J. Biochem. Biophys. Methods* **1990**, *20*, 171–180. [CrossRef]

26. Pezet, R.; Gindro, K.; Viret, O.; Richter, H. Effects of resveratrol, viniferins and pterostilbene on *Plasmopara viticola* zoospore mobility and disease development. *VITIS-GEILWEILERHOF* **2004**, *43*, 145–148.

27. Pezet, R.; Gindro, K.; Viret, O.; Spring, J.-L. Glycosylation and oxidative dimerization of resveratrol are respectively associated to sensitivity and resistance of grapevine cultivars to downy mildew. *Physiol. Mol. Plant Pathol.* **2004**, *65*, 297–303. [CrossRef]

28. Javadi Khederi, S.; Khanjani, M.; Gholami, M.; Panzarino, O.; de Lillo, E. Influence of the erineum strain of *Colomerus vitis* (Acari: *Eriophyidae*) on grape (*Vitis vinifera*) defense mechanisms. *Exp. Appl. Acarol.* **2018**, *75*, 1–24. [CrossRef]

29. Reid, K.E.; Olsson, N.; Schlosser, J.; Peng, F.; Lund, S.T. An optimized grapevine RNA isolation procedure and statistical determination of reference genes for real-time RT-PCR during berry development. *BMC Plant Biol.* **2006**, *6*, 27. [CrossRef]

30. Afzal, F.; Khurshid, R.; Ashraf, M.; Kazi, A.G. Reactive oxygen species and antioxidants in response to pathogens and wounding. In *Oxidative Damage to Plants*; Elsevier: Amsterdam, The Netherlands, 2014; pp. 397–424.

31. Zhang, L.; Yu, Y.; Chang, L.; Wang, X.; Zhang, S. Melatonin enhanced the disease resistance by regulating reactive oxygen species metabolism in postharvest jujube fruit. *J. Food Process. Preserv.* **2022**, *46*, e16363. [CrossRef]

32. Zhu, Y.; Guo, M.-J.; Song, J.-B.; Zhang, S.-Y.; Guo, R.; Hou, D.-R.; Hao, C.-Y.; An, H.-L.; Huang, X. Roles of endogenous melatonin in resistance to *Botrytis cinerea* infection in an Arabidopsis model. *Front. Plant Sci.* **2021**, *12*, 683228. [CrossRef] [PubMed]

33. Walling, L.L. The myriad plant responses to herbivores. *J. Plant Growth Regul.* **2000**, *19*, 195–216. [CrossRef] [PubMed]

34. Mittler, R.; Vanderauwera, S.; Gollery, M.; Van Breusegem, F. Reactive oxygen gene network of plants. *Trends Plant Sci.* **2004**, *9*, 490–498. [CrossRef] [PubMed]

35. Zhu, X.; Shi, X.; Yong, D.; Zhang, Y.; Li, B.; Liang, W.; Wang, C. Induction of resistance against Glomerella cingulata in apple by endophytic actinomycetes strain A-1. *Plant Physiol. Comm* **2015**, *51*, 949–954.
36. Mohammadi, M.; Kazemi, H. Changes in peroxidase and polyphenol oxidase activities in susceptible and resistant wheat heads inoculated with *Fusarium graminearum* and induced resistance. *Plant Sci.* **2002**, *162*, 491–498. [CrossRef]
37. Gao, Y.; Yin, X.; Jiang, H.; Hansen, J.; Jørgensen, B.; Moore, J.P.; Fu, P.; Wu, W.; Yang, B.; Ye, W. Comprehensive Leaf Cell Wall Analysis Using Carbohydrate Microarrays Reveals Polysaccharide-Level Variation between Vitis Species with Differing Resistance to Downy Mildew. *Polymers* **2021**, *13*, 1379. [CrossRef]
38. Khederi, S.J.; Khanjani, M.; Fayaz, B.A. Resistance of three grapevine cultivars to Grape Erineum Mite, *Colomerus vitis* (Acari: Eriophyidae), in field conditions. *Persian J. Acarol.* **2014**, *3*. [CrossRef]
39. Petanović, R.; Kielkiewicz, M. Plant–eriophyoid mite interactions: Cellular biochemistry and metabolic responses induced in mite-injured plants. Part I. *Exp. Appl. Acarol.* **2010**, *51*, 61–80. [CrossRef]
40. Akhtar, Y.; Isman, M.B.; Lee, C.-H.; Lee, S.-G.; Lee, H.-S. Toxicity of quinones against two-spotted spider mite and three species of aphids in laboratory and greenhouse conditions. *Ind. Crops Prod.* **2012**, *37*, 536–541. [CrossRef]
41. Wang, H.; Junli, S.; Baolong, Z.; Zhijun, Z.; Lianling, L.; Lizhong, P.; Xinyi, C. Effects of rootstocks on the content of resveratrol and relative enzyme activities in 'Cabernet Sauvignon' grape. *J. Fruit Sci.* **2019**, *6*, 738–747. [CrossRef]
42. Zhang, Z.; Sun, J.; Zhao, S.; Lu, Q.; Pan, L.; Zhao, B.; Yu, S. Effects of different rootstocks on phenolics in the skin of 'Cabernet Sauvignon' and widely targeted metabolome and transcriptome analysis. *Hortic. Res.* **2022**, *9*, uhac053. [CrossRef] [PubMed]
43. Samira, H.; Carolyn, A.B.; Ulrike, M. Effectors of plant parasitic nematodes that re-program root cell development. *Funct. Plant Biol.* **2010**, *37*, 933–942.
44. Jaouannet, M.; Magliano, M.; Arguel, M.J.; Gourgues, M.; Evangelisti, E.; Abad, P.; Rosso, M.-N. The root-knot nematode calreticulin Mi-CRT is a key effector in plant defense suppression. *Mol. Plant-Microbe Interact.* **2013**, *26*, 97–105. [CrossRef] [PubMed]
45. Brader, G.; Tas, E.; Palva, E.T. Jasmonate-dependent induction of indole glucosinolates in Arabidopsis by culture filtrates of the nonspecific pathogen Erwinia carotovora. *Plant Physiol.* **2001**, *126*, 849–860. [CrossRef]
46. Schnee, S.; Viret, O.; Gindro, K. Role of stilbenes in the resistance of grapevine to powdery mildew. *Physiol. Mol. Plant Pathol.* **2008**, *72*, 128–133. [CrossRef]

sustainability

Article

Peppers under Siege: Revealing the Prevalence of Viruses and Discovery of a Novel Potyvirus Species in Venezuela

Eduardo Rodríguez-Román [1], Yrvin León [1], Yearlys Perez [1], Paola Amaya [1], Alexander Mejías [1], Jose Orlando Montilla [2], Rafael Ortega [1], Karla Zambrano [3], Barlin Orlando Olivares [4,*] and Edgloris Marys [1,*]

[1] Laboratorio de Biotecnología y Virología Vegetal, Centro de Microbiologia y Biologia Celular, Instituto Venezolano de Investigaciones Científicas (IVIC), Caracas 1204, Venezuela; erodriguezroman@gmail.com (E.R.-R.); yrvinln@gmail.com (Y.L.); yearlys@gmail.com (Y.P.); amayapao2011@gmail.com (P.A.); alexandermejias@gmail.com (A.M.); rafaeskeo@gmail.com (R.O.)
[2] Decanato de Agronomía, Cátedra de Fitopatología, Universidad Centroccidental Lisandro Alvarado (UCLA), Barquisimeto 3001, Venezuela; jmontilla@ucla.edu.ve
[3] Decanato de Agronomía, Cátedra de Biología, Universidad Centroccidental Lisandro Alvarado (UCLA), Barquisimeto 3001, Venezuela; karlazambrano07@gmail.com
[4] Biodiversity Management Research Group (GESBIO-UCO), Rabanales Campus, University of Cordoba (UCO), Carretera Nacional IV, km 396, 14014 Cordoba, Spain
* Correspondence: ep2olcab@uco.es (B.O.O.); edgloris@gmail.com (E.M.)

Citation: Rodríguez-Román, E.; León, Y.; Perez, Y.; Amaya, P.; Mejías, A.; Montilla, J.O.; Ortega, R.; Zambrano, K.; Olivares, B.O.; Marys, E. Peppers under Siege: Revealing the Prevalence of Viruses and Discovery of a Novel Potyvirus Species in Venezuela. *Sustainability* **2023**, *15*, 14825. https://doi.org/10.3390/su152014825

Academic Editor: Haicheng Zhang

Received: 29 August 2023
Revised: 7 October 2023
Accepted: 10 October 2023
Published: 12 October 2023

Abstract: Many plant virus outbreaks have been recorded in the last two decades, threatening food security around the world. During pepper production seasons in 2008, 2014, and 2022, virus outbreaks were reported from Lara (western) and Miranda (central) states in Venezuela. Three hundred seventy-three plants exhibiting virus-like symptoms were collected and tested for virus infection through reverse transcription PCR (RT-PCR). The most prevalent viruses during the 2008 surveys conducted in Lara were potato virus Y (PVY, 66.25%), cucumber mosaic virus (CMV, 57.50%), pepper mild mottle virus (PMMoV, 35%), alfalfa mosaic virus (AMV, 23.75%), and tobacco rattle virus (TRV, 17.50%). This survey revealed for the first time that pepper is a natural host of AMV and TRV in Venezuela. A further, divergent potyvirus isolate was also detected in 23.75% of pepper plants from Lara state. In 2014, a follow-up survey after virus outbreaks reported in Lara and Miranda states also detected this divergent potyvirus isolate in 21.68% of pepper plants, with tomato spotted wilt virus (TSWV) and PMMoV dominating the viral landscape (62.65 and 21.68% of tested plants, respectively). By comparison, the surveys revealed significant changes in viral community composition. The complete capsid protein (CP) sequence of the putative potyvirus was obtained from two pepper samples. According to the *Potyvirus* taxonomic criteria, these results suggest that the isolate represents a distinct virus species, for which we propose the name "pepper severe mottle virus" (PepSMoV). Virus outbreaks could be attributed to agricultural and environmental factors, such as climate change, the use of wastewater, the use of uncertified seeds, misuse of agricultural chemicals, transmission with food trade networks, and the development of new viral strains due to mutations and recombination and pathogen spillover. This study demonstrates the value of knowledge of the prevalence and distribution of viral species to recommend virus-resistant cultivars to replace susceptible ones, especially in virus hotspot areas.

Keywords: diagnostic; disease; pepper; ribonucleic acid; survey; viruses

1. Introduction

Pepper (*Capsicum annuum* L.) belongs to the family *Solanaceae* and is an important crop worldwide [1,2]. Pepper production in Venezuela for 2021 was 146,817 tons, with a harvested area of 10,126 ha and a yield of 144,996 hg/ha [3]. The pepper crop faces numerous challenges due to viral infections that can severely impact its production and quality on a global scale [4,5]. In the central and western regions of Venezuela, pepper cultivation serves as a significant agricultural activity, contributing to the nation's economy and food security. However, recent reports have raised concerns about the escalating prevalence of viral diseases in these areas, which is threatening the sustainability of pepper farming [6]. Hence, it is essential to identify the most commonly occurring and damaging viruses to recommend management strategies.

To date, 11 RNA viruses belonging to the genera *Cucumovirus, Ilarvirus, Nepovirus, Orthotospovirus, Potexvirus, Potyvirus, Tobamovirus,* and *Tobravirus* have been identified infecting pepper crops in Venezuela [6,7], mainly in the horticultural Lara state (located in the western region). Lara is the country's leading pepper producer where almost 70% of the production is achieved in the Quibor Valley. Multiple infections have been documented in the area, associated with an increased synergistic effect on the crop, leading to higher yield losses [7]. In 2008, viral outbreaks were reported by pepper producers in Quibor, where previously undescribed *Alfamovirus, Ilarvirus,* and *Potyvirus* species were detected.

During the spring of 2014, another major virus outbreak emerged in Quibor Valley (Lara) and in Altos Mirandinos (Miranda), the latter state now being considered the largest development pole of greenhouses dedicated to pepper production in Venezuela [8]. Pepper plants exhibiting virus-like symptoms were observed in different greenhouses throughout the states with incidences ranging from 21 to 62%. Remarkable pepper yield losses of up to 90% were associated with the emerging tomato spotted wilt virus (TSWV, *Orthotospovirus*) [6], causing field abandonment before harvest and making the cultivation of pepper not profitable. Preliminary observations by electron microscopy of crude sap from some of the diseased plants showed long flexuous filamentous particles of about 750 nm in length suggestive of viruses belonging to the genus *Potyvirus* (Appendix A, Figure A1). Moreover, the potyvirus "core" CP gene fragment obtained by RT-PCR from these plants exhibited 97–99% sequence identity with the above-mentioned isolates from 2008 surveys.

Given recurrent complaints from farmers facing heavy crop losses, molecular diagnostic assays were initiated to assess the relative importance of viruses infecting pepper in Lara and Miranda states. This paper aims to compare our findings during 2008, 2014 and 2022 pepper virus surveys in Lara and Miranda states. Also, in this study, we characterized the entire CP coding region of the newly isolated potyvirus. Phylogenetic analyses indicated that the virus was most closely related to pepper yellow mosaic virus (PepYMV). The results suggest that the virus should be classified as a novel species within the genus Potyvirus, which we tentatively name pepper severe mottle virus (PepSMoV). By shedding light on the viral landscape and uncovering this newly emerged pathogen, we aimed to enhance the understanding of the complex interactions between viruses and pepper plants, paving the way for effective management strategies and safeguarding the future of pepper cultivation.

2. Materials and Methods

2.1. Virus Surveys

Sample collection was carried out in Lara state in September 2008 and 2014 and in Miranda state in September 2014 and September 2022 (Table 1). Leaves and fruits showing virus-like symptoms (i.e., stunt, mottle, vein yellowing, mosaic, leaf distortion, light green leaves, necrotic spots on leaves or fruits, or distorted fruit showing brown streaks) were collected. Three or four apical leaves or fruits were collected from symptomatic crops. Immediately after collection, each sample was placed in a plastic bag, transported to the laboratory on ice, and stored at $-80\ ^{\circ}\text{C}$ or $-20\ ^{\circ}\text{C}$ pending analysis. Global positioning system (GPS) data were recorded at each greenhouse site, and a subsequent map was generated using QGIS v.3.28 [9].

Table 1. A number of pepper samples were collected from different regions of Venezuela.

Sampling State	Year	Localities	Farms	Number of Samples (*n*)
Lara	2008	Tintorero	F10 to F13	80
Lara	2014	Tintorero	F1 to F9	83
Miranda	2014	Pozo de Rosas San Pedro El Jarillo	F1 to F6	108
Miranda	2022	Pozo de Rosas, El Jarillo, Hoyo de La Puerta, La Reinosa	F7 to F10	92

2.2. RNA Extraction, RT-PCR

Total RNA was extracted from 100 mg of frozen plant tissue samples using the TRIzol^{TM} reagent (Life Technologies, Carlsbad, CA, USA) according to the manufacturer's recommendations. Approximately 1 µg of total RNA was used to generate first-strand cDNA with random, degenerated, or specific primers using a SuperScript III Reverse Transcription kit (Thermo Fisher Scientific Inc., Waltham, MA, USA) as reported [10–17] or designed in this study (Appendix A, Table A1). Each RT reaction contained about 200 ng of total RNA (2 µL), 1 µL of Random Oligo-dT (N6) primer or a specific reverse primer (10 µM), and 3 µL nuclease-free ddH$_2$O. RNAs from virus-infected samples used as positive controls were obtained from the Plant Virus Collection at IVIC. The PCR reaction (12.5 µL) contained 50 ng template cDNA, 10 pmol of each amplification primer, 200 mM each dNTP, 1.25U Taq DNA polymerase (Invitrogen, Carlsbad, CA, USA), 5 mM MgCl$_2$, and 10X Taq Polymerase PCR Buffer. Amplification parameters were as reported (Appendix A, Table A1). To determine the size of the amplified PCR products, a DNA ladder (Thermo Scientific, Waltham, MA, USA) was used. The RT-PCR products were examined by electrophoresis in 1% agarose gels containing ethidium bromide and examined and recorded using a Fotodyne UV/Digital camera transilluminator system.

2.3. Cloning and DNA Sequencing

Amplified PCR products were purified using the AccuPrep PCR Purification Kit (Bioneer, Seoul, Republic of Korea) and then cloned into the pGEM-T easy Vector System (Promega, Madison, WI, USA) according to standard methods [18]. Plasmid DNA preparations were obtained using the QIAprep Spin Miniprep Kit (Qiagen, Hilden, Germany). At least three clones representing a single PCR product were sequenced by Macrogen Inc. (Seoul, Republic of Korea). Sequences of the amplified PCR products were edited and assembled using MEGA7 [19]. The resulting sequences were used to BLAST search the sequence database at the National Center for Biotechnology Information (http://blast.ncbi.nlm.nih.gov/Blast.cgi, accessed on 5 March 2017).

2.4. Molecular Characterization of Putative Potyvirus Species

Some samples that tested positive only for the presence of a potyvirus with "core" CP primers showing 85% nt sequence identity to PepYMV were selected to amplify a larger portion of the genome, including the 3$'$ terminal sequence and the complete CP gene. The cDNA was synthesized using the primer B1570 Oligo(dT) complementary to the PepYMV polyadenylated tail. PCR was performed using the primer PY10, designed based on the pepper mottle virus (PepMoV) sequence, towards a conserved region in the nuclear inclusion body b (NIb) cistron, and B1570 [11] (Appendix A, Table A1). Expected fragments (ca. 1200 bp) comprising the CP and 3$'$ untranslated region (3$'$-UTR) of the potyvirus genome were cloned using the pGEM-T vector (Promega) according to the manufacturer's recommendations. Two clones were selected and sequenced using M13F/M13R primers. The nucleotide (nt) and predicted amino acid (aa) sequences of the whole CP coding region and the nt sequence corresponding to the 3′-UTR were compared with potyvirus sequences deposited in GenBank, EMBL, DDBJ, and TrEMBL databases using the pairwise Align program. Sequence assembly and analysis were performed utilizing the DNA Dragon–Contig Sequence Assembly Software v1.7.1 [20]. Multiple sequence alignments produced by the Clustal W algorithm were used as input data for reconstructing phylogenetic trees by the neighbor-joining method using the software MEGA version 4 [21] (Table A2). Statistical significance was estimated by performing 1000 replications of bootstrap resampling of the original alignment using the bootstrap option of the phylogenetic tree menu.

2.5. Mechanical Transmission of Putative Potyvirus Isolate

To demonstrate the infectivity of the new potyvirus, frozen leaf tissues (0.3 g) from a single-infected potyvirus sample, AMPIM8, were used for mechanical inoculation of 10-day-old seedlings of *C. annuum* cv. Magistral. Sap was extracted from frozen samples by grinding tissue of leaves into a cold 0.1 M potassium phosphate buffer pH 7.0 containing 1% magnesium trisilicate, using a pre-chilled mortar. The homogenate was inoculated by gently rubbing the bottom leaves of healthy, carborundum-dusted pepper seedlings. Mock-inoculated plants were used as controls. Plants were subsequently grown in an insect-proof greenhouse until symptom expression. Four weeks after inoculation, virus infection was tested by RT-PCR using PepSMoV-specific primer pair PepSMoV-CP-F/PepSMoV-CP-R (Table A1) followed by Sanger sequencing. The virus from *C. annuum* cv. Magistral was reinoculated onto *C. annuum* cv. Magistral and cv. Acero for the fulfillment of Koch's postulates. Plants were tested by RT-PCR and Sanger sequencing for PepSMoV infection.

2.6. Analysis of Principal Coordinates (PCoA)

The analysis of principal coordinates was applied to assess the distribution of viruses across various farms (designated as F1 to F13) in the Miranda and Lara states of Venezuela through InfoStat software v.11 [22]. The methodology involved encoding the presence and absence of different viruses, including PMMoV, TSWV, PepSMoV, AMV, CMV, TRV, and PVY, as binary values (1 for presence and 0 for absence), using the distance matrix obtained from the S transformation $(1–S_{ij})^{1/2}$ with S = Dice's similarity index, to which a minimum spanning tree was superimposed to facilitate the visualization of the ordering. These binary values were then subjected to principal coordinate analysis, which aimed to uncover patterns and relationships in the viral distribution among the surveyed farms.

3. Results

3.1. Field Survey Results

Pepper fields were surveyed for virus diseases in Lara state in 2008 and 2014 and in Miranda state in 2014 and 2022 (Table 1, Figure 1). A total number of 373 pepper samples were collected during the surveys. The most common symptoms on the infected pepper plants during the 2008 surveys in Lara were stunting, mottling or mosaic, yellowing, and distortion in leaves (Figure 1b, left), whereas in 2014, the main symptoms found were ringspots on leaves and fruits and necrosis of leaves (Figure 1b, right). In Miranda state, chlorotic line patterns with necrotic spots were observed in mature plants, often showing cupped downward leaves (Figure 1d, left). Severe stunting of younger plants with chlorotic mosaic or yellow flecking of the leaves was also observed during 2022 surveys (Figure 1d, right). Based on symptomatology, an average virus incidence of 70–100% was recorded during surveys.

Surveys from Lara state in 2008 revealed the presence of five viruses; PVY was the most prevalent virus (66.25%), followed by CMV (57.50%), PMMoV (35%), AMV (23.75%), and TRV (17.50%). In this survey, AMV (GenBank accession OR420758) and TRV (GenBank accession OR420759) were found for the first time in pepper fields in Venezuela. The primer pair MJ1–MJ2 used for potyvirus diagnosis also amplified a 327 bp fragment from 40 pepper samples (10.70%) that showed 85% nucleotide (nt) sequence identity to PepYMV.

Data from surveys in Lara state (2014) indicated that out of all samples showing virus-like symptoms, 75.9% (63/83) were positive for at least one virus. *Orthotospovirus* was the most common genus identified in 52 samples (62.6%), with TSWV being the only species found. *Tobamoviruses* were detected in 18 samples (21.6%), with PMMoV being the only species found. Potyvirus isolates sharing 85% nucleotide (nt) sequence identity to PepYMV, and 99–100% nt sequence identity to potyvirus isolates from 2008 in Lara, were also detected with the primer pair MJ1-MJ2 in 11 of the samples (13.2%), indicating relatively low prevalence. Single infections were more frequent than mixed infections (96.3 and 8.4%, respectively). TSWV and potyvirus coinfections were the most frequent mixed infections for all samples (4.8%), while triple infections (*Orthotospovirus–Tobamovirus–Potyvirus*) occurred at even lower percentages (3.6%). PMMoV and potyvirus coinfections were not recorded. A similar scenario occurred in Miranda in the same year, where TSWV had the highest overall prevalence (54.62%), followed by PMMoV (34.25%) and the divergent potyvirus species (23.14%). Double and triple infections were also recorded. These virus species (TSWV, PMMoV, and the divergent potyvirus) accounted for the viral population in Miranda during recent (2022) surveys.

Figure 1. Virus detection using field-collected pepper samples in Lara Farms (F1 to F13) (**a**) and Miranda Farms (F1 to F10) (**c**), Venezuela. (**b**) Pepper surveys in Lara showed mosaic, yellowing of veinlets, and leaf deformation (**left**) during 2008 surveys and chlorotic spots on fruits and leaves (**right**)

during 2014 surveys. (**d**) Pepper surveys in Miranda show cupped downward leaves and chlorotic spots on fruits (2014, (**left**)) and severe mottling (2022, (**right**)). Pictures by Edgloris Marys (Miranda and Trujillo, Venezuela).

The circular dendrogram (Figure 2a) presents data on the distribution of various viruses in different regions and farms within the Miranda and Lara states of Venezuela. In Lara state, the virus population structure varied across different farms (F1 to F13). For instance, TSWV was detected in varying numbers in different farms, with the highest detection being 15 out of 28 samples in Miranda (F1). Similarly, PMMoV, PepSMoV, AMV, CMV, TRV, and PVY were also detected in different proportions across these farms. In Miranda state, a similar pattern emerged. PMMoV, TSWV, and PepSMoV were detected across different farms (F1 to F6), with varying numbers of positive samples. In Miranda state (2022), farms F7 and F10, for example, appear to have higher virus prevalence compared to the others, indicating potential challenges for pepper plant health in those areas. In farm F7, for PMMoV, there were 11 positive samples out of 26 plant samples tested, and for TSWV, there were 13 positive samples out of 26 plant samples tested. In F10, for TSWV, there were 10 positive samples out of 20 plant samples tested. Overall, the analysis of virus population structures in different surveyed regions within Miranda and Lara states revealed fluctuations in the prevalence of various viruses across farms and variables.

The percentages provided for each virus reflect the proportion of pepper samples in which each virus was detected (Figure 2b). Notably, TSWV had the highest detection rate, being present in 41.6% of the samples, while PMMoV followed with a detection rate of 30.7%. PepS MoV was found in 20.8% of the samples, while PVY was detected in 19.6% of the samples.

3.2. Analysis of Principal Coordinates (PCoA)

Figure 3 displays the outcomes of the analysis of principal coordinates (PCoA) and the resulting tree of minimum distance, based on the presence (1) and absence (0) values of various viruses within different surveyed farms in Miranda and Lara states of Venezuela. The graph built with the first two axes of the PCoA explains 65.5% of the total variability in the presence or absence of the studied viruses. The analysis revealed that CP1 explained 39.4% of the variance, and farms F10, F11, F13 (Lara), and F9 (Miranda) had the highest scores along this coordinate, suggesting a common pattern of virus occurrence in these farms, while CP2 accounted for 26.1% of the variance, and farms F8 (Miranda) and F6 and F9 (Lara) had the highest scores along this coordinate, indicating a distinct pattern of virus distribution compared to other farms.

These principal coordinates offer insights into the relationships between the surveyed farms based on the virus presence/absence data. The tree of minimum distance illustrates the clustering of farms that have similar virus compositions, indicating potential patterns of virus distribution within these regions. The presence or absence of viruses within farms was used to create a distance metric, and the resulting tree visually represents the similarity between farms in terms of their virus populations.

3.3. Characterization of Potyvirus Isolates

The detection of potyvirus was carried out using genus-specific primers MJ1 and MJ2, designed to amplify a short 327 nt fragment spanning conserved motifs MVWCIEN to QMKAAA in the "core" of the CP of potyviruses. The primers were chosen because they gave superior amplification signals in preliminary experiments. Following the removal of primer sequences, the resulting 324 bp fragments obtained from 11 positive plants collected in Lara in 2014 (GenBank accessions MH785274 to MH785295) showed a unique 85% nucleotide (nt) sequence identity to PepYMV isolated from *Capsicum* sp. in Brazil (AF348610); 82% to pepper severe mosaic virus (AM181350.1); 81% to potato virus V (KP849483.1); 80% to Peru tomato mosaic virus (AJ516016.1), brugmansia mosaic virus (JX867236.1), and Amazon lily mosaic virus (AB158523.1); and 79% to pepper mottle virus (EU586135.1), Ecuadorian rocoto virus (EU495234.1), mashua virus Y (MH680823.1), verbena virus Y (EU564817.1) and *Amaranthus leaf mottle virus* (AJ580095.1).

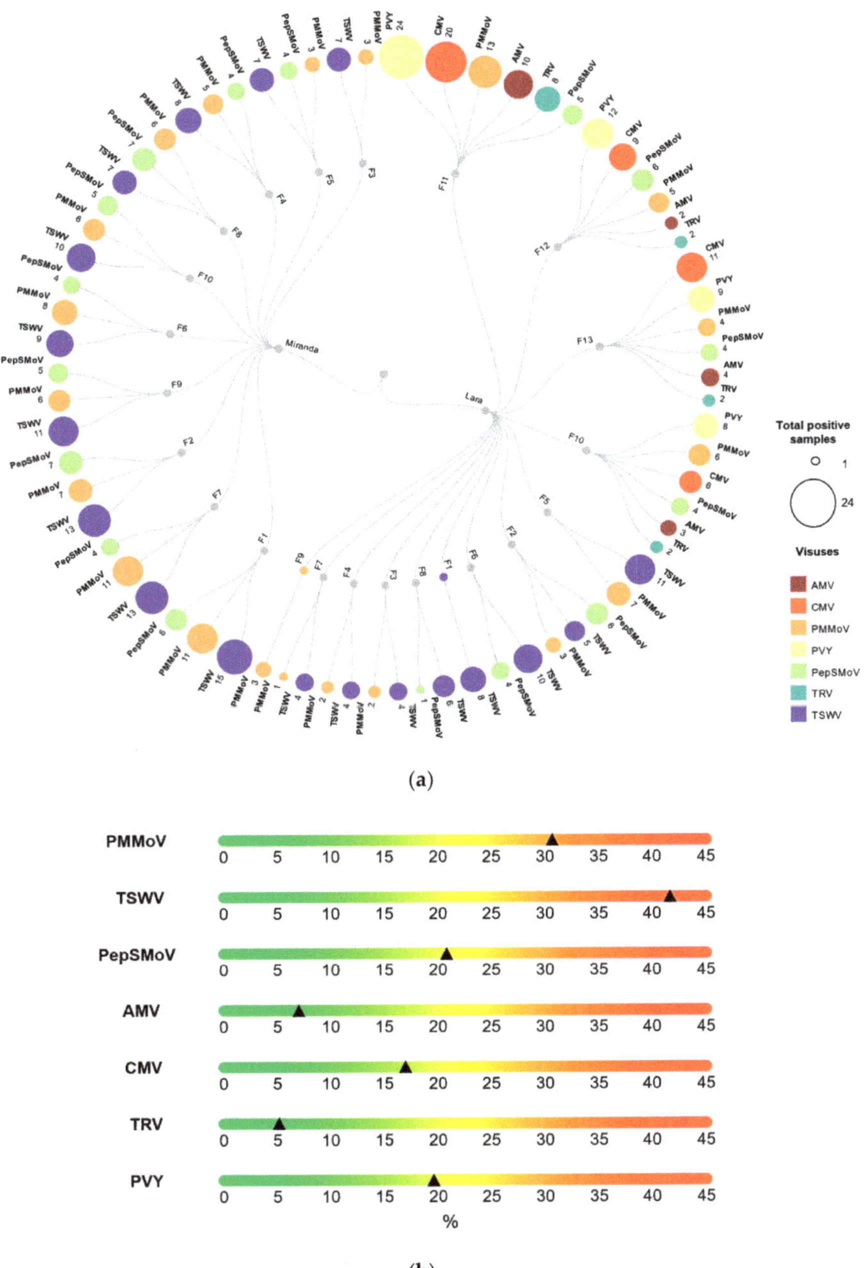

Figure 2. (**a**) Circular dendrogram of virus population structures in different regions and Farms (F1 to F13) surveyed in this study. (**b**) Detection rates of viruses (%) in pepper samples.

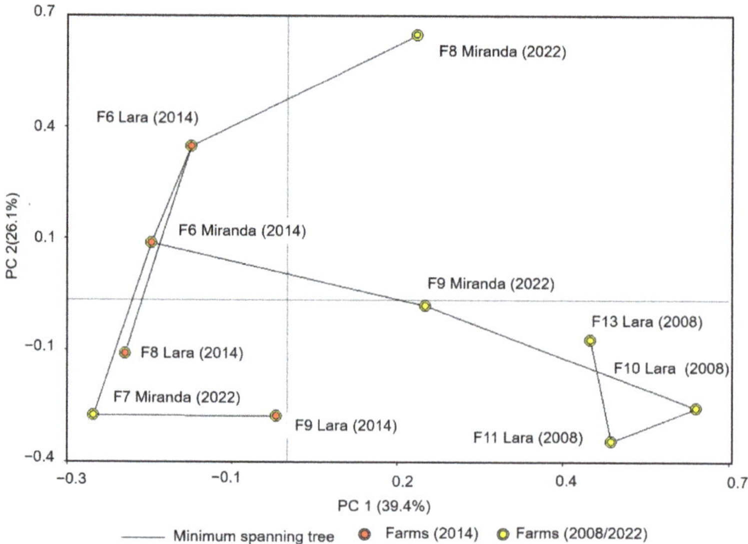

Figure 3. Scatter diagram from the main coordinates (PC1 and PC2) obtained using "Dice" distances between the farms based on the binary presence and absence of viruses. The minimum spanning tree is superimposed.

To further characterize the divergent potyvirus isolates, a larger RT-PCR fragment (ca. 1.163 bp) comprising the CP and 3' untranslated region (3'-UTR) of the potyvirus genome was obtained and sequenced (KT721567.1) using primers designed for PepYMV from single-infected samples. Then, the 3' terminal sequence was split into the NIb, CP, and 3'-UTR according to the inferred cleavage site (-VHHQ/AD-) and stop codon (TAA). The resulting 837 nt CP sequence was subjected to BLASTn and BLASTx analysis and indicated 73.9% and 76.7% nt and aa identity to PepYMV (NC_014327). These sequence identities met the current species demarcation criteria for the *Potyvirus* genus [23]. These findings, therefore, suggest that our isolate is a new potyvirus species that possesses a closer evolutionary relationship with potyvirus PepYMV. The CP is 277 aa long, and analyses showed that the well-known -DAG- motif, which is involved in aphid transmission, was absent from the CP sequence. Instead, a -DAA- motif, which is also present in PepYMV-CP [24], was identified. The deduced amino acid sequence of the CP was most similar to the CP of some potyviruses within the potyvirus supergroup (Table 2).

Table 2. Identity and similarity (%) of the deduced amino acid sequence of the divergent potyvirus to those of other potyviruses.

Virus/Genbank Accession	Identity	Similarity
PepYMV (NC_008393.1)	76.7	83.0
EcRV (EU495234.1)	71.0	82.4
PeSMV (NC_008393.1)	74.8	82.3
PTV (NC_004573.1)	73.1	83.5
PepMoV (NC_001517.1)	69.9	79.9
PVY (NC_001616.1)	73.5	81.0
TEV (NC_001555.1)	58.6	75.4
ChiRSV (NC_016044.1)	57.0	74.9
ChiVMV (NC_005778.1)	57.7	72.9
PVMV (NC_011918.1)	58.4	76.3
WTMV (NC_009744.1)	59.4	74.1

To further investigate the evolutionary relationship between our isolates and other potyviruses, we constructed phylogenetic trees at the CP protein level. According to the phylogenetic tree, our isolates were placed in a separate branch closer to PepYMV isolates within the potyvirus supergroup (Figure 4). These findings, therefore, suggest that our isolate represents a new species of *Potyvirus* that possesses a closer evolutionary relationship with PepYMV. We suggest the name pepper severe mottle virus (PepSMoV) for the novel isolate.

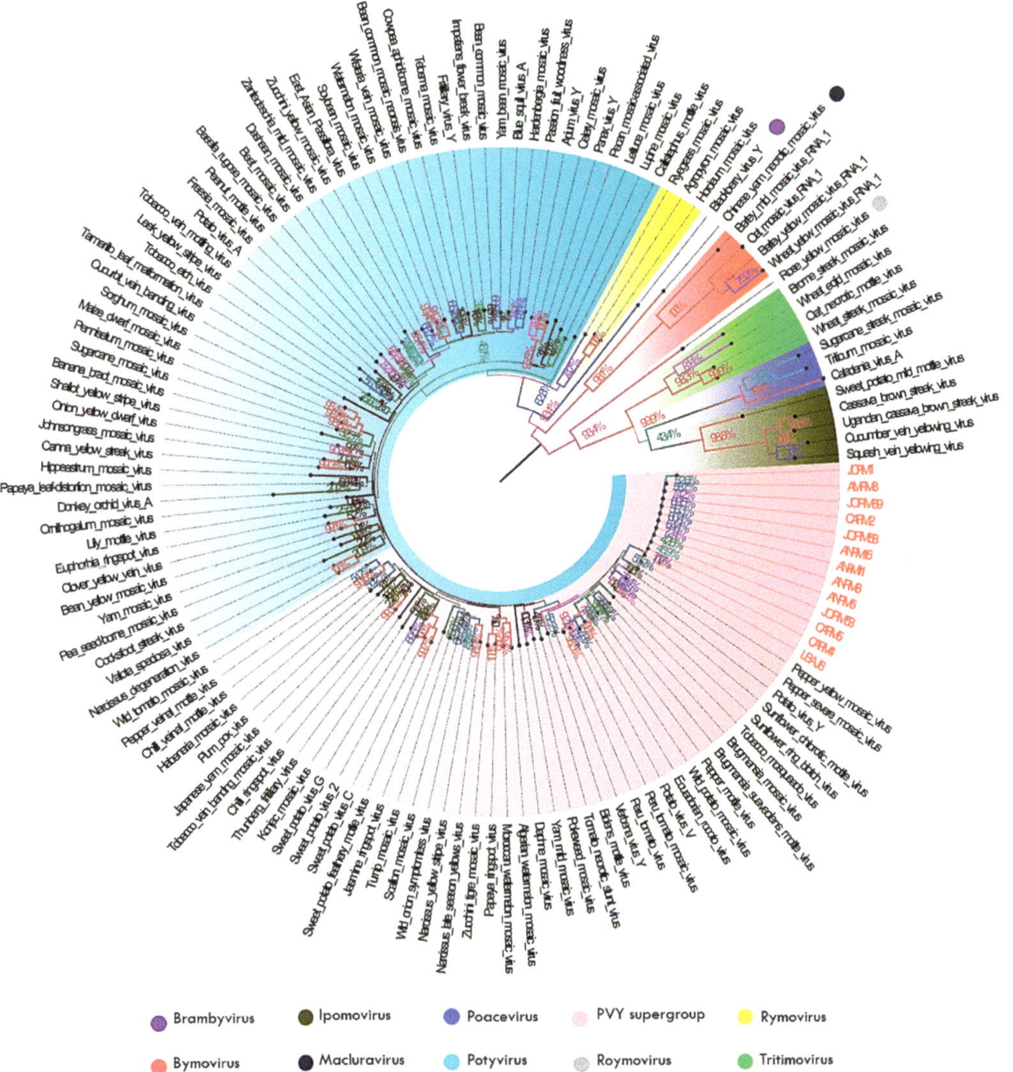

Figure 4. Phylogenetic tree illustrating the position of novel potyvirus isolates from pepper (in red) among the members of the family *Potyviridae*. The tree was constructed from multiple sequence alignments of capsid protein amino acid complete sequences obtained from Genbank, using the LG + G model. Bootstrap values (*n* = 1000) or probability estimate values larger than 70% are indicated at branch nodes for neighbor-joining/maximum-likelihood/Bayesian analysis. Each colored box indicates major phylogenetic groups.

In single virus infections, the new potyvirus isolates induced leaf distortion, severe mottling, and mosaic symptoms on systemic leaves of *C. annuum* cv. Magistral 14 days post-inoculation (dpi) (Figure 5a). The symptoms resemble those observed in single-infected PepSMoV found in pepper fields (Figure 5b). Single RT-PCR analysis of mechanically inoculated plants confirmed potyvirus infection. To fulfill Koch's postulates, a virus from *C. annuum* cv. Magistral was reinoculated onto *C. annuum* cv. Magistral and cv. Acero. Severe mottling was observed in new leaves at 14 dpi, which matched the symptoms found in the field. Reinoculated host plants were positive by RT-PCR only when tested with specific PepSMoV primers (Figure 5c), indicating that the mottling symptoms were caused by only one kind of plant virus.

Figure 5. (**a**) Pepper plants (*Capsicum annuum* L.) var. Magistral mechanically inoculated with PepSMoV showing severe mottling and yellowing on leaves. (**b**) Single-infected PepSMoV in the field. (**c**) Amplification of 837 bp PepSMoV-CP fragments from the field-infected sample AMPIM8 (1); inoculated *C. annunm* var. Magistral (2); reinoculated *C. annunm* var. Acero (3, 4). The arrow points to the 850 bp molecular weight marker. Pictures by Edgloris Marys (Miranda and Trujillo, Venezuela).

4. Discussion

Although quantification of crop losses due to pathogens in Venezuela is limited, plant disease outbreaks are causing substantial declines in major staple food and cash crops, and this impacts rural livelihoods and poses a significant and growing threat to the already complex food insecurity crisis in the country. This work highlights the importance of virus detection in pepper—one of the most popular species of vegetables in Venezuela—in Lara and Miranda states, which account for most of Venezuela's pepper production. Epidemiological knowledge on viral diseases of pepper has been accumulated poorly in Venezuela during the last few decades. Previously, the use of ELISA assays had limited our knowledge of viruses infecting pepper in the country and did not generate molecular evidence to confirm the occurrence of specific viruses [7]. In this study, most samples were infected with at least one virus (76.0%), while 24.0% of the samples remained negative. Symptoms observed in negative samples could be related to certain physiological and nutritional disorders, phytotoxicity or senescence, or possibly to other viruses not tested for. A previous survey on the presence and incidence of pepper viruses in fields located in Lara state carried out many years ago [7] showed that PMMoV and TEV were the most prevalent viruses (100%), followed by ToRSV (58.2%), TRSV (50.0%), CMV (35.29%), TMV (26.47%), PVY (17.65%), and TRS (14.72%). Thus, it was reasonable to focus on determining their incidence in this research. Surveys from 2008 revealed the presence of AMV and TRV, which constitutes—to our knowledge—the first report of this virus infecting pepper crops in Venezuela, which could pose a threat to other economic crops grown in Lara. However, AMV and TRV were not detected in the 2014 surveys, demonstrating the importance of disease surveillance and monitoring systems.

Data obtained in this studio reveals a shift in virus prevalence with the emergence of TSWV, only recently reported in the country [6], with incidence rates as high as 62.3% which could explain the nature and severity of the symptoms reported by farmers. TSWV is considered a major agricultural pest because of its worldwide distribution, wide host range, significant crop losses resulting from infection, and the difficulty in managing the thrip vector [25]. The factors driving the emergence and spread of TSWV in Lara and Miranda remain largely unknown but could be attributed in part to high thrip (*Frankliniella occidentalis*) densities recorded during sampling.

Interestingly, PMMoV remains one of the most common viruses detected in pepper. A decrease in PMMoV incidence from 2004 (100%) to 2014 (21.6%) could be attributed to the fact that—at least until 2015—a considerable number of farmers changed the practice of using their seed to buying the seed of modern hybrid cultivars. This confirmed the importance of seed origin, quality, and health status in the control of seed-transmitted *Tobamovirus* [26]. TEV (genus *Potyvirus*), whose presence in pepper fields in Jimenez has been previously established in Venezuela [7], was not detected in this work. In contrast with the 2004 survey, symptomatic samples in this study tested negative for ToRSV, TRSV, CMV, TMV, and TRS. Data on pepper variety were not collected during surveys; however, it is known that "Majestic", a highly TMV-resistant pepper variety, was the most popular variety being grown by farmers in previous years. Although there is a lack of information on the season when the 2004 survey took place, this is likely due to the different seasons in which samples were collected. In 2014, the surveys were conducted in September (end of the rainy season; moderate temperatures). As suggested by Afouda et al. [27], the abundance of virus vector populations (aphids in the case of CMV and PVY) likely correlates with seasonal variations affecting virus incidence. Nematode-transmitted virus (ToRSV, TRSV, TRS) incidence also could be influenced by seasonal fluctuations in the spatial distribution of the nematode population [28] and by the change in the pepper production system.

The varying detection rates of these viruses in the pepper samples from Miranda and Lara states can also be attributed to a combination of factors. The geographic distribution and environmental conditions of the two states may have influenced the prevalence of these viruses. Different viruses could have been favored by specific climatic conditions [29,30], temperature ranges [31], and soil types present in these regions [32]. Additionally, the population dynamics of insect vectors, such as aphids and thrips, may have contributed to the differing transmission rates of these viruses. The presence of abundant vector populations could lead to higher rates of virus transmission [33]. Furthermore, variations in host plant susceptibility might have played a role. Different pepper cultivars could have exhibited varying levels of resistance to certain viruses, leading to differences in infection rates. The movement of infected plant material, such as seeds and seedlings, could also have contributed to the spread of viruses across the two states. Additionally, cultural practices and farming strategies employed in each region could have influenced the transmission dynamics of these viruses. Interactions between viruses and possible coinfections within pepper plants might have affected detection rates as well. Some viruses could interact synergistically or competitively, influencing their ability to establish infections within the same host plant [34].

A new potyvirus was found associated with the 2014 virus outbreak. Electron microscopy analysis of infected tissues showed flexuous filamentous particles in the 800–900 nm size range typical of a potyvirus. RT-PCR using *Potyvirus* genus-specific primers and subsequent sequencing identified the infectious agent as a potentially new virus species, PepSMoV, belonging to the genus *Potyvirus*. Further coat protein sequence analysis of PepSMoV and phylogenetic analysis with other viruses confirmed that PepSMoV belongs to the genus *Potyvirus*. The molecular criteria for species discrimination within the *Potyvirus* genus have been established by the International Committee on Taxonomy of Viruses (ICTV) [24]. The species demarcation criteria, based upon the large ORF or its protein product, are generally accepted as <76% nucleotide identity and <82% amino acid identity. Pairwise homology studies of CP genes were undertaken between PepSMoV and its closest related potyvirus, PepYMV. PepSMoV has a nucleotide identity of 74% and amino acid identity of 77% with PepYMV and meets the molecular species demarcation criteria.

Based on these criteria and the results obtained from BLAST and multiple alignments of nucleotide and amino acid sequences, with high certainty, we suggest the presence of a virus that belongs to a new species of Potyvirus that is most closely related to PepYMV, with which it shares 73.9% and 76.7% nucleotide and amino acid identity in the CP.

It is important to try to reveal the origin and evolutionary history of each virus. Like its closest relative, PepSMoV did not carry the DAG motif present in most potyvirus CPs as an important factor related to aphid transmission [35], both sharing instead a DAA motif [36]. This confirms that PepYMV and PepSMV are highly similar and could have a common origin.

PepYMV is a species indigenous and confined to Brazil, and it is the prevalent virus in pepper fields [37]. It is believed that the extensive use of PVY-resistant cultivars may have contributed to the emergence of PepYMV in that country [24].

Comparing the relatively low incidence of PepSMoV in the survey samples, we speculate that it might be a minor virus of pepper. Nevertheless, to determine its significance for the genus *Capsicum*, future diagnostic surveys in Venezuela should include testing for the presence of this virus. Issues such as complete genome sequence, host range, vector transmission [38], epidemiology [39], and pathological properties [40], relevant to the proper management of viral diseases in peppers, should also be addressed. This work constitutes the first attempt to determine the role of RNA viruses in pepper production in Lara state during the 2014 epidemic. It is expected that surveillance programs aimed at diagnosing and preventing virus spread will be implemented in our country.

5. Conclusions and Prospects

Despite yield losses, very few viral emergencies or novel threats have been mentioned in pepper crops in Venezuela during the past 20 years. Our study aimed to enhance the understanding of RNA virus prevalence in pepper plants, paving the way for effective management strategies and safeguarding the future of pepper cultivation. Our key findings include the following:

(1) Five virus species (PVY, CMV, PMMoV, AMV, and TRV) were identified in samples collected during the 2008 outbreak in Lara state. Two of these species (AMV and TRV) were found infecting pepper for the first time in the country. Synergistic disease caused by mixed virus infection could account for the crop losses reported during 2008.

(2) An alarming prevalence of TSWV and PMMoV in the surveyed regions during the 2014 and 2022 virus outbreaks reaffirms the severity of their impact on pepper production.

(3) Surveys revealed the unexpected and groundbreaking discovery of an entirely new potyvirus species previously uncharacterized. This previously unknown virus represents a significant addition to the existing repertoire of viral threats to pepper crops in Venezuela and potentially beyond.

The study made important contributions to sustainability in various dimensions. The research had a social impact by addressing viral threats to pepper crops, directly benefiting the agricultural communities of Lara and Miranda in Venezuela. The study helped us understand the threats posed by these pathogens. This contributes to sustainable agriculture by safeguarding crop health and minimizing yield losses due to viral infections. The study addressed a real-world agricultural challenge by revealing the prevalence of viruses in pepper plants. It provided essential information about the existing situation and the temporal evolution (2008, 2014, and 2022), which allowed the identification and quantification of viral threats.

The research benefited local farmers and agricultural stakeholders by offering for the first time (as far as is known) the discovery of a new species of potyvirus in Venezuela. It provided them with knowledge and certainty in the identification of the pathogen to protect their crops effectively. The discovery of a new species of Potyvirus expanded the theoretical knowledge base in virology and phytopathology. It contributed to the understanding of viral diversity and evolution, supporting and enriching theoretical foundations in these fields.

The study introduced new approaches to studying and managing viral infections in crops. It offered innovative methods for detecting viruses, which can be applied in research and agricultural practices beyond Venezuela. By presenting new approaches and solutions to address the prevalence of the virus, the research justified its practical importance. From this study, practical recommendations can be established that could be implemented on farms, minimizing losses and promoting sustainable agricultural practices.

The research findings and methodologies have the potential to be applied in other institutions, communities, or organizations beyond their original context. Their insights into the prevalence and management of the virus could be adapted to benefit other regions facing similar agricultural challenges, thus extending the positive impact and sustainability beyond Venezuela.

Author Contributions: Conceptualization, E.M., B.O.O. and E.R.-R.; methodology, E.M., E.R.-R. and B.O.O.; software, E.R.-R., Y.L., K.Z. and B.O.O.; validation, E.M., E.R.-R., Y.L. and B.O.O.; formal analysis, E.M., E.R.-R., Y.L. and B.O.O.; investigation, J.O.M., A.M., Y.P., Y.L., E.R.-R., P.A., E.M., E.R.-R. and R.O.; resources, E.R.-R., E.M. and B.O.O.; data curation, E.M., E.R.-R., Y.L. and B.O.O.; writing—original draft, E.M.; writing—review and editing, B.O.O. and E.R.-R.; visualization, E.R.-R. and B.O.O.; supervision, E.M.; project administration, E.M.; funding acquisition, E.M. All authors have read and agreed to the published version of the manuscript.

Funding: This study was supported by grants from the Inter-American Development Bank (IDB), the National Fund for Science and Technology (FONACIT, Venezuela), and the Venezuelan Institute for Scientific Research (IVIC) to project CMBC-408.

Institutional Review Board Statement: Not applicable.

Informed Consent Statement: Not applicable.

Data Availability Statement: Not applicable.

Acknowledgments: The authors are grateful to all the growers for field assistance.

Conflicts of Interest: The authors declare no conflict of interest.

Appendix A

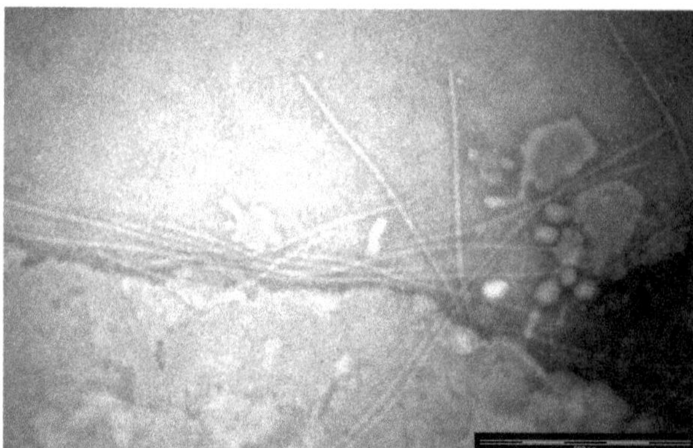

Figure A1. Electron micrograph of partially purified virions negatively stained.

Table A1. The list of primers used for the detection of pepper viruses.

Virus Genus	Primer Name	Sequence (5'-3 *	Amplicon Size (bp)	Reference
Ilarvirus	Ilar1F5 /	GCNGGWTGYGGDAARWCNAC	300	[16]
	Ilar1R7	AMDGGWAYYTGYTYNGTRTCACC		
Nepovirus	Nepo-AF	GGHDTBCAKTMYSARRARTGG	255	[12]
	Nepo-AR	TGDCCASWVARYTCYCCATA		
	Nepo-CF	TTRKDYTGGYKAAMYYCCA	640	[12]
	Nepo0CR	TMATCSWASCRHGTGSKKGCCA		
Orthotospovirus	L1 /	AATTGCCTTGCAACCAATTC	276	[10]
	L2	ATCAGTCGAAATGGTCGGCA		
Potexvirus	Potex 1RC /	TCAGTRTTDGCRTCRAARGT	584	[17]
	Potex 5	CAYCARCARGCMAARGAYGA		
Potyvirus	MJ1 /	TGGTHTGGTGYATHGARAAYGG	327	[14]
	MJ2	TGCTGCKGCYTTCATYTG		
	B1570 /	GGAGAGTCTTGGGCT	1.200	[11]
	PY10	GCAATGCTTGAGTCATGGGG		
	PepSMoV-CP-F	GCAGATGACACAAGTAAAACT	837	This study
	PepSMoV-CP-R	CATATTCTTCACCCCAAGCAA		
Tobamovirus	TobUni1 /	ATTTAAGTGGASGGAAAAVCAT	750	[15]
	TobUni2	GTYGTTGATGAGTTCRTGGA		
Tobravirus	Tobra-F3 /	GGTGGKCAATGGTCTTWTTGG	800	[13]
	Tobra-R2	GTCAGCTGYTGATCAGATAACC		

* R = A or G; Y = C or T; S = G or C; W = A or T; K = G or T; M = A or C; B = C or G or T; D = A or G or T; H = A or C or T; V = A or C or G; N = any base.

Table A2. Reference sequences used for phylogenetic analysis.

Virus/Acronym	Genbank Accession No.
agropyron mosaic virus	NC_005903.1
algerian watermelon mosaic virus	NC_010736.1
apium virus Y	NC_014905.1
arracacha mottle virus	NC_018176.1
artichoke latent virus isolate FR37	NC_026759.1
asparagus virus 1 isolate DSMZ PV-0954	NC_025821.1
banana bract mosaic virus	NC_009745.1

Table A2. *Cont.*

Virus/Acronym	Genbank Accession No.
barley mild mosaic virus RNA 1	NC_003483.1
barley mild mosaic virus RNA2	NC_003482.1
barley yellow mosaic virus RNA 1	NC_002990.1
barley yellow mosaic virus RNA 2	NC_002991.1
basella rugose mosaic virus	NC_009741.1
bean common mosaic necrosis virus	NC_004047.1
bean common mosaic virus	NC_003397.1
bean yellow mosaic virus	NC_003492.1
beet mosaic virus	NC_005304.1
bidens mosaic virus isolate SP01	NC_023014.1
bidens mottle virus	NC_014325.1
blackberry virus Y	NC_008558.1
blue squill virus A	NC_019415.1
brazilian weed virus Y isolate KLL097	NC_030847.1
brome streak mosaic virus	NC_003501.1
brugmansia mosaic virus strain SK	NC_020105.1
brugmansia suaveolens mottle virus	NC_014536.1
caladenia virus A	NC_018572.1
calla lily latent virus strain m19 polyprotein gene	NC_021196.1
callistephus mottle virus isolate DJ	NC_030794.1
canna yellow streak virus	NC_013261.1
carrot thin leaf virus isolate CTLV-Cs	NC_025254.1
cassava brown streak virus	NC_012698.2
catharanthus mosaic virus isolate Mandevilla-US	NC_027210.1
celery mosaic virus	NC_015393.1
chilli ringspot virus	NC_016044.1
chilli veinal mottle virus	NC_005778.1
chinese yam necrotic mosaic virus	NC_018455.1
clover yellow vein virus	NC_003536.1
coccinia mottle virus isolate Su12-25	NC_030840.1
cocksfoot streak virus	NC_003742.1
colombian datura virus	NC_020072.1
cowpea aphid-borne mosaic virus	NC_004013.1
cucumber vein yellowing virus	NC_006941.1
cucurbit vein banding virus isolate 3.1	NC_035134.1
daphne mosaic virus	NC_008028.1
dasheen mosaic virus	NC_003537.1
donkey orchid virus A isolate SW3.1 polyprotein gene	NC_021197.1
east asian passiflora virus	NC_007728.1

Table A2. *Cont.*

Virus/Acronym	Genbank Accession No.
endive necrotic mosaic virus strain ENMV-FR	NC_034273.1
ecuadorian rocoto virus isolate Rocoto	EU495234.1
euphorbia ringspot virus isolate PV-0902	NC_031339.1
freesia mosaic virus	NC_014064.1
fritillary virus Y	NC_010954.1
habenaria mosaic virus genomic RNA	NC_021786.1
hardenbergia mosaic virus	NC_015394.2
hippeastrum mosaic virus	NC_017967.1
hordeum mosaic virus	NC_005904.1
hubei poty-like virus 1 strain SCM51506 polyprotein gene	NC_032912.1
impatiens flower break potyvirus isolate Asan	NC_030236.1
iranian johnsongrass mosaic virus	NC_018833.1
iris severe mosaic virus isolate BJ	NC_029076.1
japanese yam mosaic virus	NC_000947.1
jasmine ringspot virus	NC_029051.1
johnsongrass mosaic virus	NC_003606.1
keunjorong mosaic virus isolate Cheongwon	NC_016159.1
konjac mosaic virus	NC_007913.1
leek yellow stripe virus	NC_004011.1
lettuce italian necrotic virus	NC_027706.1
lettuce mosaic virus	NC_003605.1
lily mottle virus	NC_005288.1
longan witches broom-associated virus isolate Han1	NC_034835.1
lupine mosaic virus	NC_014898.1
maize dwarf mosaic virus	NC_003377.1
moroccan watermelon mosaic virus	NC_009995.1
narcissus degeneration virus	NC_008824.1
narcissus late season yellows virus isolate Marijiiup8	NC_023628.1
narcissus yellow stripe virus	NC_011541.1
oat mosaic virus RNA 1	NC_004016.1
oat mosaic virus RNA 2	NC_004017.1
oat necrotic mottle virus	NC_005136.1
onion yellow dwarf virus	NC_005029.1
ornithogalum mosaic virus	NC_019409.1
panax virus Y	NC_014252.1
papaya leaf distortion mosaic virus	NC_005028.1
papaya ringspot virus	NC_001785.1
passion fruit woodiness virus	NC_014790.2

Table A2. *Cont.*

Virus/Acronym	Genbank Accession No.
pea seed-borne mosaic virus	NC_001671.1
peanut mottle virus	NC_002600.1
pecan mosaic-associated virus isolate LA	NC_030293.1
pennisetum mosaic virus	NC_007147.1
pepper mottle virus	NC_001517.1
pepper severe mosaic virus	NC_008393.1
pepper veinal mottle virus	NC_011918.1
pepper yellow mosaic virus	NC_014327.1
peru tomato mosaic virus	NC_004573.1
plum pox virus	NC_001445.1
pokeweed mosaic virus isolate PkMV-PA	NC_018872.2
potato virus A	NC_004039.1
potato virus V	NC_004010.1
potato virus Y	NC_001616.1
rice necrosis mosaic virus RNA 1	NC_028144.1
rice necrosis mosaic virus RNA 2	NC_028145.1
rose yellow mosaic virus	NC_019031.1
ryegrass mosaic virus	NC_001814.1
scallion mosaic virus	NC_003399.1
shallot yellow stripe virus	NC_007433.1
sorghum mosaic virus	NC_004035.1
soybean mosaic virus	NC_002634.1
squash vein yellowing virus	NC_010521.1
sugarcane mosaic virus	NC_003398.1
sugarcane streak mosaic virus	NC_014037.1
sunflower chlorotic mottle virus	NC_014038.1
sunflower mild mosaic virus isolate Entre Rios	NC_021065.1
sunflower ring blotch virus isolate Chaco	NC_034208.1
sweet potato feathery mottle virus	NC_001841.1
sweet potato latent virus	NC_020896.1
sweet potato mild mottle virus	NC_003797.1
sweet potato virus 2	NC_017970.1
sweet potato virus C	NC_014742.1
sweet potato virus G isolate Jesus Maria	NC_018093.1
tall oatgrass mosaic virus isolate Benesov	NC_022745.1
tamarillo leaf malformation virus isolate A	NC_026615.1
telosma mosaic virus	NC_009742.1
thunberg fritillary virus	NC_007180.1
tobacco etch virus	NC_001555.1
tobacco mosqueado virus isolate RS-01	NC_030118.1

References

1. Tandukar, S.; Sherchan, S.P.; Haramoto, E. Applicability of crAssphage, pepper mild mottle virus, and tobacco mosaic virus as indicators of reduction of enteric viruses during wastewater treatment. *Sci. Rep.* **2020**, *10*, 3616. [CrossRef] [PubMed]
2. Zayed, M.S.; Taha, E.-K.A.; Hassan, M.M.; Elnabawy, E.-S.M. Enhance Systemic Resistance Significantly Reduces the Silverleaf Whitefly Population and Increases the Yield of Sweet Pepper, *Capsicum annuum* L. var. annuum. *Sustainability* **2022**, *14*, 6583. [CrossRef]
3. FAO. The FAO Statistical Database (FAOSTAT): Food and Agriculture Organization of the United Nations. 2023. Available online: http://faostat.fao.org/DesktopDefault.aspx?PageID=567#ancor (accessed on 7 August 2023).
4. Rivera-Toro, D.M.; López-López, K.; Vaca-Vaca, J.C. First molecular characterization of pepper severe mottle virus infecting chili pepper crops in Colombia. *J. Plant Pathol.* **2021**, *103*, 321–325. [CrossRef]
5. Nava, A.; Londoño, A.; Polston, J.E. Characterization and distribution of tomato yellow margin leaf curl virus, a begomovirus from Venezuela. *Arch. Virol.* **2013**, *158*, 399–406. [CrossRef]
6. Pérez-Colmenares, Y.; Mejías, A.; Rodríguez-Román, E.; Avilán, D.; Gómez, J.C.; Olaechea, J.E.; Zambrano, K.; Marys, E. Identification of Tomato spotted wilt virus Associated with Fruit Damage During a Recent Virus Outbreak in Pepper in Venezuela. *Plant Dis.* **2015**, *99*, 896. [CrossRef]
7. Rodríguez, Y.; Rangel, E.; Centeno, F.; Mendoza, O.; Parra, A. Detection of viral diseases affecting sweet pepper in Counties Iribarren, Jimenez and Torres of Lara State, Venezuela, using ELISA technique. *Rev. Fac. Agron. (LUZ)* **2004**, *21*, 93–103.
8. Jaimez, R.; Costa, M.; Araque, O.; Palha, M.; Salazar, R.; Portugal, L. Greenhouses in Venezuela: Current status and development prospects. *Rev. Fac. Agron. (LUZ)* **2015**, *32*, 145–174.
9. QGIS.org. QGIS Geographic Information System. QGIS Association. 2023. Available online: http://www.qgis.org (accessed on 7 August 2023).
10. Chatzivassiliou, E.K.; Weekes, R.; Morris, J.; Wood, K.R.; Barker, I.; Katis, N.I. Tomato spotted wilt virus (TSWV) in Greece: Its incidence following the expansion of *Frankliniella occidentalis*, and characterisation of isolates collected from various hosts. *Ann. Appl. Biol.* **2000**, *137*, 127–134. [CrossRef]
11. da Cunha, L.C.; Resende, R.D.O.; Nagata, T.; Inoue-Nagata, A.K. Distinct features of Pepper yellow mosaic virus isolates from tomato and sweetpepper. *Fitol. Bras.* **2004**, *29*, 663–667. [CrossRef]
12. Digiaro, M.; Elbeaino, T.; Martelli, G. Development of degenerate and species specific primers for the differential and simultaneous RT-PCR detection of grapevine infecting nepoviruses of subgroups A, B and C. *J. Virol. Methods* **2007**, *141*, 34–40. [CrossRef]
13. Jones, D.; Farreyrol, K.; Clover, G.R.G.; Pearson, M.N. Development of a generic PCR detection method for tobraviruses. *Australas. Plant Pathol.* **2008**, *37*, 132–136. [CrossRef]
14. Marie-Jeanne, V.; Ioos, R.; Peyre, J.; Alliot, B.; Signoret, P. Differentiation of Poaceae potyviruses by reverse transcription-polymerase chain reaction and restriction analysis. *J. Phytopathol.* **2000**, *148*, 141–151. [CrossRef]
15. Letschert, B.; Adam, G.; Lesemann, D.E.; Willingmann, P.; Heinze, C. Detection and differentiation of serologically cross-reacting tobamoviruses of economical importance by RT-PCR and RT-PCR-RFLP. *J. Virol. Methods* **2002**, *106*, 1–10. [CrossRef]
16. Untiveros, M.; Perez-Egusquiza, Z.; Clover, G. PCR assays for the detection of members of the genus Ilarvirus and family Bromoviridae. *J. Virol. Methods* **2010**, *165*, 97–104. [CrossRef] [PubMed]
17. van der Vlugt, R.A.; Berendsen, M. Development of a general potexvirus detection method. *Eur. J. Plant Pathol.* **2002**, *108*, 367. [CrossRef]
18. Sambrook, J.; Russell, D.W. *Molecular Cloning: A Laboratory Manual*; Cold Spring Harbor Laboratory Press: Austin, TX, USA, 2011; p. 1546.
19. Kumar, S.; Stecher, G.; Tamura, K. MEGA7: Molecular evolutionary genetics analysis version 7.0 for bigger datasets. *Mol. Biol. Evol.* **2016**, *33*, 1870–1874. [CrossRef]
20. Hepperle, D. DNA Dragon 1.5.2—DNA Sequence Contig Assembler Software. Version 2012. Available online: https://www.dna-dragon.com/ (accessed on 7 August 2023).
21. Tamura, K.; Peterson, D.; Peterson, N.; Stecher, G.; Nei, M.; Kumar, S. MEGA5: Molecular evolutionary genetics analysis using maximum likelihood, evolutionary distance, and maximum parsimony methods. *Mol. Biol. Evol.* **2011**, *28*, 2731–2739. [CrossRef]
22. Di Rienzo, J.A.; Casanoves, F.; Balzarini, M.G.; González, L.; Tablada, M.; Robledo, Y.C. InfoStat Versión 2011. Grupo InfoStat, FCA, Universidad Nacional de Córdoba, Argentina. 2011. Available online: http://www.infostat.com (accessed on 7 August 2023).
23. Inoue-Nagata, A.; Fonseca, M.; Resende, R.; Boiteux, L.; Monte1, D.; Dusi, A.; de Avila, A.; van der Vlugt, A. Pepper yellow mosaic virus, a new potyvirus in sweetpepper, Capsicum annuum. *Arch. Virol.* **2001**, *147*, 849–855. [CrossRef] [PubMed]
24. Inoue-Nagata, A.K.; Jordan, R.; Kreuze, J.; Li, F.; López-Moya, J.J.; Mäkinen, K.; ICTV Report Consortium. ICTV virus taxonomy profile: Potyviridae 2022. *J. Gen. Virol.* **2022**, *103*, 001738. [CrossRef]
25. Scholthof, K.; Adkins, S.; Czosnek, H.; Palukaitis, P.; Jacquot, E.; Hohn, T.; Hohn, B.; Saunders, K.; Candresse, T.; Ahlquist, P.; et al. Top 10 plant viruses in molecular plant pathology. *Mol. Plant Pathol.* **2011**, *12*, 938–954. [CrossRef]
26. Dombrovsky, A.; Smith, E. Seed transmission of Tobamoviruses: Aspects of global disease distribution. *Adv. Seed Biol.* **2017**, *12*, 233–260. [CrossRef]
27. Afouda, L.A.; Kotchofa, R.; Sare, R.; Zinsou, V.; Winter, S. Occurrence and distribution of viruses infecting tomato and pepper in Alibori in northern Benin. *Phytoparasitica* **2013**, *1*, 271–276. [CrossRef]

28. Karuri, H.; Olago, D.; Neilson, R.; Neri, E.; Opere, A.; Ndegwa, P. Plant parasitic nematode assemblages associated with sweet potato in Kenya and their relationship with environmental variables. *Trop. Plant Pathol.* **2017**, *42*, 1–12. [CrossRef]
29. Viloria, J.A.; Olivares, B.O.; García, P.; Paredes-Trejo, F.; Rosales, A. Mapping Projected Variations of Temperature and Precipitation Due to Climate Change in Venezuela. *Hydrology* **2023**, *10*, 96. [CrossRef]
30. Khaled-Gasmi, W.; Souissi, R.; Boukhris-Bouhachem, S. Temporal distribution of three pepper viruses and molecular characterization of two cucumber mosaic virus isolates in Tunisia. *Tunis. J. Plant Prot.* **2020**, *15*, 1–17.
31. Olivares, B.; Hernandez, R.; Arias, A.; Molina, J.C.; Pereira, Y. Eco-territorial adaptability of tomato crops for sustainable agricultural production in Carabobo, Venezuela. *Idesia* **2020**, *38*, 95–102. [CrossRef]
32. Olivares, B.O.; Vega, A.; Calderón, M.A.R.; Rey, J.C.; Lobo, D.; Gómez, J.A.; Landa, B.B. Identification of Soil Properties Associated with the Incidence of Banana Wilt Using Supervised Methods. *Plants* **2022**, *11*, 2070. [CrossRef]
33. Claflin, S.B.; Jones, L.E.; Thaler, J.S.; Power, A.G. Crop-dominated landscapes have higher vector-borne plant virus prevalence. *J. Appl. Ecol.* **2017**, *54*, 1190–1198. [CrossRef]
34. Singhal, P.; Nabi, S.U.; Yadav, M.K.; Dubey, A. Mixed infection of plant viruses: Diagnostics, interactions and impact on host. *J. Plant Dis. Prot.* **2021**, *128*, 353–368. [CrossRef]
35. Atreya, C.D.; Raccah, B.; Pirone, T.P. A point mutation in the coat protein abolishes aphid transmissibility of a potyvirus. *Virology* **1990**, *178*, 161–165. [CrossRef]
36. Lucinda, N.; Da Rocha, W.B.; Inoue-Nagata, A.K.; Nagata, T. Complete genome sequence of pepper yellow mosaic virus, a potyvirus, occurring in Brazil. *Arch. Virol.* **2012**, *157*, 1397–1401. [CrossRef]
37. Moura, M.F.; Mituti, T.; Marubayashi, J.M.; Gioria, R.; Kobori, R.; Pavan, M.; da Silva, N.; Krause-Sacate, R. A classification of Pepper yellow mosaic virus isolates into pathotypes. *Eur. J. Plant Pathol.* **2001**, *131*, 549. [CrossRef]
38. Ma, Y.; Xing, F.; Che, H.; Gao, S.; Lin, Y.; Li, S. The virome of Piper nigrum: Identification, genomic characterization, prevalence, and transmission of three new viruses of black pepper in China. *Plant Dis.* **2022**, *106*, 2082–2089. [CrossRef] [PubMed]
39. Khaled-Gasmi, W.; Souissi, R.; Boukhris-Bouhachem, S. Cucumber mosaic virus epidemiology in pepper: Aphid dispersal, transmission efficiency and vector pressure. *Ann. Appl. Biol.* **2023**, *182*, 101–111. [CrossRef]
40. Giesbers, A.K.; Roenhorst, A.; Schenk, M.F.; Westenberg, M.; Botermans, M. African eggplant-associated virus: Characterization of a novel tobamovirus identified from Solanum macrocarpon and assessment of its potential impact on tomato and pepper crops. *PLoS ONE* **2023**, *18*, e0277840. [CrossRef] [PubMed]

 *sustainability*

Article

Cupressus sempervirens Essential Oil, Nanoemulsion, and Major Terpenes as Sustainable Green Pesticides against the Rice Weevil

Abdulrhman A. Almadiy [1], Gomah E. Nenaah [1,2,*], Bader Z. Albogami [1], Dalia M. Shawer [3] and Saeed Alasmari [1]

[1] Biology Department, College of Science and Arts, Najran University, Najran 1988, Saudi Arabia;
aaalmady@nu.edu.sa (A.A.); bzalmarzoky@nu.edu.sa (B.Z.A.); smalasmari@nu.edu.sa (S.A.)
[2] Zoology Department, Faculty of Science, Kafrelsheikh University, Kafrelsheikh 33516, Egypt
[3] Economic Entomology Department, Faculty of Agriculture, Kafrelsheikh University, Kafrelsheikh 33516, Egypt
* Correspondence: dr_nenaah1972@yahoo.com

Abstract: In order to find effective, biorational, and eco-friendly pest control tools, *Cupressus sempervirens* var. *horizontalis* essential oil (EO) was produced using hydrodistillation, before being analyzed with gas chromatography, specifically, using flame ionization detection. The monoterpene components α-pinene (46.3%), δ-3-carene (22.7%), and α-cedrol, a sesquiterpene hydrocarbon, (5.8%), were the main fractions. An oil-in-water nanoemulsion was obtained following a green protocol. The EO, its nanoemulsion, and its terpenes each exhibited both insecticidal and insect repellent activities against the rice weevil, *Sitophilus oryzae*. In a contact bioassay, the nanoemulsion induced a 100% adult mortality rate in a concentration of 10.0 μL/cm^2 after 4 days of treatment, whereas 40 μL/cm^2 of EO and α-cedrol was required to kill 100% of weevils. Using fumigation, nanoemulsion and EO at 10 μL/L air caused a 100% adult mortality rate after 4 days of treatment. The LC$_{50}$ values of botanicals ranged between 5.8 and 53.4 μL/cm^2 for contact, and between 4.1 and 19.6 μL/L for fumigation. The phytochemicals strongly repelled the weevil at concentrations between 0.11 and 0.88 μL/cm^2, as well as considerably inhibiting AChE bioactivity. They were found to be safe for earthworms (*Eisenia fetida*) at 200 mg/kg, which also caused no significant alteration in wheat grain viability. This study provides evidence for the potential of using the EO of *C. sempervirens* and its nanoemulsion as natural, eco-friendly grain protectants against *S. oryzae*.

Keywords: cypress oil; nanoemulsion; terpenes; *Sitophilus oryzae*; bioactivity; biosafety

Citation: Almadiy, A.A.; Nenaah, G.E.; Albogami, B.Z.; Shawer, D.M.; Alasmari, S. *Cupressus sempervirens* Essential Oil, Nanoemulsion, and Major Terpenes as Sustainable Green Pesticides against the Rice Weevil. *Sustainability* **2023**, *15*, 8021. https://doi.org/10.3390/su15108021

Academic Editors: Barlin Orlando Olivares Campos, Miguel Araya-Alman and Edgloris E. Marys

Received: 6 April 2023
Revised: 10 May 2023
Accepted: 11 May 2023
Published: 15 May 2023

1. Introduction

The world's population is expected to grow to more than 10 billion by 2050, which will boost the global food demand; particularly, for food crops and cereal products. Therefore, agricultural production should be doubled if we are to secure adequate food sources for this huge number of people [1]. There is no doubt that many countries of the world will face problems in this regard, especially those developing and underdeveloped countries with poverty and inadequate technologies for modern agriculture. Furthermore, harmful insects and other pathogens cause the loss of more than 30% of the world's food production, both in the field and during storage as well [2,3]. Coleopteran arthropod insect pests are the major causal agents of grain loss during storage, worldwide. Grain weevils of the genus *Sitophilus* are among the world's most damaging and widespread pests for stored grains. Infestations of grains by these weevils cause significant grain losses and promote the growth of molds, including harmful toxigenic species, by increasing temperature and moisture levels [3].

The rice weevil, *Sitophilus oryzae* L. (Coleoptera: Curculionidae), is a serious primary internal feeder pest, able to infest intact grains, causing both quantitative and qualitative damage in the grains and altering seed viability [4]. Chemical insecticides have widely been applied to prevent insect pests from attacking a variety of crop plants, both in the field and in stores. The overuse of chemical insecticides negatively affects not only the health of consumers and farmers, but also nontarget beneficial organisms and the quality of foodstuffs, in addition to bringing environmental problems as well [5,6]. Populations of *S. oryzae* have also developed resistance against many insecticides and fumigants, such as pirimiphos-methyl and phosphine [7]. These problems make the intensive use of chemical pesticides a point of renewed public debate, especially in light of increasing public awareness of the risks of insecticides. This popular awareness of the adverse effects of conventional pesticides has promoted researchers to seek novel agrochemical pesticides that can meet the increasing grower, consumer, environmental, and regulatory requirements [8]. These new pesticides should be environmentally safe, with novel action mechanisms and low negative impacts on the ecosystem. Pest control using phytochemicals, especially plant essential oils (PEOs), seems a promising alternative strategy for decreasing the intensive reliance on conventional insecticides [6,9]. Because of their wide spectrum bioactivity as pest control tools against several pest insects (including those of stored grains), PEOs are increasingly being considered a credible eco-friendly natural alternative to conventional insecticides [3,6,9–14]. This is especially significant for the protection of stored products, whose confinement promotes the action of molecules which are naturally highly volatile, in order to avoid the toxic residue from chemical insecticides contaminating food products. In this context, we can observe major shortcomings related to the procedures adopted for the extraction, formulation, application, and performance of plant products (including EOs) in pest control protocols. For strengthening and improving the inclusion of plant-based products in pest control programs, nanotechnology has emerged as a promising field across multidisciplinary research, opening up many application opportunities in such various fields as medicine, electronics, drug delivery, and agriculture [15,16]. In agriculture, the potential benefits of nanotechnology in pest control programs include the fabrication of novel plant-based nanomaterial formulations with enhanced insecticidal activities, which could diminish the need for repeated application of chemical pesticides [6,15–18]. Because of their high reactivity and solubility, in addition to their novel physical and chemical characteristics, nanopesticide materials show enhanced bioactivities as pest control tools relative to their bulk counterparts [6,15,17].

The genus Cupressus (Cupressaceae) comprises about 12 species, spread across North America, Mexico, the Mediterranean basin, southern Europe, and subtropical west Asian countries, including the Kingdom of Saudi Arabia [19]. Members of Cupressaceae are common essential-oil-bearing plants, of which an important species is *Cupressus sempervirens* L. var. *horizontalis* (The Mediterranean cypress). It is an aromatic evergreen tree, traditionally used as an expectorant in anticough and antibronchitis medications; for stomach pain; as an antidiabetic medicine; as an antiseptic; for antiulcer and anti-inflammatory purposes; and for treatment of toothaches, flu, coughs, and laryngitis [20]. This plant species has been screened for different bioactivities, including antimicrobial, antiviral, antihelmenthic, antiseptic, cytotoxic, antioxidant, anti-inflammatory, antirheumatic, antihyperlipidemic, anticancer, antispasmodic, antidiuretic, and hepatoprotective activities [20,21]. Insecticidal bioactivities of the cypress tree have also been recorded [9,22,23]. However, there have been no thorough investigations into the insecticidal bioactivity of the EO, nanoemulsion, or bioactive terpenes of *C. sempervirens* against stored-grain insects. In this study, we aimed to investigate not only the composition, but also the contact, fumigant, and repellence bioactivities of the oil, nanoemulsion, and bioactive terpenes of cypress trees growing in Saudi Arabia against *S. oryzae*. The effect of EO materials on acetylcholinesterase (AChE) bioactivity—being a common target enzyme for insect control agents—was investigated. The impact of EO products on earthworms (*E. fetida*), as well as their phytotoxic activity on wheat plant in terms of the basic growth parameters (%germination, root, and shoot growth) were evaluated.

2. Materials and Methods

2.1. Chemicals

Analytical grade monoterpenes, sesquiterpenes, and oxygenated monoterpenes, (Sigma-Aldrich Co. Ltd., St Louis, MO, USA; label purity 99.0–99.8%) were used for comparisons. To calculate retention indices, the series of hydrocarbons (C_5–C_{40}) known as triacontane (Supelco, Bellefonte, CA, USA) was used. Dimethyl sulfoxide (DMSO) and its solvents, all of an analytical grade (Carlo Erba Milan, Milan, Italy), were used in experiments.

2.2. Test Insect

A laboratory strain of the rice weevil, the *S. oryzae* were reared in a pesticide-free environment in our laboratory for more than twenty generations. Weevils were kept in 2 L glass bottles, each containing 200–250 adults and 250 g sterilized wheat grains (moisture content 14 ± 2%). The jars were covered using muslin cloth that was held in place with a rubber band, and then kept in laboratory conditions of 30 ± 2 °C and 68.5% relative humidity, in complete darkness, until the emergence of the adult specimens.

2.3. EO Extraction

On September 2020, the whole aerial parts of the cypress trees were collected from random gardens in Abha, Kingdom of Saudi Arabia; latitude 18° 19′ 45.7824″ (N), longitude 42° 45′ 33.7140″ (E), and 718 m altitude. Specimens were identified by the botanists of Najran University, Saudi Arabia. A plant specimen (No. Cs 02) was laboratory-deposited, for reference. Plant leaves were air dried in the shade, powdered mechanically using a high-speed blender, before being subjected to hydrodistillation using a Clevenger apparatus to obtain the EO. In each run, 150 g powders were hydrodistilled for 3 h with 250 mL distilled water, and in three replicates. The oil/water mixture was extracted with hexane, which had been washed with anhydrous sodium sulfate before the oil had dried, before then being concentrated under reduced pressure. For bioassays, the oil yield (% *wt./wt.*) was calculated on a dry weight basis and stored at 4 °C, where triplicates were considered in calculating the oil yield.

2.4. Analysis of EO and Identification of Constituents

Cypress oil was analyzed using an Agilent 6890 N gas chromatograph (Agilent Technologies, Palo Alto, CA, USA), coupled with a flame ionization detection (FID) and an HP-5 capillary column (30 m 0.32 mm; thickness 0.25 m). The following conditions were met by the GC–FID: one liter of EO; split mode; 50:1 split ratio; and an injector temperature of 250 °C. Oven temperature was initially set at 40 °C for 3 min, then increased to 80 °C at a rate of 5 °C/min, held at that temperature for 3 min, then increased to 250 °C at 10 °C/min, which was held for 10 min. The injector and detector were set to 250 °C, and the carrier gas was helium, at a 1.0 mL/min flow rate. The gas chromatograph was then connected to the silica gel capillary column (HP-5 MS). At a split ratio of 1:100, 0.1 μL of EO was injected onto the column; the carrier gas was helium (1.0 mL/min flow rate). Operation of the mass detector was set at 70 eV ionization voltage. The mass range was taken at 45~550 AMU. Temperatures of the ion source, transfer-line, and the quadrupole were 230, 250, and 150 °C, respectively. Temperature of the oven was programmed as described for the GC. Retention indices of terpenes were calculated depending on *n*-alkanes (C_5–C_{40}) co-injected into the column, in accordance with Van Den Dool and Kratz's equation. Identification of EO profile was accomplished by comparing terpene retention indices and mass spectra to those recorded by Adams [24] and the data stored in the database NIST Standard Reference Database Number 69, [25]. The oil terpenes were quantified as percentages by integrating their peak areas, calibrating, and comparing them to internal standards without the use of a response factor correction. The remaining terpenes were likewise quantified as percentages, calculated by integrating their peak areas, calibrating, and then comparing to standards without using a response factor correction.

2.5. Isolation of Main Terpenes

Ten milliliters of cypress oil was fractionated on a silica gel capillary column (Kieslgel 60, 230–400 mesh, Merck). Trials were conducted in order to determine the best eluent. To accomplish this, several solutions of both n-hexane: ethyl acetate and toluene: ethyl acetate were prepared, which were then tested using Thin-layer Chromatographic plates (TLC), with toluene: ethyl acetate (90:10, then 93:7) being chosen as the best eluent [26]. According to TLC data, developed fractions were divided into 3 main fractions: Main fraction 1 (fractions 8–14, 274.3 mg) was fractionated on a silica gel column, affording 41 mg of α-pinene. The main fraction 2 (fractions 21–27, 193.1 mg) yielded 23 mg of δ-3-carene. By contrast, when fraction 3 (fractions 31–37, 82.5 mg) was developed it provided 9.2 mg of α-cedrol. Terpene fractions were visualized using an UV lamp (254 and 365 nm), and the R_f values of terpenes were calculated and then compared against standards. The structures of the terpene fractions were elucidated using spectroscopic instruments.

2.6. Nanoemulsion Formulation and Characterization

An oil/water nanoemulsion was made following a low-energy emulsification protocol at a constant temperature, with the following proportions: deionized H_2O (90%), EO 5% (*wt./wt.*), and Tween 80 (5%) as a nonionic surfactant [17]. The EO/emulsifier mixture was stirred at 800 rpm for 30 min in a water bath at 35 ± 5 °C. After reaching an oily phase, deionized H_2O was gradually added (2.5 mL/min). After stirring for 45 min at 800 rpm, the temperature was gradually reduced to room temperature, and the nanoemulsion was formed and then preserved in dark screw-capped vials at 24 ± 2 °C. At 0, 1, 10, 20, 30, and 45 days following its preparation, the nanoemulsion was examined for thermodynamic changes. Nanoemulsion was characterized using a polydispersity index (PDI), mean droplet size, and thermodynamic stability measurements (centrifugation, heating, cooling, freezing cycles, and viscosity) [17]. The oil emulsion was first centrifuged (5000 rpm at 25 °C for 25 min), and checked for phase separation, turbidity, and cracking, if any. Heating/cooling tests were performed on the stable formulations for 6 cycles (4–40 °C, each cycle lasting 48 h). Emulsions that demonstrated stability were subjected to a freeze–thaw stress test by alternately storing them at 2 different temperatures (20 °C and 20 °C, 24 h each). Nanoemulsions that demonstrated stability were stored in dark, tightly closed vials at room temperature for one month to observe any creaming, phase separation, or flocculation. The emulsion's pH was measured at 25 ± 0.2 °C, and its viscosity (μ) was elucidated at 200 rpm. Samples were left to stand for about 2 min to reach an equilibrium; thus, readings and experiments were each taken and performed in triplicate. The Z-average diameter of droplets and PDI were measured using a nanoparticle analyzer apparatus (Zetasizer, Nano ZS, Malvern Instruments, Worcestershire, UK) that operates on a dynamic light-scattering basis. Prior to measurements, test formulations were diluted to 10% with deionized H_2O to avoid multiple scattering. The droplets' size and their PDI were calculated using DLS data. The measurements were taken at a scattering angle of 90°, and trials were repeated three times. A Scanning Electron Microscope, or SEM (JEOL, JFC-1600, Tokyo, Japan), was used to determine the morphology of droplets; for this, 15 µL of emulsion that had been dissolved in deionized H_2O was placed onto a carbon-coated copper grid that had been stained with 2% phosphotungstic acid (pH of 6.8). The test samples were dried at 26 °C before being imaged at 80 kV.

2.7. Contact Insecticidal Activity

The contact bioactivity of the EO, nanoemulsion, and terpenes from cypress trees against *S. oryzae* adults were determined using the dipping filter paper technique [6]. Test concentrations of 0.398, 0.795, 1.59 and 3.18 mL of EO, nanoemulsion, δ-3-carene, α-cedrol, and α-pinene were dissolved in 5 mL acetone to obtain the test solutions. Each test concentration was uniformly dropped onto a filter paper (Whatman No. 1, 9 cm d, 63.6 cm^2) to achieve serial test concentrations of 5.0, 10.0, 20.0, and 40.0 µL/cm^2, respectively. Acetone was evaporated from the treated papers; then a treated paper was placed into the bottom of a Petri dish (9 cm d) and twenty adult weevils (of mixed sexes, and 15–20 days old) were introduced. Control groups (adults exposed to acetone-treated filter papers) were included. Treatment and control dishes were kept in the dark at 30 ± 2 °C and $68 \pm 5\%$ r.h. After 24 h, insects were placed into clean Petri dishes, enriched with wheat grains, and kept in rearing conditions. Experiments were performed six times alongside control, and mortality was recorded 1, 2, 4 and 7 days after the treatment, whereafter the end-point mortality was reached, and the resulting contact toxicity was expressed in $\mu L/cm^2$.

2.8. Fumigation

Insecticidal bioactivity using fumigation was investigated as follows: a filter paper (7.0 cm diameter) was dipped in 25 μL of an appropriate concentration of each terpene dissolved in acetone, and control sets were made using acetone, only [6]. Botanicals were screened at 4 dose rates (2.5, 5.0, 10.0, and 20.0 μL/L air). After evaporating the acetone, each treated paper was attached to the undersurface of the screw cap of 250 mL volume glass bottles, which served as fumigant chambers. Twenty weevils were placed into each bottle as adults (15–20 days old), and the bottle was covered with a tape-fixed fine gauze. Experiments were undertaken in six replicates, alongside control groups. After 24 h, the weevils were transferred back to food-enriched clean vials and kept in the rearing conditions described before; whereupon mortality was measured after 1, 2, 4 and 7 days had elapsed from treatment, and the resulting fumigant bioactivity was expressed in μL/L air.

2.9. Repellence Bioactivity

The repellence bioactivity of *C. sempervirens* oil materials against adult weevils was studied by adopting a chosen (area preference) bioassay [22]. A piece of filter paper (Whatman No. 1, 9 cm diameter) was divided into two halves. Test solutions of oil materials were prepared, with 3.5, 7.0, 14.0, and 28 μL of each material dissolved in 0.5 mL n-hexane. Each test concentration was uniformly dropped onto a half filter paper disc, which served as a test area, to obtain bioassay concentrations of 0.11, 0.22, 0.44, and 0.88 $\mu L/cm^2$. The second half was treated only with n-hexane, representing a control. The treated and untreated paper discs were then air dried for 5 min, and thereafter attached to their corresponding opposite surface with adhesive tape, and put in the bottom of a Petri dish (9 cm). Twenty (15–20 days old) unsexed adult weevils of *S. oryzae* were released at the center of each disc, then the lid was covered using a parafilm. Five replicates (100 adults) were considered for each concentration, and the experiments were achieved in rearing conditions. The number of weevils that were observed across both the treated and control halves were counted after 2, 6, 12, and 24 h. The Repellency percentage (RP) was calculated using the following formula: RP = $(C - T)/(C + T) \times 100$, where C is the No. of weevils on untreated zone, and T is the No. of weevils on control zone.

2.10. AChE Inhibition and Estimation of IC_{50}

Anticholinesterase (AChE) enzymatic activity was measured in accordance with Ellman et al. [27]. One gram's worth of the adult weevils were homogenized in 20 mL of an ice-cold phosphate buffer (50 mM and pH 7.4). Acetylthiocholine iodide (25 μL of 15 mM) was dropped as a substrate. The inhibition in AChE activity was measured calorimetrically using a supernatant as an enzyme source [17]. Botanicals were formulated initially in acetone, then in Triton-X 100 (0.01%), and were then tested at 2.5~100 mM. Test and control solutions were corrected using blanks for the nonenzymatic hydrolysis. Trials were performed in triplicate. Absorbance of the solution reflecting AChE specific activity ($\Delta OD/mg$ protein/min) was monitored at a wavelength of 412 nm.

2.11. Phytotoxicity

The phytotoxic impact of the oil terpene components (as indicated by basic growth parameters (%germination, root, and shoot growth)) was evaluated on wheat plants (*Triticum aestivum* L.). Wheat seeds were sterilized using a solution of sodium hypochlorite (15%) for about 40 s, followed by rinsing in sterile deionized water. The grains were placed in clean 9 cm diameter Petri plates, each containing five layers of Whatman filter paper, onto which 1 mL of each botanical (at concentrations of 50, 100, and 150 µL/mL) was dropped. 2 mL of methanol was sprayed on the control. After the evaporation of methanol, ten healthy grains (~0.3–0.36 g) were deposited in each dish. Dishes were maintained at $20 \pm 2\,°C$; $65 \pm 5\%$ R.H., with a natural photoperiod (optimized environmental conditions for wheat germination). Additionally, 10 mL of water was given daily. Each concentration had five replicates, as well as a control. Germination and the seedling growth were noticed after 10 days of planting. The length of shoot and number of leaves were counted 2 weeks later, and seed germination was indicated by the emergence of radicles.

2.12. Toxicity on Earthworm

The acute toxicity of the botanicals against *E. fetida* earthworms was investigated, according to the guidelines of OECD (Organization for Economic Co-Operation and Development [28]. The animals were reared on artificial diet, as detailed by Pavela [29]. Terpenes were admixed with the soil at concentrations equaling 50, 100, and 200 mg kg^{-1}. The positive control was α-cypermethrin at 10 and 20 mg kg^{-1} soil alongside, with deionized water as a negative one. In 1 L glass pots containing either treated or untreated (control) soil, the earthworms were confined as ten adults, and triplicates were made for each run. Pots were covered with a fine gauze, then incubated at $22 \pm 2\,°C$, $75 \pm 5\%$ R.H., and 16:8 h light/dark photoperiod. Mortality was recorded after five and ten days of treatment.

2.13. Statistical Analysis

Data of mortality were adjusted for control mortality and corrected using Abbott's formula [30] when mortality in control exceeded (5%), and data were expressed as % means ($\pm$S.E.). A one-way analysis of variance (ANOVA) at the probability level (=0.05) was adopted on transformed data to compare significance differences between means in both the treatment samples and the controls, followed by individual pairwise comparisons, adopting Tukey's HSD test. Dose–response mortality was analyzed using Finney's Probit analysis to estimate the LC$_{50}$ and LC$_{95}$ and their limits across 48 exposure periods [31]. Probit analysis was adopted to calculate the concentrations that inhibited AChE bioactivity by 50% (IC$_{50}$). The Statistical Package for Social Sciences was used for data analysis (version 23.0; SPSS, Chicago, IL, USA).

3. Results

3.1. Composition of EO

A pale yellowish EO with a strong odor (yield 0.74% *w/w*) was obtained from *C. sempervirens* var. *horizontalis* using hydrodistillation. A total of 62 terpenes amounting 99.7% (*wt./wt.*) were identified in the oil (Table 1 and Figure 1a), and then listed according to their retention indices. The main oil terpenes were (1*S*,5*S*)-2,6,6-trimethylbicyclo [3.1.1] hept-2-ene (α-pinene, 46.3%), 3,7,7-trimethylbicyclo [4.1.0] hept-3-ene (δ-3-carene, 22.7%), and (1*S*,2*R*,5*S*,7*R*,8*R*)-2,6,6,8-tetramethyltricyclo [5.3.1.0] undecan-8-ol (α-cedrol, 5.8%) (Figure 1b). The structure of the terpenes was confirmed using physical and spectroscopic methods, which corroborates with published data:

α-pinene

Colorless, $C_{10}H_{16}$. ^{1}H NMR (CDCl$_3$, 300 MHz): δ 1.92 (m, 1H), δ 1.4 (s, 3H), δ 1.6 (s, 3H), δ 1.8 (s, 3H), δ 1.9 (m, 2H), δ 2.3 (m, 1H), δ 2.4 (m, 1H), δ 4.2 (s, 1H), δ 5.6 (t, 1H), 20.67, 22.35, 22.47, 26.36, 32.62, 36.08, 68.03, 94.15, 123.87, 134.38; ^{13}C NMR (125 MHz, CHCl$_3$): δ 46.99 (C-1), 144.6 (C-2), 116.0 (C-3), δ 31.3 (C-4), 40.69 (C-5), 37.97 (C-6),31.5 (C-7), δ 26.3 (C-8), 20.8 (C-9), 23.01 (C-10) [32,33].

δ-3-carene

Colorless, $C_{10}H_{16}$. ^{1}H-NMR (600 MHz, CDCl$_3$) δ ppm: 5.23 (2H, t, H-2), 2.62 (2H, t, H-6), 1.60 (4H, m, H-3, 5), 1.022 (1H, m, H-4), 0.761 (6H, s, 9, 10-CH$_3$), 0.49 (3H, s, 7-CH$_3$);

^{13}C-NMR (125 MHz, CDCl$_3$) δ ppm: 131.30 (C-6), 119.56 (C-1), 28.42 (C-7), 24.93 (C-5), 23.63 (C-4), 20.89 (C-2), 18.71 (CH$_3$-8), 16.90 (C-3),16.78 (CH$_3$-9), 13.20 (CH$_3$-l0) [34,35].

α-cedrol

Colorless, C$_{15}$H$_{26}$O. ν$_{max}$/cm^{-1} 3380, 1460, 1030, and 1000; δ_{H}(300 MH$_z$) 0.74 (3H, s, Me), 0.78 (3H, d, *J* 7.1, Me), 0.85 (3H, s, Me), 0.88 (3H, s, Me), 0.85–1.61 (10H, m), 1.87–1.98 (1H, m), 2.1–2.2 (1H, m), and 3.94 (1H, ddd, *J* 2.2, 5.6 and 9.7, CHOH); *m/z* 222 (M1, 44%), 206 (15), 178 (100), and 123 (40) (Found: M1, 222.1992. C$_{15}$H$_{26}$O requires *M*, 222.1985) [36–38].

Table 1. Chemical profile of *Cupressus sempervirens* essential oil.

[a,b] Components	[c] RI exp.	[d] RI lit.	Concentration (%)
2-Hexanal	860	862	0.2
Tricyclene	918	916	0.1
a-Thujene	920	921	0.4
α-Pinene	928	930	46.3
Camphene	930	932	1.2
a-Fenchene	941	942,	0.1
Sabinene	966	967	0.6
ß-Pinene	980	980	0.9
ß-Myrcene	988	988	0.1
α-Phellandrene	1006	1008	0.2
δ-3-Carene	1010	1010	22.7
α-Terpinene	1016	1018	1.3
p-Cymene	1021	1020	0.6
Limonene	1032	1029	1.6
β-Phellandrene	1034	1032	0.2
Z-*ß*-Ocimene	1038	1037	0.2
E-*ß*-Ocimene	1044	1044	0.1
γ-Terpinene	1054	1055	0.3
cis-Sabinene hydrate	1067	1066	0.4
p-Cymenene	1070	1072	0.9
α-Terpinolene	1085	1086	1.3
Linalool	1096	1095	0.4
trans-Sabinene hydrate	1099	1097	1.1
Pinocarveol	1138	1140	0.1
Camphor	1142	1144	0.2
Pinocarvone	1158	1162	1.6
Borneol	1162	1165	0.2
Terpinen-4-ol	1176	1174	1.1
p-Cymen-8-ol	1180	1181	0.4
trans-Pinocarveol	1182	1184	0.6
α-Terpineol	1188	1186	0.2
Myrtenol	1192	1195	0.2
Pulegone	1235	1233	0.3
Carvacrol methyl ether	1241	1241	0,2
cis-Chrysanthenyl acetate	1244	1242	0.1
cis-Piperitone epoxide	1247	1248	0.4
trans-Piperitone epoxide	1251	1252	0.7
Carvone	1254	1258	1.1
Carvenone oxide	1260	1260	0.2
Bornyl acetate	1285	1186	0.4
Thymol	1289	1288	0.3
trans-Sabinyl acetate	1290	1292	0.2
Carvacrol	1296	1298	0.3
α-Cedrene	1295	1294	0.1
α-Copaene	1372	1374	0.2
β-Bourbonene	1382	1384	0.1
α-Gurjunene	1408	1408	0.3
β-Caryophyllene	1414	1417	0.1

Table 1. *Cont.*

[a,b] Components	[c] RI exp.	[d] RI lit.	Concentration (%)
β-Gurjunene	1430	1432	0.2
α-Humulene	1450	1452	0.2
Alloaromadendrene	1472	1474	0.3
Germacrene D	1480	1478	0.3
Bicyclogermacrene	1496	1495	0.1
β-Bisabolene	1508	1510	0.2
cis-Calamenene	1544	1443	0.2
Spathulenol	1574	1576	0.6
α-Cedrol	1596	1591	5.8
α-Acorenol	1632	1630	0.3
β-Acorenol	1635	1637	0.2
γ-Cadinol	1648	1649	0.1
Cadalene	1674	1674	0.1
Manool	1990	1989	0.3
Grouped compounds (%)	-	-	
Monoterpene hydrocarbons	-	-	77.1%
Oxygenated monoterpenes	-	-	12.9%
Sesquiterpene hydrocarbons	-	-	9.7%
% peaks identified	-	-	99.7
Total yield % (mL/100 g)	-	-	0.74

[a] Compounds are listed in the order of their elution from a HP-5MS column. [b] Identification methods: a, based on comparison of RT, RI, and MS with those of authentic compounds; b, based on comparison of mass spectrum with those reported in Wiley, Adams [24], and NIST 69 MS libraries [25]. [c] Linear retention index on the HP-5MS column, experimentally determined using homologous (C_5–C_{40}) *n*-alkane series. [d] Linear retention index based on Adams [24] or NIST 69 [25], and literature.

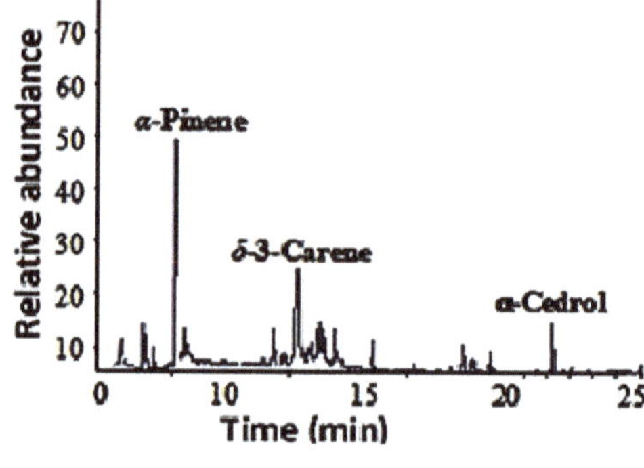

(a)

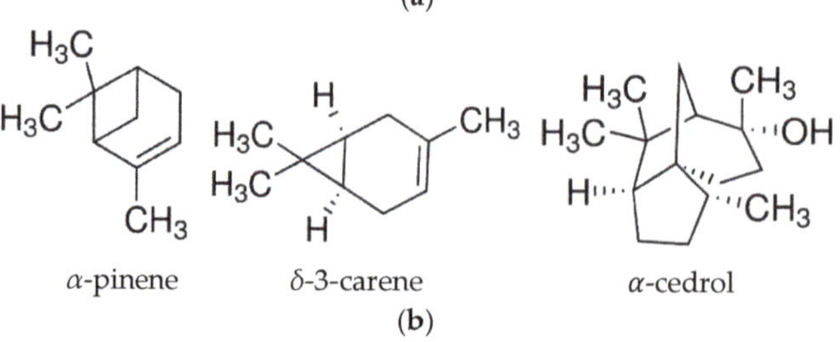

(b)

Figure 1. (**a**) GC–FID chromatogram of cypress EO. Major terpenes are highlighted; (**b**) Major terpene components of cypress EO.

3.2. Nanoemulsion Characterization

The developed emulsion (droplet size 67.8 ± 3.1 nm) showed stability during extreme conditions of centrifugation, temperature, heating–cooling cycle (4–40 °C), and a freezing cycle at −4 °C. The optimum conditions of nanoemulsion preparation, droplet size, and their PDI, are listed in Table 2 and illustrated in Figure 2. The SEM revealed that a transparent nanoemulsion consisting of dispersed and spherical-shaped nanoparticles had been developed (Figure 3).

Table 2. Characterization of *Cupressus sempervirens* oil nanoemulsion.

Storage Period (Days)	Viscosity (mPa·s)	pH	PDI	Size (nm ± S.E.)
0	4.1	6.1 ± 0.04 [c]	0.18 ± 0.03 [a]	67.8 ± 3.1 [a]
1	4.1	5.8 ± 0.06 [b]	0.20 ± 0.05 [a]	69.2 ± 3.6 [a]
10	4.4	5.6 ± 0.04 [b]	0.20 ± 0.02 [a]	73.4 ± 4.2 [b]
20	4.8	5.1 ± 0.08 [a]	0.23 ± 0.03 [b]	78.6 ± 5.7 [bc]
30	5.1	4.9 ± 0.16 [a]	0.25 ± 0.02 [bc]	86.1 ± 6.3 [c]
45	5.5	4.7 ± 0.14 [a]	0.28 ± 0.02 [c]	92.1 ± 6.1 [d]

Each experiment is the mean of three replicates. Within a column, means followed by same letter(s) are not significantly different ($p \leq 0.05$).

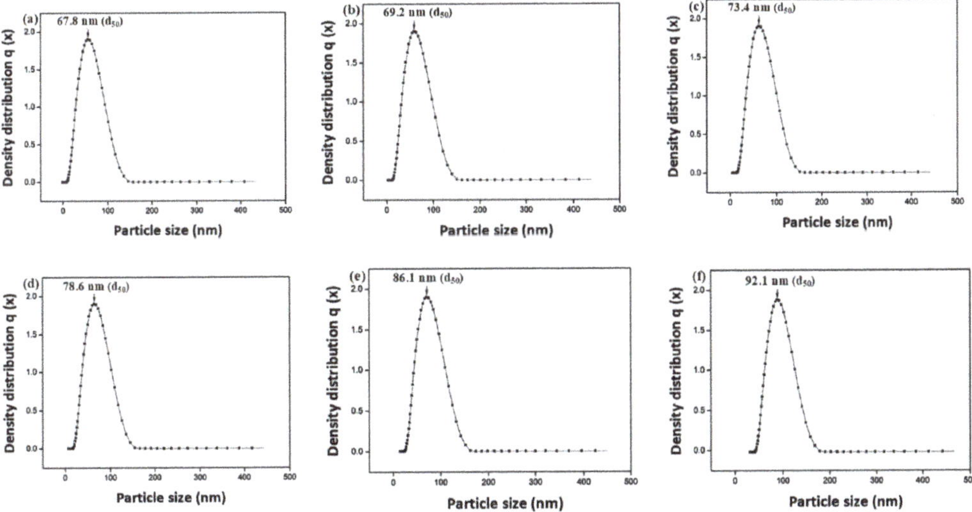

Figure 2. Particle size of nanoemulsion from cypress oil after: (**a**) 0 day, (**b**) 1 day, (**c**) 10 days, (**d**) 20 days, (**e**) 30 days, and (**f**) 45 days.

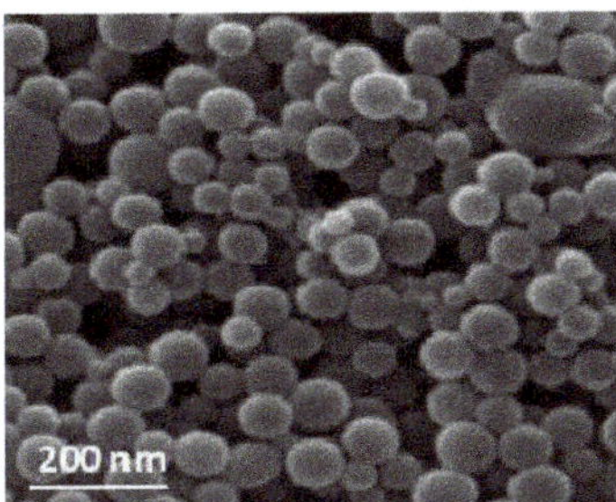

Figure 3. SEM of cypress oil nanoemulsion.

3.3. Contact Bioactivity

Contact bioactivity of EO materials was both dose- and time-dependent (Table 3). The oil/water nanoemulsion caused the strongest activity, with which 100% adult mortality of *S. oryzae* was reached at 10.0 μL/cm^2 at 4 days' following exposure. Under these conditions, the percentage mortality was 71.1, 60.3, 37.1, and 33.6% for EO, α-cedrol, δ-3-carene and α-pinene, respectively. After 7 days from exposing the weevils to 40 μL/cm^2 of botanicals, the mortality of the adults ranged between 77.5 and 100%.

Table 3. Contact insecticidal bioactivity of *S. oryzae* exposed to essential oil, nanoemulsion, and terpene fractions of *C. sempervirens*.

Test Material	Concentration (μL/cm^2)	Mortality (% Mean $\pm$ S.E.) after Exposure Period			
		Day 1	**Day 2**	**Day 4**	**Day 7**
Crude oil	5.0	18.1 $\pm$ 2.3 [ghi]	25.3 $\pm$ 1.1 [fg]	44.3 $\pm$ 2.6 [g]	70.3 $\pm$ 2.1 [d]
	10.0	27.6 $\pm$ 2.1 [f]	46.6 $\pm$ 2.3 [e]	71.1 $\pm$ 2.3 [d]	84.1 $\pm$ 2.2 [c]
	20.0	39.9 $\pm$ 2.1 [e]	63.6 $\pm$ 3.1 [d]	88.0 $\pm$ 3.2 [b]	100.0 $\pm$ 0.0 [a]
	40.0	71.3 $\pm$ 3.0 [b]	93.8 $\pm$ 2.6 [a]	100.0 $\pm$ 0.0 [a]	100.0 $\pm$ 0.0 [a]
Nanoemulsion	5.0	36.3 $\pm$ 3.1 [e]	48.1 $\pm$ 2.1 [e]	83.6 $\pm$ 3.1 [bc]	92.4 $\pm$ 2.1 b [b]
	10.0	54.4 $\pm$ 3.1 [d]	70.9 $\pm$ 2.1 [c]	100.0 $\pm$ 0.0 [a]	100.0 $\pm$ 0.0 [a]
	20.0	67.8 $\pm$ 2.6 [bc]	81.3 $\pm$ 1.9 [b]	100.0 $\pm$ 0.0 [a]	100.0 $\pm$ 0.0 [a]
	40.0	92.1 $\pm$ 3.3 [a]	100.0 $\pm$ 0.0 [a]	100.0 $\pm$ 0.0 [a]	100.0 $\pm$ 0.0 [a]
α-Cedrol	5.0	15.8 $\pm$ 1.1 [ijk]	29.3 $\pm$ 2.1 [f]	36.1 $\pm$ 2.1 [h]	44.8 $\pm$ 2.1 [h]
	10.0	25.5 $\pm$ 2.1 [fg]	41.9 $\pm$ 2.3 [e]	60.0 $\pm$ 2.1 [e]	71.0 $\pm$ 2.1 [d]
	20.0	36.0 $\pm$ 2.3 [e]	60.8 $\pm$ 2.6 [d]	70.9 $\pm$ 1.9 [d]	83.3 $\pm$ 1.7 [c]
	40.0	61.3 $\pm$ 3.1 [cd]	83.1 $\pm$ 3.1 [b]	100.0 $\pm$ 0.0 [a]	100.0 $\pm$ 0.0 [a]
δ-3-Carene	5.0	11.3 $\pm$ 2.0 [jk]	16.6 $\pm$ 2.1 [h]	22.6 $\pm$ 3.1 [i]	29.3 $\pm$ 4.2 [j]
	10.0	19.0 $\pm$ 3.0 [ghi]	30.3 $\pm$ 1.9 [f]	37.1 $\pm$ 2.1 [h]	46.7 $\pm$ 2.4 [h]
	20.0	27.3 $\pm$ 2.3 [f]	41.9 $\pm$ 1.6 [e]	54.3 $\pm$ 2.6 [ef]	65.9 $\pm$ 2.0 [f]
	40.0	38.2 $\pm$ 3.3 [e]	60.4 $\pm$ 3.1 [d]	77.3 $\pm$ 1.9 [cd]	84.3 $\pm$ 1.9 [c]
α-Pinene	5.0	8.3 $\pm$ 1.1 [k]	14.3 $\pm$ 1.1 [h]	19.4 $\pm$ 1.3 [i]	25.4 $\pm$ 2.1 [k]
	10.0	13.1 $\pm$ 1.1 [jk]	20.8 $\pm$ 1.3 [gh]	33.6 $\pm$ 1.3 [h]	42.6 $\pm$ 1.8 [hi]
	20.0	19.6 $\pm$ 1.6 [ghi]	28.5 $\pm$ 2.3 [f]	51.3 $\pm$ 1.9 [f]	56.3 $\pm$ 1.7 [g]
	40.0	23.0 $\pm$ 2.3 [fgh]	44.0 $\pm$ 2.9 [e]	70.5 $\pm$ 2.3 [d]	77.5 $\pm$ 2.0 [e]
* *F*-value	-	167.90	278.00	323.96	436.45

Each result is the mean of 6 replicates, each made with 20 adults (n = 120). Means within a column followed by same letters are not significantly different ($p \leq 0.05$) (Tukey's HSD test). * All *F*-values are significant, at $p \leq 0.001$.

3.4. Fumigation Bioactivity

Results of fumigation bioactivity (Table 4) tests demonstrated that the nanoemulsion and the crude oil exhibited the strongest fumigant bioactivity (10 μL/L air of both botanicals caused 100% adult mortality after 4 days). At these conditions, (%) mortality was 68.6, 53.3, and 51.6% for α-cedrol, α-pinene, and δ-3-carene, respectively. After 4 days of exposing weevils to 20 μL/L air, α-cedrol caused 100% adult mortality. After 7 days of exposing insects to 20 μL/L air of botanicals, the mortality of adult weevils ranged between 81.5 and 100%.

Table 4. Fumigant insecticidal bioactivity of *S. oryzae* exposed to essential oil, nanoemulsion, and terpene fractions of *C. sempervirens*.

Test Material	Concentration (μL/L Air)	Mortality (% Mean $\pm$ S.E.) after Exposure Period			
		Day 1	**Day 2**	**Day 4**	**Day 7**
Crude oil	2.5	15.6 $\pm$ 1.3 [ijk]	23.1 $\pm$ 1.6 [fg]	39.3 $\pm$ 2.4 [e]	53.8 $\pm$ 2.1 [h]
	5.0	24.1 $\pm$ 1.3 [hi]	36.0 $\pm$ 2.3 [e]	63.6 $\pm$ 3.2 [c]	72.7 $\pm$ 2.4 [e]
	10.0	33.6 $\pm$ 2.1 [efg]	55.3 $\pm$ 3.6 [cd]	100.0 $\pm$ 0.0 [a]	100.0 $\pm$ 0.0 [a]
	20.0	55.3 $\pm$ 2.3 [c]	94.6 $\pm$ 2.2 [a]	100.0 $\pm$ 0.0 [a]	100.0 $\pm$ 0.0 [a]

Table 4. *Cont.*

Test Material	Concentration (µL/L Air)	Mortality (% Mean ±S.E.) after Exposure Period			
		Day1	**Day 2**	**Day 4**	**Day 7**
Nanoemulsion	2.5	27.5 ± 1.1 [fgh]	36.1 ± 2.3 [e]	66.6 ± 2.1 [c]	74.1 ± 2.4 [e]
	5.0	49.9 ± 2.3 [cd]	60.3 ± 3.6 [c]	92.1 ± 2.1 [a]	100.0 ± 0.0 [a]
	10.0	68.4 ± 3.1 [b]	94.3 ± 3.0 [a]	100.0 ± 0.0 [a]	100.0 ± 0.0 [a]
	20.0	91.1 ± 3.3 [a]	100.0 ± 0.0 [a]	100.0 ± 0.0 [a]	100.0 ± 0.0 [a]
α-Cedrol	2.5	13.3 ± 1.6 [jk]	20.6 ± 3.3 [fg]	34.3 ± 4.5 [ef]	49.5 ± 3.4 [i]
	5.0	20.3 ± 2.1 [hij]	32.3 ± 3.3 [e]	49.3 ± 4.5 [d]	58.9 ± 2.7 [g]
	10.0	34.8 ± 2.1 [ef]	47.9 ± 3.6 [d]	68.6 ± 3.3 [c]	76.1 ± 2.3 [e]
	20.0	46.4 ± 2.3 [d]	70.3 ± 4.1 [b]	100.0 ± 0.0 [a]	100.0 ± 0.0 [a]
δ-3-Carene	2.5	9.3 ± 1.3 [k]	15.3 ± 3.3 [fg]	30.3 ± 4.5 [fg]	40.8 ± 3.4 [j]
	5.0	14.3 ± 2.3 [jk]	23.0 ± 3.3 [f]	40.0 ± 4.5 [e]	52.5 ± 2.8 [h]
	10.0	25.0 ± 2.0 [gh]	37.3 ± 3.6 [e]	53.3 ± 3.3 [d]	64.0 ± 3.2 [f]
	20.0	37.3 ± 2.9 [e]	56.3 ± 4.1 [c]	77.0 ± 3.9 [b]	89.9 ± 2.1 [c]
α-Pinene	2.5	8.9 ± 1.1 [k]	12.9 ± 3.3 [g]	26.0 ± 3.2 [g]	33.0 ± 3.3 [g]
	5.0	12.1 ± 1.1 [jk]	19.1 ± 1.8 [fg]	36.3 ± 4.5 [ef]	47.4 ± 3.6 [ef]
	10.0	19.6 ± 2.1 [hij]	30.9 ± 1.5 [e]	51.6 ± 3.1 [d]	59.7± 3.1 [g]
	20.0	33.1 ± 2.3 [efg]	53.3 ± 2.3 [cd]	70.1 ±2.9 [c]	81.2 ±2.6 [d]
* *F*-value	-	134.04	265.03	317.22	302.08

Each result is the mean of 6 replicates, each made with 20 adults (n = 120). Means within a column followed by same letters are not significantly different ($p \leq 0.05$) (Tukey's HSD test). * All *F*-values are significant, at $p \leq 0.00$.

3.5. The Dose-Response Mortality

The LC_{50} and LC_{90} and their confidence limits are illustrated in (Table 5). For the contact bioassay, LC_{50} values of the botanicals after 48 h of treatment were: Nanoemulsion (LC_{50} = 5.8 µL/cm^2, χ^2 = 0.94, df = 4), crude oil (LC_{50} = 13.3 µL/cm^2, χ^2 = 1.33, df = 4), α-cedrol (LC_{50} = 15.1 µL/cm^2, χ^2 = 2.04, df = 4), δ-3-carene (LC_{50} = 30.7 µL/cm^2, χ^2 = 2.77, df = 4), and α-pinene (LC_{50} = 53.4 µL/cm^2, χ^2 = 3.12, df = 4). The $LC_{50's}$ of the phytochemicals after 48 h fumigation were as follows: Nanoemulsion (LC_{50} = 4.1 µL/L air, χ^2 = 0.91, df = 54), crude oil (LC_{50} = 8.7 µL/L air, χ^2 = 1.08, df = 4), α-cedrol (LC_{50} = 12.2 µL/L air, χ^2 = 2.16, df = 4), δ-3-carene (LC_{50} = 17.2 µL/L air, χ^2 = 3.06, df = 54), and α-pinene (LC_{50} = 19.6 µL/L air, χ^2 = 3.22, df = 4).

Table 5. * LC_{50} and LC_{95} and their fiducial limits of EO materials against *S. oryzae* 48 h post treatment.

Test Material	Bioassay	LC_{50} ** (95% fl)	LC_{95} ** (95% fl)	Slope (±S.E.)	*** χ^2 (df = 4)
Crude oil	Contact (µL/cm^2)	13.3 (11.1–16.3)	25.9 (19.3–32.2)	2.1 ± 0.20	1.33
	Fumigation (µL/L)	8.7 (7.5–10.1)	16.3 (13.8–21.3)	2.0 ± 0.24	1.08
Nanoemulsion	Contact (µL/cm^2)	5.8 (5.3–7.2)	10.2 (8.6–13.3)	1.5 ± 0.18	0.94
	Fumigation (µL/L)	4.1 (3.7–4.9)	7.3 (6.1–8.6)	1.6 ± 0.14	0.91
α-Cedrol	Contact (µL/cm^2)	15.1 (13.4–18.9)	27.5 (22.9–35.3)	2.1 ± 0.28	2.04
	Fumigation (µL/L)	12.2 (10.5–15.8)	22.9 (18.6–27.1)	2.6 ± 0.26	2.18
δ-3-Carene	Contact (µL/cm^2)	30.7 (27.6–36.6)	55.3 (48.4–64.8)	2.9 ± 0.32	2.77
	Fumigation (µL/L)	17.2 (15.4–21.3)	39.6 (34.7–47.6)	2.8 ± 0.40	3.06
α-Pinene	Contact (µL/cm^2)	53.4 (46.3–63.1)	114.8 (101.8–119.1)	3.1 ± 0.30	3.12
	Fumigation (µL/L)	19.6 (17.3–24.5)	42.2 (37.0–50.3)	2.7 ± 0.41	3.22

Each result is the mean of 6 replicates, each including 20 individuals (n = 120). * LC_{50} and LC_{95} are considered significantly different when the 95% fiducial limits (f.l.) fail to overlap. ** fl = fiducial limits. *** *Chi*-square value, significant at $p \leq 0.05$ level; df = degree of freedom.

3.6. Repellence Bioactivity

As illustrated in Table 6 and Figure 4, the EO materials strongly repelled the adult weevils, and the repellent bioactivity was both time- and dose-dependent. The crude oil of *C. sempervirens* was the strongest insect repellent, even at low concentrations, followed by nanoemulsion, α-cedrol, and δ-3-carene; by contrast, the monoterpene α-pinene showed a weak-to-moderate repelling efficacy. The crude oil completely repelled the adult weevils at 0.44 μL/cm^2 after 12 h. The crude oil, nanoemulsion, and α-cedrol caused 100% repellency when the weevils were treated with a concentration equaling 0.88 μL/cm^2 of these products after 24 h. At this concentration, the remaining monoterpenes caused moderate repelling activities. At the lowest concentration tested (0.22 μL/cm^2), the percentage repellency was 73.9, 61.3, 43.9, 33.3, and 22.1% for the EO, the nanoemulsion, α-cedrol, δ-3-carene, and α-pinene, respectively.

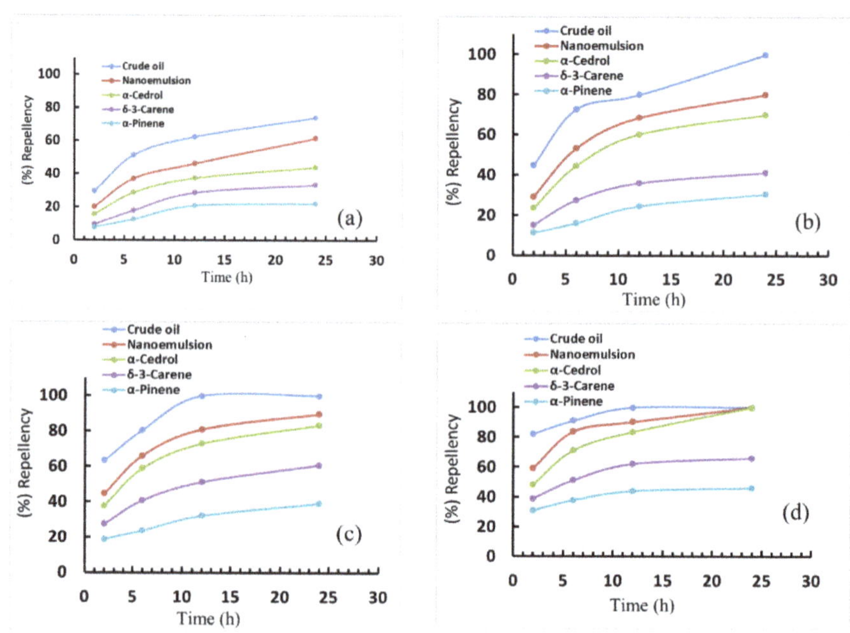

Figure 4. Repellent activity of cypress EO products against *S. oryzae* at (**a**) 0.11 μL/cm^2; (**b**) 0.22 μL/cm^2; (**c**) 0.44 μL/cm^2; and (**d**) 0.88 μL/cm^2. (%) (Repellency in control was nil).

Table 6. Repellence activity of EO, nanoemulsion, and terpenes of *C. sempervirens* against *S. oryzae* adult weevils.

Test Material	Concentration (μL/cm^2)	Repellency (% Mean ± S.E.) after Period (h)			
		2	6	12	24
Crude oil	0.11	29.6 ± 1.1 [gh]	51.3 ± 2.3 [efg]	62.3 ± 3.1 [f]	73.9 ± 2.6 [e]
	0.22	44.9 ± 2.3 [e]	72.9 ± 2.3 [bcd]	80.1 ± 2.9 [c]	100.0 ± 0.0 [a]
	0.44	63.3 ± 3.1 [b]	80.6 ± 1.9 [abc]	100.0 ± 0.0 [a]	100.0 ± 0.0 [a]
	0.88	82.3 ± 3.1 [a]	91.1 ± 2.6 [a]	100.0 ± 0.0 [a]	100.0 ± 0.0 [a]
Nanoemulsion	0.11	20.1 ± 1.3 [j]	37.1 ± 1.3 [gh]	46.1 ± 2.3 [h]	61.3 ± 2.1 [h]
	0.22	29.3 ± 1.6 [gh]	53.3 ± 2.1 [ef]	68.6 ± 2.6 [e]	80.1 ± 2.2 [d]
	0.44	44.6 ± 2.1 [e]	66.1 ± 3.6 [cde]	80.9 ± 3.3 [c]	89.6 ± 0.0 [b]
	0.88	59.3 ± 2.6 [c]	83.9 ± 2.6 [ab]	90.3 ± 3.1 [b]	100.0 ± 0.0 [a]
α-Cedrol	0.11	15.6 ± 1.3 [k]	28.6 ± 2.1 [hi]	37.3 ± 2.3 [i]	43.9 ± 2.3 [i]
	0.22	23.6 ± 2.1 [i]	44.6 ± 2.3 [fg]	60.3 ± 2.1 [f]	70.1 ± 2.1 [f]
	0.44	37.9 ± 2.1 [f]	58.9 ± 2.6 [def]	72.9 ± 1.9 [d]	83.3 ± 1.6 [c]
	0.88	48.3 ± 2.3 [d]	71.3 ± 3.1 [bcd]	83.6 ± 2.1 [c]	100.0 ± 0.0 [a]

Table 6. *Cont.*

Test Material	Concentration (μL/cm^2)	Repellency (% Mean $\pm$ S.E.) after Period (h)			
		2	**6**	**12**	**24**
δ-3-Carene	0.11	9.6 $\pm$ 1.3 [lm]	17.9 $\pm$ 1.9 [ij]	28.6 $\pm$ 3.1 [k]	33.3 $\pm$ 3.1 [k]
	0.22	15.3 $\pm$ 2.1 [k]	27.6 $\pm$ 1.6 [hij]	36.1 $\pm$ 2.1 [i]	41.3 $\pm$ 2.6 [j]
	0.44	27.6 $\pm$ 2.1 [h]	43.9 $\pm$ 2.6 [fg]	51.3 $\pm$ 2.6 [g]	60.9 $\pm$ 2.3 [h]
	0.88	39.1 $\pm$ 2.3 [f]	51.3 $\pm$ 3.3 [efg]	62.3 $\pm$ 1.9 [f]	66.1 $\pm$ 1.9 [g]
α-Pinene	0.11	7.9 $\pm$ 1.1 [m]	12.6 $\pm$ 1.1 [k]	20.9 $\pm$ 1.6 [m]	22.1 $\pm$ 2.3 [m]
	0.22	11.3 $\pm$ 1.3 [l]	16.1 $\pm$ 1.3 [ij]	24.6 $\pm$ 1.9 [l]	30.6 $\pm$ 1.9 [l]
	0.44	18.9 $\pm$ 1.6 [j]	23.9 $\pm$ 2.3 [hij]	32.3 $\pm$ 1.9 [j]	39.3 $\pm$ 2.6 [j]
	0.88	31.1 $\pm$ 2.1 [g]	37.9 $\pm$ 2.9 [gh]	44.1 $\pm$ 2.1 [h]	46.1 $\pm$ 2.3 [i]
Control	-	0.0 $\pm$ 0.0	0.0 $\pm$ 0.0	0.0 $\pm$ 0.0	0.0 $\pm$ 0.0
* *F*-value	-	961.75	61.65	1237.03	2644.87

Each result is the mean of 5 repeats, each including 20 individuals (n = 100). Means within a column followed by same letter(s) are not significantly different. ($p \leq 0.05$) (Tukey's HSD test). * All *F*-values are significant at $p \leq 0.001$.

3.7. AChE Inhibition

All test EO materials caused a remarkable inhibition of AChE activity in *S. oryzae* (Table 7). The nanoemulsion (IC$_{50}$ = 9.88 mM, χ^2 = 1.68, df = 5, p = 0.201) was the superior AChE inhibitor, followed by EO (IC$_{50}$ = 14.03 mM, χ^2 = 2.41, df = 5, p = 0.243), and α-cedrol (IC$_{50}$ = 17.21 mM, χ^2 = 2.88, df = 5, p = 0.311). By contrast, δ-3-carene and α-pinene caused moderate effects. The IC$_{50}$ of methomyl was 2.44 $\times$ 10^{-3} mM.

Table 7. Inhibition of acetylcholinesterase (AChE) of *S. oryzae* larvae by EO materials of *C. sempervirens*.

Plant Material	* IC$_{50}$ (mM)	(95% Fiducial Limits)	Slope ($\pm$S.E.)	** χ^2 (df = 5)	p
Crude oil	14.03	(12.21–16.28)	1.44 $\pm$ 0.22	2.41	0.243
Nanoemulsion	9.88	(7.94–11.71)	1.09 $\pm$ 0.14	2.68	0.201
α-Cedrol	17.21	(15.20–20.07)	1.30 $\pm$ 0.19	2.88	0.311
δ-3-Carene	34.54	(30.00–39.33)	1.66 $\pm$ 0.25	3.82	0.355
α-Pinene	39.83	(34.44–46.12)	2.08 $\pm$ 0.34	4.05	0.512
Methomyl	2.17 $\times$ 10^{-3}	(1.73 $\times$ 10^{-3}–3.66 $\times$ 10^{-3})	1.03 $\pm$ 0.18	2.62	0.377

* The concentration causing 50% enzyme inhibition. ** *Chi*-square value, not significant at $p \leq 0.05$ level; df = degree of freedom.

3.8. Phytotoxicity Assessment

Phytotoxicity testing revealed that the botanicals were not phytotoxic to wheat plants, where the agronomical parameters of wheat (%) germination, and the growth of radicals and shoots were unaffected after treatment with botanicals at concentrations ranging between 50.0 and 150 μL/mL (Table 8 and Figure 5). Percentage germination and growth of shoots are slightly affected at 150 μL/mL, especially with EO, nanoemulsion, and α-cedrol. By contrast, the remaining compounds were nonphytotoxic, even at high test concentrations.

Table 8. * Phytotoxic activities of essential oil, nanoemulsion, and major fractions of *C. sempervirens* against wheat plants.

Plant Material	Concentration (μL/mL)	Germination (%)	RL	SL
Crude oil	50	90.6 $\pm$ 1.4 [a]	9.08 $\pm$ 0.31 [a]	3.32 $\pm$ 0.14 [ab]
	100	80.3 $\pm$ 1.5 [ab]	8.89 $\pm$ 0.23 [a]	3.12 $\pm$ 0.11 [ab]
	150	70.1 $\pm$ 1.3 [b]	8.07 $\pm$ 0.18 [a]	2.26 $\pm$ 0.15 [c]

Table 8. *Cont.*

Plant Material	Concentration (μL/mL)	Germination (%)	RL	SL
Nanoemulsion	50	88.9 ± 1.3 [a]	9.03 ± 0.19 [a]	3.16 ± 0.12 [ab]
	100	76.3 ± 1.2 [ab]	8.70 ± 0.20 [a]	3.03 ± 0.14 [ab]
	150	65.2 ± 1.2 [bc]	7.08 ± 0.28 [a]	2.05 ± 0.12 [c]
α-Cedrol	50	88.2 ± 1.3 [a]	9.11 ± 0.41 [a]	3.28 ± 0.15 [ab]
	100	83.4 ± 1.9 [ab]	9.02 ± 0.26 [a]	3.20 ± 0.13 [ab]
	150	74.2 ± 1.4 [b]	8.24 ± 0.12 [a]	2.59 ± 0.17 [bc]
δ-3-Carene	50	90.6 ± 1.7 [a]	9.20 ± 0.18 [a]	3.39 ± 0.16 [a]
	100	88.7 ± 1.4 [a]	9.12 ± 0.31 [a]	3.34 ± 0.11 [a]
	150	88.3 ± 1.9 [a]	9.01 ± 0.22 [a]	3.09 ± 0.11 [ab]
α-Pinene	50	91.8 ± 1.1 [a]	9.24 ± 0.19 [a]	3.41 ± 0.13 [a]
	100	91.0 ± 1.3 [a]	9.23 ± 0.08 [a]	3.31 ± 0.11 [ab]
	150	90.2 ± 1.3 [a]	9.19 ± 0.22 [a]	3.16 ± 0.12 [ab]
Control	-	91.7 ± 1.6 [a]	9.22 ± 0.32 [a]	3.43 ± 0.18 [a]
F-value	-	4.28	1.16	7.94

* Each value is the mean ± S.E. of 4 trials; RL = Radicle growth (length of seeds, cm); SL = Shoot length (cm). In a column, means followed by same letter (s) are not significantly different ($p \leq 0.05$). All *F*-values are significant at $p \leq 0.001$.

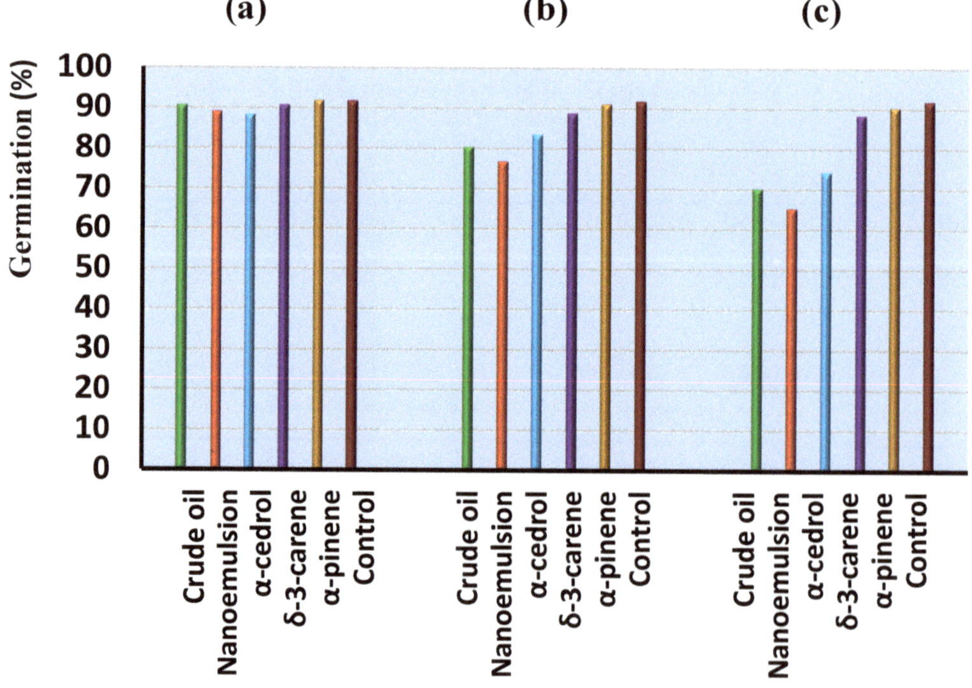

Figure 5. Viability of wheat grains as (%) germination treated with cypress EO products at (**a**) 50 μL/mL; (**b**) 100 μL/mL; and (**c**) 150 μL/mL.

3.9. Toxicity against Earthworms

The test botanicals showed relative safety toward *E. fetida*. Neither mortality nor toxicity signs were recorded in treated animals, even at 200 mg per kg^{-1} soil. On the other hand, the chemical pesticide α-cypermethrin at 20.0 mg per kg^{-1} caused 100% mortality of earthworms after 10 days.

4. Discussion

The yield and composition of *C. sempervirens* var. *horizontalis* oil are in a good accordance with previous studies of the same Saudi species or those of other similar flora, where α-pinene was the main component of cypress EO [21,39–42]. Variations both in the yield and the abundant oil terpenes of plant oils have been recorded in previous reports [22,43–45]. However, β-thujene was presented as the main terpene component (31.4%) in the EO of Brazilian *C. sempervirens* [43]. These variations are mainly dependent on many factors, including genetic and geographic factors. The soil status, the method of cultivation, water availability, seasonality, the extracted parts, and extraction techniques are also of major influence [6,46–48].

According to our findings, the EO of cypress, its nanoemulsion, and individual terpenes exhibited remarkable insecticidal, repellent, and AChE effects against *S. oryzae*. To our knowledge, bioactivity of the EO of cypress belonging to Saudi flora, particularly its nanoemulsion and individual terpenes, had not been investigated against insects of stored grain; hence our study is considered a first report. A remarkable fumigation bioactivity of the EO extracted from Egyptian cypress was recorded against adults of *S. oryzae* with $LC_{50} = 17.2$ mg/L air.

The EO of *C. sempervirens* was reported to possess toxic and repellent bioactivities against *Sitophilus zeamais* and *Tribolium confusum* using the impregnated filter paper bioassay, as well as treated grains [22]; a remarkable repellent potential against the codling moth, *Cydia pomonella*; and moderate toxicity and repellent activities against the mosquitoes, *Aedes albopictus* and *Ae. aegypti* [49–51]. The oil materials tested herein strongly repelled *S. oryzae* adult weevils. There are many factors affecting the repellent bioactivity of the plant-based products against harmful insects, which depend mainly on the nature of products under investigation; the respiratory system upon which the plant bioactive substances act, and the insect's olfactory receptors are of a major influence [52]. The high volatile nature of EOs play a main role in this phenomenon, where they can be inhaled, ingested, or easily absorbed through the insect's skin [53,54]. The repellence bioactivity of cypress EO products toward *S. oryzae* was in accordance with studies and reports that investigated the bioinsecticidal and repellant potential of plant EOs against insects of stored grain, including *S. oryzae* [10,11,55–57].

In this study, a green approach was followed to prepare an oil-in-water nanoemulsion (droplet size 67.8 ± 3.1 nm) from cypress oil using fewer toxic chemicals at acceptable concentrations, in the proportions 5:90:5% (EO:H_2O:Tween 80 as an emulsifier). Tween surfactants, especially Tween 80 and Tween 20, are frequently utilized as emulsifiers in the preparations of oil/water nanoemulsions as they can produce stable formulations without using cosurfactants [58]; However, Tween 80 was selected herein as a nonanionic emulsifier, due to its miscibility with water and good solubility for EOs. It is characterized by a high hydrophilic–lipophilic balance (HLB = 15); hence, decreasing the tension between the oil and aqueous phases, and resulting in the formation of stable emulsions [17]. In most cases, Tween 80 also appeared to perform better than Tween 20 in terms of droplet size distribution and the stability of the nanoemulsion, which may be due to the structural differences in the nonpolar tail of the two molecules [59,60]. In both micro- and nanoemulsion preparations, the surfactant functions to reduce the interfacial energy by providing a mechanical barrier to coalescence [17]. The nanoemulsion of cypress oil exhibited a good stability up to 45 days after preparation when exposed to stress conditions during storage. Meanwhile, nonequilibrium emulsion formulations may undergo a breakdown, resulting in sedimentation, flocculation, and coalescence, resulting in many shortcomings in their biological activities. Alternatively, because of their novel properties, such as subcellular size, nanoemulsions have good stability under extreme conditions [17,61]. The pH of the nanoemulsion stabilized around 6.5 during storage. The pH of an emulsion is critical to its stability because changes in pH affect the surface charge of the globules, disrupting their stability. Furthermore, increases in the surface charge of globules cause electrostatic repulsion, which reduces flocculation and leads to the dissolution of micro- and nanoemulsions [17]. In a nanoemulsion formulation, the PDI determines droplet size stability and uniformity; a low PDI ensures high droplet size uniformity. Over 30 days of storage, the PDI of cypress oil nanoemulsion ranged between 0.18 ± 0.03 and 0.24 ± 0.02. Many authors have reported that a PDI of less than 0.25 indicates a narrow distribution of particle size, providing stability and homogeneity due to a reduced Ostwald ripening [17,61].

The nanoemulsion of cypress oil exhibited superior bioactivity against the target weevil. When materials are formulated at the nanoscale, they acquire novel chemical and physical properties, such as increased surface area, solubility, and high affinity to the targeted biosystems, which promotes their biological activities [17]. Because of these novel criteria, nanomaterials are promising candidates for developing effective eco-friendly insecticides. To avoid the overuse of the toxic solvents or high-energy inputs that are commonly used in pesticide synthesis, the "green synthesis" concept has been proposed, outlining the potential use of animal, microbial, and plant-borne compounds as stabilizing agents for the production of bioactive nanomaterials [15,62]. As a result, nanotechnology is being considered as an alternative strategy to improve the stability and bioactivity of pesticide materials that rely on various nanocarriers, such as plant-oil-based nanoemulsions [6,15,17,63]. In the literature, the reported insecticidal activity of plant-based nanopesticides, including oil nanoemulsions against serious insects, such as those infesting stored grains, has been reported [6,15,16,18]. Nenaah reported that nanoemulsions made from the EOs of three Achillea species, *A. biebersteinii, A. santolina, and A. millefolium,* outperformed their bulk counterparts in adulticidal activity against *T. castaneum* [6]. Similar results have been reported for *T. confusum* and *Cryptolestes ferrugineus* [18,63].

The bioactivity of plant oils are attributed to several components, with demonstrable insecticidal activity contained in the plant EO, especially in monoterpenes such as α-pinene, α-terpinene, limonene, camphor, carvacrol, thymol, δ-3-carene, α-thujone, 1,8-cineol (eucalyptol), eugenol, and ascaridole [3,6,9–11,17,57]. Although a synergism with other minor constituents is common where each oil component participates in penetration, fixation, and distribution into biomembranes [6,9,17], synergy between components of an EO might be occur between several components contained in the same oil, or between different essential oils with known biological activities [17,64,65].

Essential oils, particularly monoterpenes, are volatile and lipophilic, allowing them to quickly penetrate the integument of insects, interfering with physiological parameters and causing alteration in all vital functions. [6,9,17,53]. The EO materials caused a considerable inhibition in the AChE bioactivity of *S. oryzae*, indicating a neurotoxic mechanism of action. As mixtures, the toxicity of EOs is not yet fully understood. Nevertheless, the rapid action against some pests is major evidence of a neurotoxic action, which is attributed to AChE inhibition, as described herein [9–12,17]. Comparing our results with previous reports, the dichloromethane, acetone, ethyl acetate, and methanol extracts of the cones and leaves of *Cupressus sempervirens* var. *horizantalis* displayed a moderate inhibition of butyrylcholinesterase, AChE, and tyrosinase bioactivities at 200 µg/mL [66]. Aazza reported the acetylcholinesterase inhibitory effect of cypress oil (where IC_{50} was 0.2837 mg/mL) using bovine acetylcholine [67]. Recently, Alimi et al. found that the EO of cypress displayed a significant inhibition in AChE activity of *Hyalomma scupense* (Acari: Ixodidae) [68]. The plant EOs can interfere with other protein targets, which disrupts the insect's nervous system, such as the nicotinic acetylcholine receptors (nAChR), and the octopamine or the neurotransmitter inhibitor γ-aminobutyric acid (GABA). EOs were found to inhibit enzymatic biosystems (superoxide dismutase (SOD), catalase (CAT), glutathione-S-transferase (GST), and glutathione reductase (GR)), peroxidases (POx), and the nonenzymatic (glutathione (GSH)) antioxidant defense biosystems [9,11].

According to our findings, the EO materials showed a relative safety within the limit of the test concentrations when tested on *E. fetida* (a common earthworm) and wheat plants. Plant products with pesticidal activities are often wrongly considered safe with no negative effects on nontargets, including humans, without this being experimentally verified. Nevertheless, many authors stated that EOs, nanoemulsion preparations, and oil terpenes were relatively safe when assessed against several nontarget species [3,17,28,69]. In that regard, most reports have focussed on assessing the acute toxicity, whereas both subchronic and chronic evaluations have not been fully undertaken [5,17,69–71]. Regarding cypress oil, health risks or side effects following administration of designated therapeutic dosages are not recorded. Nevertheless, kidney irritation was recorded with the intake of large doses [72]. However, not all natural products are free of risk, therefore deep investigations are always required to explore the biosafety of the plant-based pesticides before practical use in stored-product insect control programs. The authors should discuss the findings and how they can be interpreted in light of previous research and the working hypotheses. The findings and implications should be discussed in the broadest possible context. Future research directions may be highlighted as well.

5. Conclusions

According to the results of the present study, cypress EO, the oil nanoemulsion, and its individual terpenes showed remarkable insecticidal, repellence, and acetylcholinesterase inhibitory bioactivities against the rice weevil, *S. oryzae*. There were no significant adverse effects on the earthworms, nor the agronomical parameters of wheat plant. When properly prepared, cypress oil, its nanoemulsion, and its main terpenes could be applied as novel ecofriendly natural pest-control options against *S. oryzae*, being more appropriate than the chemical insecticides. However, deep toxicological evaluations should be carried out to substantiate the relevant concentrations and adverse effects of the test products against mammals and other nontarget organisms.

Author Contributions: Conceptualization, G.E.N.; methods, investigation, and validation, G.E.N., B.Z.A., A.A.A., D.M.S. and S.A.; analysis, G.E.N.; resources, curation, G.E.N., B.Z.A., S.A., D.M.S. and A.A.A.; writing original draft, G.E.N.; review and editing, G.E.N. and A.A.A.; supervision, G.E.N. All authors have read and agreed to the published version of the manuscript.

Funding: This research received no external funding.

Institutional Review Board Statement: The study was conducted in accordance with the Declaration of Helsinki, and approved by the Institutional Ethics Committee.

Informed Consent Statement: Not applicable.

Data Availability Statement: Data supporting the conclusions of this article are presented in the main manuscript.

Conflicts of Interest: The authors declare no conflict of interest.

References

1. FAO. *The Future of Food and Agriculture—Trends and Challenges*; Food and Agriculture Organization of the United Nations: Rome, Italy, 2017.
2. Sparks, T.C.; Bryant, R.J. Impact of natural products on discovery of, and innovation in, crop protection compounds. *Pest Manag. Sci.* **2021**, *78*, 399–408. [CrossRef] [PubMed]
3. Prakash, B.; Kumar, A.; Singh, P.; Das, S.; Dubey, N.K. Prospects of plant products in the management of insect pests of food grains: Current status and future perspectives. In *Natural Bioactive Compounds*; Sinha, R.P., Häder, D.-P., Eds.; Academic Press: Cambridge, MA, USA, 2021; pp. 317–335. [CrossRef]
4. Hossain, F.; Boddupalli, P.M.; Sharma, R.K.; Pradyumn Kumar, P.; Bahadur Singh, B. Evaluation of quality protein maize genotypes for resistance to stored grain weevil *Sitophilus oryzae* (Coleoptera: Curculionidae). *Int. J. Trop. Insect Sci.* **2007**, *27*, 114–121. [CrossRef]
5. Desneux, N.; Decourtye, A.; Delpuech, J.M. The sublethal effects of pesticides on beneficial arthropods. *Annu. Rev. Entomol.* **2007**, *52*, 81–106. [CrossRef] [PubMed]
6. Nenaah, G.E. Chemical composition, toxicity and growth inhibitory activities of essential oils of three Achillea species and their nano-emulsions against *Tribolium castaneum* (Herbst). *Ind. Crops Prod.* **2014**, *53*, 252–260. [CrossRef]
7. Okpile, C.; Zakka, U.; Nwosu, L.C. Susceptibility of ten rice brands to weevil, *Sitophilus oryzae* L. (Coleoptera: Curculionidae), and their influence on the insect and infestation rate. *Bull. Natl. Res. Cent.* **2021**, *45*, 2. [CrossRef]
8. Sparks, T.C.; Bryant, R.J. Innovation in insecticide discovery: Approaches to the discovery of new classes of insecticides. *Pest Manag. Sci.* **2022**, *78*, 3226–3247. [CrossRef]

9.   Isman, M.B. Bioinsecticides based on plant essential oils: A short overview. *Z. Naturforsch.* **2020**, *75*, 179–182. [CrossRef]
10. Abdelgaleil, S.A.M.; Mohamed, M.I.E.; Shawir, M.S.; Abou-Taleb, H.K. Chemical composition, insecticidal and biochemical effects of essential oils of different plant species from Northern Egypt on the rice weevil, *Sitophilus oryzae* L. *J. Pest Sci.* **2016**, *89*, 219–229. [CrossRef]
11. Chaudhari, A.K.; Singh, V.K.; Kedia, A.; Das, S.; Dubey, N.K. Essential oils and their bioactive compounds as eco-friendly novel green pesticides for management of storage insect pests: Prospects and retrospects. *Environ. Sci. Pollut. Res.* **2021**, *28*, 18918–18940. [CrossRef]
12. Pavela, R.; Benelli, G. Essential oils as ecofriendly biopesticides? Challenges and constraints. *Trends Plant Sci.* **2016**, *21*, 1000–1007. [CrossRef]
13. Spinozzi, E.; Pavela, R.; Bonacucina, G.; Perinelli, D.R.; Cespi, M.; Petrelli, R.; Cappellacci, L.; Fiorini, D.; Scortichini, S.; Garzoli, S.; et al. Spilanthol-rich essential oil obtained by microwave-assisted extraction from *Acmella oleracea* (L.) R.K. Jansen and its nanoemulsion: Insecticidal, cytotoxic and anti-inflammatory activities. *Ind. Crops Prod.* **2021**, *172*, 114027. [CrossRef]
14. Kavallieratos, N.G.; Bonacucina, G.; Nika, E.P.; Skourti, A.; Georgakopoulou, S.K.C.; Filintas, C.S.; Panariti, A.M.E.; Maggi, F.; Petrelli, R.; Ferrati, M.; et al. The Type of Grain Counts: Effectiveness of Three Essential Oil-Based Nanoemulsions against Sitophilus oryzae. *Plants* **2023**, *12*, 813. [CrossRef] [PubMed]
15. Benelli, G. Green synthesis of nanomaterials and their biological applications. *Nanomaterials* **2021**, *11*, 2842. [CrossRef] [PubMed]
16. Srivastava, S.; Bhargava, A. Applications of biosynthesized nanoparticles. In *Green Nanoparticles: The Future of Nanobiotechnology*; Springer: Singapore, 2022. [CrossRef]
17. Nenaah, G.E.; Almadiy, A.A.; Al-Assiuty, B.A.; Mahnashi, M.H. The essential oil of *Schinus terebinthifolius* and its nanoemulsion and isolated monoterpenes: Investigation of their activity against *Culex pipiens* with insights into the adverse effects on non-target organisms. *Pest Manag. Sci.* **2022**, *78*, 1035–1047. [CrossRef]
18. Giunti, G.; Palermo, D.; Laudani, F.; Algeri, G.M.; Campolo, O.; Palmeri, V. Repellence and acute toxicity of a nano-emulsion of sweet orange essential oil toward two major stored grain insect pests. *Ind. Crops Prod.* **2019**, *142*, 111869. [CrossRef]
19. Rawat, P.; Khan, M.F.; Kumar, M.; Tamarkar, A.K.; Srivastava, A.K.; Arya, K.R.; Maurya, R. Constituents from fruits of *Cupressus sempervirens*. *Fitoterapia* **2010**, *81*, 162–166. [CrossRef]
20. Grunewald, J.; Brendler, T.; Jaenicke, C. *PDR for Herbal Medicine*, 4th ed.; Medical Economics Co.: Montreal, QC, Canada, 2007; pp. 248–249.
21. Galovičová, L.; Čmiková, N.; Schwarzová, M.; Vukic, M.D.; Vukovic, N.L.; Kowalczewski, P.Ł.; Bakay, L.; Kluz, M.I.; Puchalski, C.; Obradovic, A.D.; et al. Biological Activity of *Cupressus sempervirens* Essential Oil. *Plants* **2023**, *12*, 1097. [CrossRef]
22. Tapondjou, A.L.; Adler, C.; Fontem, D.A.; Bouda, H.; Rechmuth, C. Bioactivities of cymol and essential oils of *Cupressus sempervirens* and *Eucalyptus saligna* against *Sitophilus zeamais* Motschulsky and *Tribolium confusum* du Val. *J. Stored Prod. Res.* **2005**, *41*, 91–102. [CrossRef]
23. Hasnaoui, F.; Zouaoui, I.; Dallali, S.; Dibouba, F. Insecticidal effects of essential oils from six aromatic and medicinal plants on the pine processionary caterpillar (*Thaumetopoea pityocampa* Schiff). *Adv. Med. Plant Res.* **2020**, *8*, 81–88.
24. Adams, R.P. *Identification of Essential Oil Components by Gas Chromatography/Mass Spectrometry*, 4th ed.; Allured Publishing Corporation: Carol Stream, IL, USA, 2007; ISBN 9781932633214.
25. Linstorm, P. NIST Chemistry Webbook, NIST Standard Reference Database Number 69. *J. Phys. Chem. Ref. Data Monogr.* **1998**, *9*, 1–1951.
26. Matysik, E.; Wofniak, A.; Paduch, R.; Rejdak, R.; Polak, B.; Donica, H. The new TLC method for separation and determination of multicomponent mixtures of plant extracts. *J. Anal. Methods Chem.* **2016**, *6*, 1813581. [CrossRef] [PubMed]
27. Ellman, G.L.; Courtney, K.D.; Andres, V.; Featherstone, R.M., Jr. A new and rapid colorimetric determination of acetylcholinesterase activity. *Biochem. Pharmacol.* **1961**, *7*, 88–95. [CrossRef] [PubMed]
28. Organization for Economic Development (OECD). *Guideline for Testing of Chemicals No. 207. Earthworm, Acute Toxicity Tests*; OECD—Guideline for Testing Chemicals; OECD: Paris, France, 1984.
29. Pavela, R. Essential oils from *Foeniculum vulgare* Miller as a safe environmental insecticide against the aphid *Myzus persicae* Sulzer. *Environ. Sci. Pollut. Res.* **2018**, *25*, 10904–10910. [CrossRef] [PubMed]
30. Abbott, W.S. A method of computing the effectiveness of an insecticide. *J. Econ. Entomol.* **1925**, *18*, 265–267. [CrossRef]
31. Finney, D.J. *Probit Analysis*, 3rd ed.; Cambridge University: London, UK, 1971; pp. 68–78.
32. Lee, S.-G. α-Pinene and myrtenol: Complete ^{1}H NMR assignment. *Magn. Res. Chem.* **2002**, *40*, 311–312. [CrossRef]
33. Matsuo, A.L.; Figueiredo, C.R.; Arruda, D.C.; Pereira, F.V.; Borin Scutti, J.A.; Massaoka, M.H.; Travassos, L.R.; Sartorelli, P.; Largo, J.H.G. α-Pinene isolated from *Schinus terebinthifolius* Raddi (Anacardiaceae) induces apoptosis and confers antimetastatic protection in a melanoma model. *Biochem. Biophys. Res. Commun.* **2011**, *411*, 449–454. [CrossRef]
34. Baldwin, S.J.; Clarke, S.E.; Chenery, R.J. Characterization of the cytochrome P_{450} enzymes involved in the in vitro metabolism of rosiglitazone. *Br. J. Clin. Pharmacol.* **1999**, *48*, 424–432. [CrossRef]
35. Dvorakova, M.; Valterova, I.; Saman, D.; Vanek, T. Biotransformation of (1S)-2-carene and (1S)-3-carene by *Picea abies* suspension culture. *Molecules* **2011**, *16*, 10541–10555. [CrossRef]
36. Shanker, P.S.; Rao, G.S. Synthesis based on cyclohexadienes. Part 25. Total synthesis of (+/−)-allo-cedrol (khusiol). *J. Chem. Soc. Perkin Trans.* **1998**, *1*, 539–548.

37. Kim, K.H.; Moon, E.; Kim, S.Y.; Choi, S.U.; Son, M.W.; Choi, S.Z.; Lee, K.R. Bioactive sesquiterpenes from the essential oil of *Thuja orientalis*. *Planta Med.* **2013**, *79*, 1680–1684. [CrossRef]

38. Perez-Hernandez, N.; Gordillo-Roman, B.; Arrieta-Baez, D.; Cerda-Garcia-Rojas, C.M.; Joseph-Nathan, P. Complete ^{1}H NMR assignment of cedranolides. *Magn. Res. Chem.* **2017**, *55*, 169–176. [CrossRef] [PubMed]

39. Asgary, S.; Naderi, G.A.; Shams Ardekani, M.R.; Sahebkar, A.; Airin, A.; Aslani, S.; Kasher, T.; Emami, S.A. Chemical analysis and biological activities of *Cupressus sempervirens* var. *horizontalis* essential oils. *Pharm. Biol.* **2013**, *51*, 137–144. [CrossRef]

40. Selim, S.A.; Adam, M.E.; Hassan, S.M.; Albalawi, A.R. Chemical composition, antimicrobial and antibiofilm activity of the essential oil and methanol extract of the Mediterranean cypress (*Cupressus sempervirens* L.). *BMC Complement. Altern. Med.* **2014**, *14*, 179. [CrossRef] [PubMed]

41. Jahani, M.; Akaberi, M.; Khayyat, M.H.; Emami, S.A. Chemical Composition and Antioxidant Activity of Essential Oils from *Cupressus sempervirens*. var. sempervirens, C. sempervirens. cv. Cereiformis and C. sempervirens var. horizentalis. *J. Essent. Oil Bear. Plants* **2019**, *22*, 917–931.

42. Argui, H.; Ben Youchret-Zalleza, O.; Can Suner, S.; Periz, C.D.; Türker, G.; Ulusoy, S.; Ben-Attia, M.; Büyükkaya, F.; Oral, Y.; Said, H. Isolation, Chemical Composition, Physicochemical Properties, and Antibacterial Activity of *Cupressus sempervirens* L. Essential Oil. *J. Essent. Oil Bear. Plants* **2021**, *24*, 439–452. [CrossRef]

43. Pansera, M.R.; Silvestre, W.P.; Sartori, V.C. Bioactivity of *Cupressus sempervirens* and *Cupressus lusitanica* leaf essential oils on *Colletotrichum fructicola*. *J. Essent. Oil Res.* **2023**, *35*, 51–59. [CrossRef]

44. M'barek, K. Chemical composition and phytotoxicity of *Cupressus sempervirens* leaves against crops. *J. Essent. Oil Bear. Plants* **2016**, *19*, 1582–1599. [CrossRef]

45. Fadel, H.; Benayache, F.; Chalchat, J.C.; Figueredo, G.; Chalard, P.; Hazmoune, H.; Benayache, S. Essential oil constituents of *Juniperus oxycedrus* L. and *Cupressus sempervirens* L. (Cupressaceae) growing in Aures region of Algeria. *Nat. Prod. Res.* **2021**, *35*, 2616–2620. [CrossRef]

46. Miloš, M.; Radonić, A.; Mastelić, J. Seasonal variation in essential oil compositions of *Cupressus sempervirens* L. *J. Essent. Oil Res.* **2002**, *14*, 222–223. [CrossRef]

47. Nejia, H.; Séverine, C.; Jalloul, B.; Mehrez, R.; Stéphane, C.J. Extraction of essential oil from *Cupressus sempervirens*: Comparison of global yields, chemical composition and antioxidant activity obtained by hydrodistillation and supercritical extraction. *Nat. Prod. Res.* **2013**, *27*, 1795–1799. [CrossRef]

48. Jain, P.L.B.; Patel, S.R.; Desai, M.A. Patchouli oil: An overview on extraction method, composition and biological activities. *J. Essent. Oil Res.* **2022**, *34*, 1–11. [CrossRef]

49. Landolt, P.J.; Hofstetter, R.W.; Biddick, L.L. Plant essential oils as arrestants and repellents for neonate larvae of the codling moth (Lepidoptera: Tortricidae). *Environ. Entomol.* **1999**, *28*, 954–960. [CrossRef]

50. Giatropoulos, A.; Pitarokili, D.; Papaioannou, F.; Papachristos, D.P.; Koliopoulos, G.; Emmanouel, N.; Tzakou, O.; Michaelakis, A. Essential oil composition, adult repellency and larvicidal activity of eight Cupressaceae species from Greece against *Aedes albopictus* (Diptera: Culicidae). *Parasitol. Res.* **2013**, *112*, 1113–1123. [CrossRef] [PubMed]

51. Drapeau, J.; Fröhler, C.; Touraud, D.; Kröckel, U.; Geier, M.; Rose, A.; Kunz, W. Repellent studies with *Aedes aegypti* mosquitoes and human olfactory tests on 19 essential oils from Corsica, France. *Flavour Fragr. J.* **2009**, *24*, 160–169. [CrossRef]

52. Moore, S.J. Plant-based insect repellents. In *Insect Repellents Handbook*, 2nd ed.; Debboun, M., Frances, S.P., Strickman, D., Eds.; CRC Press: Boca Raton, FL, USA, 2014; pp. 179–213.

53. Regnault-Roger, C. The potential of botanical essential oils for insect pest control. *Integr. Pest Manag. Rev.* **1997**, *2*, 25–34. [CrossRef]

54. Nerio, L.S.; Olivero-Verbel, J.; Stashenko, E. Repellent activity of essential oils: A review. *Bioresour. Technol.* **2010**, *101*, 372–378. [CrossRef]

55. Khani, M.; Marouf, A.; Amini, S.; Yazdani, D.; Farashiani, M.E.; Ahvazi, M.; Khalighi-Sigaroodi, F.; Hosseini-Gharalari, A. Efficacy of three herbal essential oils against rice weevil, *Sitophilus oryzae* (Coleoptera: Curculionidae). *J. Essent. Oil Bear. Plants* **2017**, *20*, 4937–4950. [CrossRef]

56. Bincy, K.; Remesh, A.V.; Prabhakar, P.R.; Babu, C.S.V. Chemical composition and insecticidal activity of Ocimum basilicum (Lamiaceae) essential oil and its major constituent, estragole against *Sitophilus oryzae* (Coleoptera: Curculionidae). *J. Plant. Dis. Prot.* **2022**, *130*, 529–541. [CrossRef]

57. Fouad, H.A.; Tavares, W.; Zanuncio, J.C. Toxicity and repellent activity of monoterpene enantiomers to rice weevils (*Sitophilus oryzae*). *Pest. Manag. Sci.* **2021**, *77*, 3500–3507. [CrossRef]

58. Balasubramani, S.; Rajendhiran, T.; Moola, A.K.; Kumari, R.; Diana, B. Development of nanoemulsion from *Vitex negundo* L. essential oil and their efficacy of antioxidant, antimicrobial and larvicidal activities (*Aedes aegypti* L.). *Environ. Sci. Pollut. Res.* **2017**, *24*, 15125–15133. [CrossRef]

59. Dai, L.; Li, W.; Hou, X. Effect of the molecular structure of mixed non-ionic surfactants on the temperature of miniemulsion formation. *Colloids Surf. A Physicochem. Eng. Asp.* **1997**, *125*, 27–32. [CrossRef]

60. Mondal, P.; Laishram, R.; Sarkar, P.; Kumar, R.; Karmakar, R.; Hazra, D.K.; Banerjee, K.; Pal, K.; Choudhury, A. Plant essential oil-based nanoemulsions: A novel asset in the crop protection arsenal. In *Agricultural Nanobiotechnology, Biogenic Nanoparticles, Nanofertilizers and Nanoscale Biocontrol Agents*; Ghosh, S., Thongmee, S., Kumar, A., Eds.; Woodhead Publishing Series in Food Science, Technology and Nutrition; Woodhead Publishing: Sawston, UK, 2022; pp. 325–353. [CrossRef]

61. Reddy, S.R.; Fogler, H.S. Emulsion stability: Determination from turbidity. *J. Colloid Interface Sci.* **1981**, *79*, 101–104. [CrossRef]
62. Hano, C.; Abbasi, B.H. Plant-based green synthesis of nanoparticles: Production, characterization and applications. *Biomolecules* **2022**, *12*, 31. [CrossRef] [PubMed]
63. Palermo, D.; Giunti, G.; Laudani, F.; Palmeri, V.; Campolo, O. Essential Oil-Based Nano-Biopesticides: Formulation and Bioactivity Against the Confused Flour Beetle *Tribolium confusum*. *Sustainability* **2021**, *13*, 9746. [CrossRef]
64. Santana, A.S.; Baldin, E.L.L.; dos Santos, T.L.B.; Baptista, Y.A.; dos Santos, M.C.; Lima, A.P.S.; Tanajura, L.S.; Vieira, T.M.; Crotti, A.E.M. Synergism between essential oils: A promising alternative to control *Sitophilus zeamais* (Coleoptera: Curculionidae). *Crop Prot.* **2022**, *153*, 105882. [CrossRef]
65. Arokiyaraj, C.; Bhattacharyya, K.; Reddy, S. Toxicity and synergistic activity of compounds from essential oils and their effect on detoxification enzymes against *Planococcus lilacinus*. *Front. Plant Sci.* **2022**, *13*, 1016737. [CrossRef] [PubMed]
66. Tumen, I.; Senol, F.S.; Orhan, I.E. Evaluation of possible in vitro neurobiological effects of two varieties of *Cupressus sempervirens* (Mediterranean cypress) through their antioxidant and enzyme inhibition actions. *Türk. Biyokimya. Dergisi. (Turk. J. Biochem.)* **2012**, *37*, 5–13. [CrossRef]
67. Aazza, S.; Lyoussi, B.; Miguel, M.G. Antioxidant and antiacetylcholinesterase activities of some commercial essential oils and their major compounds. *Molecules* **2011**, *16*, 7672–7690. [CrossRef] [PubMed]
68. Alimi, D.; Hajri, A.; Jallouli, S.; Sebai, H. Phytochemistry, anti-tick, repellency and anti-cholinesterase activities of *Cupressus sempervirens* L. and *Mentha pulegium* L. combinations against *Hyalomma scupense* (Acari: Ixodidae). *Vet. Parasitol.* **2022**, *303*, 109665. [CrossRef]
69. Almadiy, A.A.; Nenaah, G.E. Bioactivity and safety evaluations of *Cupressus sempervirens* essential oil, its nanoemulsion and main terpenes against *Culex quinquefasciatus* Say. *Environ. Sci. Pollut. Res.* **2022**, *29*, 13417–13430. [CrossRef]
70. Giunti, G.; Benelli, G.; Palmeri, V.; Laudani, F.; Ricupero, M.; Ricciardi, R.; Maggi, F.; Lucchi, A.; Guedes, R.N.C.; Desneux, N.; et al. Non-target effects of essential oil-based biopesticides for crop protection: Impact on natural enemies, pollinators, and soil invertebrates. *Biol. Control* **2022**, *176*, 105071. [CrossRef]
71. Song, S.Y.; Chang, H.J.; Kim, S.D.; Kwag, E.B.; Park, S.J.; Yoo, H.S. Acute and sub-chronic toxicological evaluation of the herbal product HAD-B1 in Beagle dogs. *Toxicol. Rep.* **2021**, *8*, 1819–1829. [CrossRef]
72. IUCN Centre for Mediterranean Cooperation. *A Guide to Medicinal Plants in North Africa.* C. sempervirens; IUCN Centre for Mediterranean Cooperation: Malaga, Spain, 2005; p. 106.

 *sustainability*

Article

Control of White Rot Caused by *Sclerotinia sclerotiorum* in Strawberry Using Arbuscular Mycorrhizae and Plant-Growth-Promoting Bacteria

Andrea Delgado [1]**, Marcia Toro** [1,2,*]**, Miriam Memenza-Zegarra** [3] **and Doris Zúñiga-Dávila** [1,*]

[1] Laboratorio de Ecología Microbiana y Biotecnología, Departamento de Biología, Facultad de Ciencias, Universidad Nacional Agraria La Molina, Lima 15024, Peru

[2] Laboratorio de Ecología de Agroecosistemas, Instituto de Zoología y Ecología Tropical, Facultad de Ciencias, Universidad Central de Venezuela, Caracas 1040, Venezuela

[3] Laboratorio de Biotecnología Microbiana, Departamento de Química Orgánica, Facultad de Química e Ingeniería Química, Universidad Nacional Mayor de San Marcos, Lima 15081, Peru

* Correspondence: mtoro@lamolina.edu.pe or marcia.toro@ucv.ve (M.T.); dzuniga@lamolina.edu.pe (D.Z.-D.); Tel.: +58-414-3056459 (M.T.); +51-975286657 (D.Z.-D.)

Abstract: *Sclerotinia sclerotiorum* is a phytopathogenic fungus that causes wilting and white rot in several species such as strawberry. The overuse of agrochemicals has caused environmental pollution and plant resistance to phytopathogens. Inoculation of crops with beneficial microorganisms such as arbuscular mycorrhizae (AM), plant-growth-promoting rhizobacteria (PGPR), and their metabolites is considered as an alternative to agrochemicals. *B.halotolerans* IcBac2.1 (BM) and *Bacillus* TrujBac2.32 (B), native from Peruvian soils, produce antifungal compounds and are plant-growth-promoting rhizobacteria (PGPR). *B. halotolerans* IcBac2.1 and *Bacillus* TrujBac2 with or without *G. intraradices* mycorrhizal fungi (M) are capable of controlling *S. sclerotiorum* disease in strawberries. Inoculation of mycorrhiza alone decreases disease incidence as well. Treatments with chitosan (Ch), which is used to elicit plant defense responses against fungal pathogens, were used for comparison, as well as non-inoculated plants (C). Co-inoculation of mycorrhiza and bacteria increases plant shoot and root biomass. Our results show that the inoculation of arbuscular mycorrhiza and antifungal *Bacillus* are good biocontrols of *S. sclerotiorum* in strawberry.

Keywords: biological control; beneficial microorganisms; microbial metabolite; *Fragaria annanasa*

Citation: Delgado, A.; Toro, M.; Memenza-Zegarra, M.; Zúñiga-Dávila, D. Control of White Rot Caused by *Sclerotinia sclerotiorum* in Strawberry Using Arbuscular Mycorrhizae and Plant-Growth-Promoting Bacteria. *Sustainability* **2023**, *15*, 2901. https://doi.org/10.3390/su15042901

Academic Editors: Barlin Orlando Olivares Campos, Miguel Araya-Alman, Edgloris E. Marys and Shibao Chen

Received: 26 November 2022
Revised: 14 January 2023
Accepted: 1 February 2023
Published: 6 February 2023

1. Introduction

Agrochemicals are used for a fast and easy control of pests. However, they may be very toxic, can cause pest resistance, accumulation in the food chain, and soil and water contamination. Therefore, an alternative is the use of biological control [1], using predators, parasites, bacteria or fungi, or natural resources that cause less contamination, long-term results, and less risk of resistance.

Strawberries (*Fragaria annanasa* var Aromas) are popular fruits consumed in many countries [2]. In Peru, strawberry crops have increased due to favorable weather and soil conditions. Exports of frozen fruit generated profits of 5.5 million USD in 2010 with Europe and North America as the main markets. Harvest is labor-intensive and generates jobs [3]. However, annual losses caused by pathogens exceeded 200 million USD in the United States [4].

Strawberry is vulnerable to several pathogens such as *Sclerotinia sclerotiorum* fungi that attacks the plant crown causing wilt and decreasing production [5,6]. The attack of phytopathogens is generally treated with pesticides [7]. It is important to consider that foreign markets restrict the use of pesticides in fruits [3]. Thus, we considered testing biological control in strawberries [8]. Plant-growth-promoting bacteria (PGPR) produce growth substances and metabolites with phytopathogenic controlling effects and have been successfully used in field and greenhouse assays in Peru [9,10]. In addition, arbuscular mycorrhizae (AM), a mutualistic symbiotic association present in most plants, confers on the plant a higher nutrient content, tolerance to abiotic stresses, and resistance to different diseases or pests [11–13]. Thus, AM is widely used in agriculture [14]. The objective of this work was to analyze the effect of arbuscular mycorrhizae, plant-growth-promoting bacteria, and their metabolites in the control of white rot caused by the fungus *S. sclerotiorum* in strawberry plants, comparing its effect with chitosan, an agrochemical widely used to control phytopathogens.

2. Materials and Methods

2.1. Plant and Soil Materials

Strawberry seedlings (Aromas variety), from the province of Huaral, Lima, were used. Seedlings were previously grown in the greenhouse for 40 days in a sterile mixture of rice husk, peat, and sand. Seedlings were transplanted into 800 g pots with a sterile substrate comprised of agricultural soil, peat, and sand as substrate (2:1:1). Physicochemical characteristics of the mixture used as substrate were neutral pH (7.0); slightly saline (electrical conductivity: 1.21 dS/m); high organic matter content (5.13%); low available phosphorus content (6.1 ppm); and medium available potassium content (146 ppm).

2.2. Inoculation of Plant-Growth-Promoting Bacteria

Bacillus TrujBac2.32 (B) isolated from the rhizosphere of common beans [15], was grown in 100 mL of tryptone soy broth (TSB) at 28 °C, at constant shaking (150 rpm) for 24 h. At the end of incubation, bacterial biomass was diluted with sterile TSB broth (1:1) to an optical density (at 600 nm) equivalent to 10^8 cfu/mL.

For the production of bacterial biomass and metabolites with antifungal activity, *B. halotolerans* IcBac2.1 (BM) isolated from the rhizosphere of common beans [15] was used. The inoculum was cultured in 20 mL of TSB, incubated at 28 °C with constant shaking (150 rpm) for 24 h. Biomass obtained was added to 200 mL of minimum mineral medium (pH 7) and incubated at 28 °C, shaking constantly (150 rpm) for 72 h in order to obtain antifungal metabolites. All the fermentative broth was used, which included the biomass and all the synthesis products of the bacteria [8].

2.3. Glomeromycota Fungus Inoculum (M)

The fungus *Rhizophagus intraradices* was obtained from *Brachiaria decumbens* (grown in pots), in which spores and propagules were reproduced in 2 cycles of 4 months each. Afterwards, a sample of rhizospheric soil was extracted, the spores were quantified (wet sieving and decanting), and separated in sucrose gradient according to [16]. Rootlets were used to quantify mycorrhizal colonization [17] with trypan blue [18]. An inoculum was prepared with propagules richness of 2950 spores/mL and rootlets with 90% colonization. A total of 2 mL of this inoculum was added, per plant, 15 days before infecting the plant with *S. sclerotiorum* in order to achieve colonization of arbuscular mycorrhizal fungi in the roots of strawberries seedlings, previous to exposure to the pathogen.

2.4. Preparation of the Phytopathogen Fungus (P)

S. sclerotiorum, (P), isolated from common bean with symptoms of stem rot [9] was used to infect plants. It is a highly virulent pathogen with a wide host range. In pathogenicity plant tests, the characteristic symptoms of rot and mycelium proliferation were observed in the green organs of strawberry plants [5,6]. This fungus was maintained in potato dextrose agar (PDA) medium at 25 °C in the dark for 5 days. For the infection of strawberry plants, wheat seeds were used as substrate for the growth of *S. sclerotiorum*. Seeds were hydrated by placing them in a beaker with sterile water, and pre-cooked in the microwave for 5 min. After this, the water was decanted and the seeds dried under sterile conditions. Subsequently, wheat seeds were sterilized in small portions in polyethylene bags for 15 min at 121 °C for three successive periods of 24 h. PDA discs colonized with the fungus were placed inside the bags with sterile wheat and incubated at 25 °C for 5 days.

2.5. Substrate Sterilization

The substrate that was used in pots to grow plants was sterilized (fluent steam) in an autoclave for 1 h, 100 °C. After 24 h, this process was repeated twice to remove Glomeromycota fungal spores and propagules from the substrate as described by [19].

2.6. Plant Cultures

Strawberries were grown in a greenhouse with 12 h photoperiod, maximum (32 °C) and minimum (17 °C) temperature, and 89% of relative humidity. A completely randomized design with the following inoculation treatments was performed: Control without microorganisms (C), *S. sclerotiorum* (P), arbuscular mycorrhiza (M), and the following combinations: P+M, P+B, P+BM, P+Ch, P+M+BM, and P+M+B. The experiment consisted of 9 treatments with 5 repetitions and a total of 45 pots. First, 2 plants per pot were set; bacterial inoculum and mycorrhizae were added to the seedling neck. M inoculated at sowing, B and BM 10 days after sowing. Subsequently, 15 days after installation, 4 wheat seeds covered with the mycelium of *S. sclerotiorum* were placed in the base of the plant. The appearance of the symptoms of stem wilt caused by *S. sclerotiorum* in strawberries were observed after seven days of inoculation. The disease index analysis was calculated weekly for 6 weeks, according to the scale explained below, in Table 1.

Table 1. Scale for measuring the degree of rot in strawberries.

Disease Index	Percentage Damage	Plant Characteristic
1	0%	Healthy plant
2	0–25%	Plant tissue damage
3	26–50%	Plant tissue damage
4	51–75%	Plant tissue damage
5	76–100%	Plant tissue damage
6	100%	Dead plant

2.7. Disease Evaluation

Damage caused by *S. sclerotiorum* in strawberry plants was evaluated using a disease index (1–6) in the scale described in [20] that considers the largest percentage damage as 6 and healthy plants as 1.

2.8. Evaluation of Plant Growth Parameters

At harvest (six and a half weeks), fresh and dry weight (g) of shoot and root were recorded. Dry weight of the plants was recorded after drying the plants at 60 °C for 3 days. To evaluate the efficiency of disease control of the microbial inoculations, the following indexes were calculated: % I PSA C, (percentage of increase in aerial dry weight compared to the negative control without microorganisms, C) and % I PSA P, percentage of increase in aerial dry weight compared to the positive control (P, with *S. sclerotiorum*).

2.9. Quantification of Sclerotinia sclerotiorum at the End of the Experiment

A total of 10 g of soil was weighed and placed in a flask with 90 mL of 0.85% saline solution and shaken vigorously. Subsequently, with a sterile pipette, 1 mL of the suspension was taken and transferred to a test tube with 9 mL of saline solution until reaching the 10^{-4} dilution. Then, 1 mL aliquots of each dilution was placed in sterile Petri dishes (3 per dilution), and incorporated into PDA culture medium. A total of 30 mg/L of streptomycin was added to inhibit the bacterial population. The plates were incubated at room temperature (22 °C) for 5 days, after which the count of *S. sclerotiorum* colonies was made in the plates that contained between 8–80 colonies. The results were expressed in cfu/g of dry soil.

2.10. Disease Progress Curve

The area under the disease progress curve (AUDPC) evaluates the progress of the disease using the severity records, according to the following formula proposed by [21]:

$$\text{ABCPE} = \sum_{i=1}^{n-1} \left[\frac{X_{i+1} + X_i}{2} \right] (t_{i+1} - t_i)$$

where:

X_i: Percentage of tissue severity due to *S. sclerotiorum* at time *I*;
t_i: Time elapsed in days in the evaluation *I*;
n: Total number of evaluations.

This method determines the severity of disease accumulated during the time of the study.

2.11. Statistical Analysis

A completely randomized design was performed. Least significant difference (LSD) of Fisher was applied and standard deviation of each treatment was calculated. For the analysis of aerial, root, total plant biomass, and AUDPC curve, after the ANOVA analysis, the posteriori Tukey test was applied for the comparison between averages of treatments ($p \leq 0.05$). The analyses were carried out with Statgraphics program.

3. Results

Antifungal capacity and growth promotion of *Bacillus* TrujBac2.32 and *B. halotolerans* IcBac2.1 were tested in previous studies carried out by [15] and their results are shown in Table 2. Both bacteria are able to control *R.solani*, *S. sclerotiorum* and *F. oxysporum* fungi. *B. halotolerans* ICBac2.1 produces antifungal lipopeptides, antibiotics, and siderophores. *Bacillus* TrujBac2.32 produces volatile metabolites and has the ability to solubilize phosphates. Both bacteria produce hydrolytic enzymes, as proteases and cellulases, and indoleacetic acid (Table 2).

Control plants without the phytopathogen have an aerial dry weight of 1.31 g, while infected plants with the pathogen (P) weigh 0.65 g. Strawberry plants inoculated with mycorrhiza (M) and *Bacillus* (B) have an increase in dry weight over 24% and 23%, respectively, compared to the infected plant. All treatments inoculated with promoting microorganisms increase the dry weight of the plant between 55 and 101% compared to (C), except the interaction of (BM) and (M) (Table 3). The dry weight of the root has a similar tendency to the shoot dry weight, except in (P) whose weight is greater (1.6 g) than C (1.46 g).

Table 4 shows the disease index of the plants evaluated for six weeks. It is observed in all cases that the damage rate increases in time. After six weeks, the plant inoculated with the fungus (P) has an index of almost 6. It should be noted that chitosan does not diminish plant damage.

The treatments that best control the disease are mycorrhizae (M), BM, and BM+M inoculations, which reduce the damage to 2.3, 4.1, and 4.3, respectively, according to the disease index (Table 1). The infection begins with the proliferation of the mycelium of the phytopathogenic fungus around the neck of the plant causing a necrosis of the plant, then the disease gradually extends to the foliage of the plant. Figure 1a shows a healthy plant, while Figure 1b shows the mycelium of *S. sclerotiorum* starting to cover the neck of the plant.

Table 2. Plant-growth-promoting and antifungal activities of *Bacillus* TrujBac2.32 and *B. halotolerans* IcBac2.1. The fungal growth inhibition was quantified using the percentage inhibition. Enzymatic activity: − (no activity) to + (activity). Each value represents the mean ± SE (n = 2) [15].

Strain	Antifungal Activity by						Antibiotics Production	AIA Production (mg/mL)	P Solubilizing Ability	Siderophore Production Halo (mm)	Hydrolytic Enzymes			
	Dual Culture Technique			Volatile Metabolites							Proteases	Cellulases	Lipases	Chitinases
	R. solani	S. sclerotiorum	F. oxisporum	R. solani	S. sclerotiorum	F. oxisporum								
Bacillus TrujBac2.32	61 ± 0.01	65 ± 0.01	38 ± 0.01	22 ± 7.07	14 ± 2.12	0	-	32.6 ± 0.02	Bi and tricalcic phosphates	0	+	+	-	-
B. halotolerans IcBac2.1	85 ± 0.01	71 ± 0.01	69 ± 0.02	0	0	0	+	8.7	-	20 ± 0.01	+	+	-	-

Table 3. Dry weight of shoot and root (g) of strawberry plants inoculated with *S. sclerotiorum*, growth-promoting microorganisms and their metabolites.

Inoculation Treatments	Shoot Dry Weight (g)	Root Dry Weight (g)	Plant Total Dry Weight (g)	Percentage of Increase in Aerial Dry Weight Compared to Control C (%)	Percentage of Increase in Aerial Dry Weight Compared to P (%)
P	0.65 ab	1.6 ab	2.25 ± 0.05 abc	−50.38	0
M	1.94 b	1.35 b	3.29 ± 0.66 c	48.09	198.46
P+M	1.63 b	1.30 b	2.93 ± 0.57 c	24.43	150.77
P+B	1.62 b	1.24 ab	2.86 ± 0.40 bc	23.66	149.23
P+BM	1.01 ab	0.56 a	1.57 ± 0.64 ab	−22.9	55.38
P+Ch	0.49 ab	0.42 ab	0.91 ± 0.05 abc	−62.6	−24.62
P+BM+M	0.64 a	0.63 ab	1.27 ± 0.27 a	−51.15	−1.54
P+B+M	1.23 ab	1.33 ab	2.56 ± 1.28 abc	−6.11	89.23
Control (C)	1.31 ab	1.46 ab	2.77 ± 1.39 abc	0	101.54

Values followed by the same letter are not significantly different according to Tukey test (*p* < 0.05). n = 5. Abbreviations: Control (without microorganisms), (P) *S. sclerotiorum*, (M) arbuscular mycorrhiza, (B) *Bacillus* TrujBac2.32 inoculation, (BM) *B. halotolerans* IcBac2.1 and its antimicrobial metabolites, and (Ch) chitosan.

Table 4. Disease index of wilt caused by *S. sclerotiorum* in strawberry plants inoculated with growth-promoting microorganisms and its metabolites.

Inoculation Treatments	Weeks					
	1	2	3	4	5	6
P	3.75 ± 2.05	4.25 ± 1.91	4.88 ± 1.36	5.63 ± 0.74	5.75 ± 0.71	5.75 ± 0.71
M	-	-	-	-	-	-
P+M	1.62 ± 0.52	1.75 ± 0.46	2 ± 0.53	2 ± 0.53	2.13 ± 0.35	2.38 ± 0.52
P+B	3.75 ± 2.05	3.88 ± 1.89	4.25 ± 1.67	4.38 ± 1.77	4.63 ± 1.60	4.63 ± 1.60
P+BM	2.38 ± 1.19	2.62 ± 1.41	3.75 ± 1.98	4 ± 1.51	4 ± 1.51	4.13 ± 1.64
P+Ch	3.63 ± 1.60	4.38 ± 1.19	5.38 ± 1.06	5.5 ± 0.93	5.5 ± 0.93	5.75 ± 0.71
P+BM+M	3.38 ± 2.20	3.50 ± 2.14	4 ± 1.93	4.13 ± 1.89	4.25 ± 1.98	4.38 ± 2.00
Control (C)	-	-	-	-	-	-

The values were compared using least significance difference (LSD) test of Fisher, and show means ± standard deviation per treatment. n = 5. Abbreviations: Control (without microorganisms), (P) *S. sclerotiorum*, (M) arbuscular mycorrhiza, (B) *Bacillus* TrujBac2.32 inoculation, (BM) *B. halotolerans* IcBac2.1 and its antimicrobial metabolites, and (Ch) chitosan. Treatments M and Control show no symptoms since they are not inoculated with the phytopathogenic fungus.

(a)   (b)

Figure 1. (a) Healthy plant without disease symptoms. (b) Appearance of the mycelium of the *S. sclerotiorum* developing in the neck of strawberry.

Table 5 shows that plants treated with BM+M have a higher percentage of AM root colonization (20.4%), followed by B+MI (12%). The bacteria significantly stimulates mycorrhizal roots colonization of the plant, compared to that plant that is only inoculated with the mycorrhizal fungi, while the pathogen diminishes mycorrhizal colonization.

Table 5. Arbuscular mycorrhizal colonization (%) in strawberry plants inoculated with *S. sclerotiorum*, growth-promoting bacteria and their metabolites.

Inoculation Treatments	% AM Root Colonization
P+BM+M	20.46 ± 1.65 c
P+M	8.94 ± 2.06 a
P+B+ M	14.96 ± 0.95 b
M	17.49 ± 3.39 bc

Values followed by the same letter are not significantly different, according to Tukey test ($p < 0.05$). n = 5.

At harvest, the population of *S. sclerotiorum* that was in the soil around the crown of the plant was quantified. Table 6 shows that there were 80×10^3 cfu of *S. sclerotiorum*/g soil. All treatments reduce the population of the phytopathogenic fungus, with (BM) and (B+M) (50×10^3 cfu/ g dry soil) being the most effective treatments. Chitosan has no effect.

Table 6. Number of colonies of the phytopathogen *S. sclerotiorum* in rhizospheric soil (cfu/g dry soil) after the harvest of strawberries inoculated with growth-promoting microorganisms and its metabolites.

Inoculation with Beneficial Microorganisms in Addition to *S. sclerotiorum*	10^4 cfu *S. sclerotiorum*/g Dry Soil
P	80 ± 35
M	60 ± 30
B	60 ± 32
BM	50 ± 25
Ch	80 ± 15
BM+M	60 ± 35
B+M	50 ± 35

The values were compared using least significance difference (LSD) test of Fisher, and show means ± standard deviation per treatment. n = 3. Abbreviations: (P) *S. sclerotiorum*, (M) arbuscular mycorrhiza, (B) *Bacillus* TrujBac2.32, (BM) *B. halotolerans* IcBac2.1 and its antimicrobial metabolites, and (Ch) chitosan.

Figure 2 shows the number of colonies of *S. sclerotiorum* grown in PDA medium isolated from soil in pots after plant harvest.

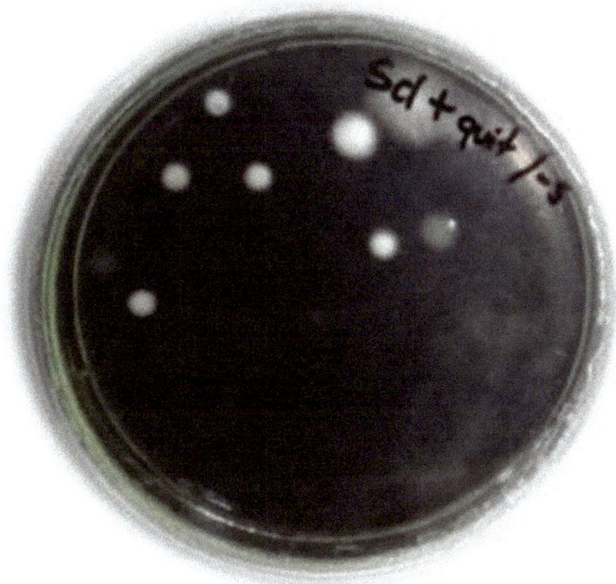

Figure 2. Colonies of *S. sclerotiorum* in PDA medium.

In relation to AUDPC (Figure 3), the values of each bar correspond to the average of eight repetitions/treatment, representing the progress of the disease on plants. According to the ANOVA analysis, P+M is the treatment with the lowest ABCE value (31), however, it is statistically similar to P+B, P+BM, and P-BM+M. P is statistically to similar P+Ch, indicating that chitosan does not diminish the disease. The analysis indicates that beneficial microorganisms alone (M, B, or BM) or in combination, BM+M, control *S. sclerotiorum*.

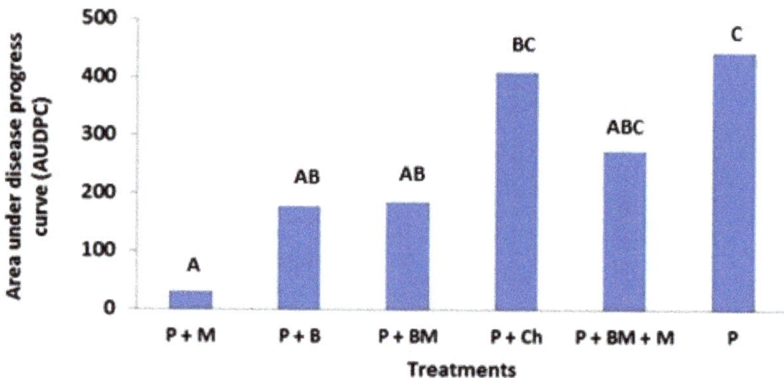

Figure 3. Area under disease progress curve (AUDPC) of the disease caused by *S. sclerotiorum* in strawberry plants inoculated with arbuscular mycorrhiza, growth-promoting bacteria and their metabolites. Abbreviations: (P) *S. sclerotiorum*, (M) arbuscular mycorrhiza, (B) *Bacillus* TrujBac2.32, (BM) *B. halotolerans* IcBac2.1and its antimicrobial metabolites, and (Ch) chitosan. Different letters over the bars indicate significant differences between treatments (Tukey $p \leq 0.05$). n = 8.

4. Discussion

Bacillus are commonly used to control phytopathogens [22,23]. Different mechanisms have been proposed to explain the ability to inhibit mycelial growth of fungal pathogens, such as the production of antimicrobial metabolites, competition for nutrients or space, or a combination of them [24]. Ref. [25] found that *B. amyloliquefaciens* protected tomato, squash, and eggplant from *S. sclerotiorum*. The protection is related to the antibiosis capacity of *B. amyloliquefaciens*. This bacterium also controls other fungi through the production of a-1,3-glucanase. In addition, due to its capacity to produce enzymes such as chitinases, proteases, amylases, and celluloses, *B. amyloliquefaciens* and other *Bacillus* species cause leakage of fungal cell contents when hydrolyzing cell walls of the phytopathogen. Here, we found that both *Bacillus* strains produce hydrolyzing enzymes. Ref. [26] also observed that cell suspensions, cell-free culture filtrates, and broth cultures of *B. subtilis* control *S. sclerotiorum* in soybean, through the reduction in its mycelial growth (by 50 to 75%) and suppressing sclerotial formation by > 90%. Antibiotic peptides present in its cell-free supernatant may be responsible for this effect [10,15]. Antimicrobial metabolites produced by bacteria include iron-chelating siderophores, antibiotics, volatile biocides, and non-volatile antimicrobial compounds such as cyclic lipopeptides (CLPs) such as iturins, fengycins, and surfactins involved in the inhibition of mycelial growth of phytopathogens [27,28].

Here, we found that BM reduces the infection rate of the phytopathogenic fungus to 50%, while the disease index of P is greater than 81%. In our group we reported previously a reduction in the incidence of *S. sclerotiorum* wilt in beans by 96% due to the application of the strain *B. halotolerans* IcBac2.1 (BM) [9]. There, the bacterium was inoculated at different stages of growth, seed, flowering, formation, and filling of pods. Notably, the same strain was able to reduce the population of the phytopathogen in the soil, similar to our results with strawberry.

Bacteria may promote plant growth by phytohormones and solubilize phosphate [29]. Several species of *Bacillus* produce indole-3-acetic acid (IAA), which increases root length and the number of secondary roots [30]. Ref. [31] demonstrates that *B. velezensis* and *B. amyloliquefaciens* can produce phytohormones, fix nitrogen, and solubilize phosphate, in addition to reducing disease against phytopathogens such as *Helicobasidium purpureum*, *F. oxysporum*, and *Rhizoctonia solani* on pepper. Both bacteria used in this work have the capacity of producing AIA; *Bacillus* TrujBac2.32 (B) produces large amounts of AIA (32.6 µg mL^{-1}) [15,32]. In this work, this strain was able to improve shoot dry weight of the plant by 149% (Table 2).

Chitosan is considered an environment-friendly alternative to the use of chemicals for controlling phytopathogens such as *Sclerotinia*, among others, and post-harvest diseases in fruits [33]. In our case, chitosan is not able to diminish disease or the fungal population in the soil. Results obtained in different studies with chitosan are variable and depend on several factors, such as the concentration used, the application method, and the experimental temperature of the greenhouse, among others [34].

Regarding arbuscular mycorrhizae, the inoculation of the fungus *R. intraradices* (M) is able to significantly reduce the disease index by effectively controlling infection. In this treatment the highest shoot dry weight of the plant increased by 150% (Table 2). However, the percentage of AM root colonization is the lowest (Table 3), which could indicate that even with 8% root colonization, the promoter effect exerted by *R. intraradices* is sufficient to achieve such biomass. Similar to our results, [35], mycorrhiza achieves adequate control of the disease caused by *Fusarium* sp. in maize, with a low number of Glomeromycota spores or root colonization, which could also be associated with the effectiveness and affinity of the AM fungus with the host in question. Arbuscular mycorrhizae can induce systemic resistance in the plant to soil and/or aerial pathogens [36]. The production of jasmonic and/or salicylic acids, among others, as phytopathogenic resistance mechanisms are stimulated in the plants by mycorrhizal inoculation [37]. In mycorrhized plants, the expression of different genes involved in plant defense against foliar and root pathogen attack may play a role locally and systemically. There are mechanisms in which acid endochitinase PR4 and β1, 3-endoglycanase EG488 may also be involved [38]. Moreover, AM fungi can improve the resistance of host plants to phytopathogens by changing the anatomical structure of plant roots and producing substances involved in plant defense against pathogenic attack [39]. Some authors [40] described the fact that mycorrhizal association causes a chemical balance in the plant that inhibits the growth and reproduction of phytopathogenic fungi. Ref. [41] observes that pre-inoculation with the AMF *Glomus mosseae* prevents the decrease in antioxidant compounds, along with the decrease in SOD activity and ascorbate content in strawberry roots after inoculation with *F. oxysporum* f. sp. *fragariae*. These mechanisms, although not evaluated in our case, could be occurring for the control of *S. sclerotiorum*.

In cotton, inoculation with PGPR stimulates root AM colonization [42], which is in agreement with the results presented here where the strains of *Bacillus* TrujBac2.32 (B) and *B. halotolerans* IcBac2.1 and its metabolites (BM) increase root mycorrhizal colonization significantly, 12 and 20.4 %, respectively (Table 3). The interaction between PGPR and AM can favor the plant both by promoting mycorrhiza activity, which, in turn, favors the development of plants (mycorrhiza helper bacteria) [43,44], and by contributing to the control of phytopathogens [36].

BM+M is the most effective treatment in the control of the disease and better mycorrhizal colonization, which could, in turn, influence the decrease in *Sclerotinia* spores in the rhizosphere of the plant. Although this treatment does not produce more dry biomass of the plant, it is effective in the parameters that indicate control of the phytopathogen. The antifungal effects of *B. halotolerans* ICBac2.1 may be due to the production of lipopeptides [15] and could explain our results.

Similarly, other authors found that the interaction of AMF and PGPR controlled plant diseases [27]. AM and PGPR favor the tolerance of strawberry plants to wilt induced by *P. capsici*. [39], indicating that the combined treatment of AMF and PGPR as *Bacillus* exerts a positive control on plant root diseases. The mixed inoculation of *G. mosseae* and *Bacillus subtilis* not only reduces the severity of *Fusarium*-caused disease by 85.0–93.4%, but also improves plant nutrient content, total soluble sugar, total soluble protein, and total free amino acid content [45,46]. Our results show that the joint use of beneficial microorganisms, AMF and PGPR, can control plant diseases, and constitute an alternative for sustainable agriculture, contributing to the reduction in the application of agrochemicals. Ref. [47] proposes that the use of biological control, together with crop rotation and the use of resistant cultivars, will allow for effective sustainable management of agriculture.

5. Conclusions

The interaction between PGPR and AM favors the development of plants and contributes to the control of phytopathogens, such is the case of BM +M, which is the most effective treatment in controlling the disease caused by *S. sclerotiorum* and promotes mycorrhiza activity. The presence of beneficial microorganisms in the rhizosphere with influence on the state of health of the plant is complex and manifests itself in our results, so it could be considered that the microorganisms used influence the defense responses of strawberry plants to *S. sclerotiorum*. Our results constitute a first approach to the joint use of native strains of both PGPR and AMF for the control of the disease caused by *S. sclerotiorum* in strawberries plantations in Peru. We suggest further research of these studies.

Author Contributions: Conceptualization, M.T. and D.Z.-D.; methodology, M.T., M.M.-Z. and D.Z.-D.; validation, A.D. and M.M.-Z.; formal analysis, M.T. and D.Z.-D.; investigation, A.D. and M.M.-Z.; resources, D.Z.-D.; writing—original draft preparation, M.T. and D.Z.-D.; writing—review and editing, M.T.; supervision, M.T., M.M.-Z. and D.Z.-D.; project administration, D.Z.-D.; funding acquisition, D.Z.-D. All authors have read and agreed to the published version of the manuscript.

Funding: This work was supported by grants 009-2016 Fondecyt from Fondo Nacional de Desarrollo Científico, Tecnológico y de Innovación Tecnológica and Innovate Perú Proyect 158-PNCIP-PIAP-2015 from Programa Nacional de Innovación para la Competitividad y Productividad del Ministerio de la Producción.

Acknowledgments: The authors express their gratitude to Wilson Castañeda for the preparation of the Glomeromycota inoculum used in the experiment and appreciate the collaboration of Auxi Toro García and Esperanza Martínez-Romero (Centro de Ciencias Genómicas, Cuernavaca, Mexico) in their support as linguistic advisors.

Conflicts of Interest: The authors declare no conflict of interest.

References

1. RedAgrícola. Las Exportaciones de Fresa Congelada le Dan una Segunda Vida a Este Cultivo. 2020. Available online: https://www.redagricola.com/pe/las-exportaciones-fresa-congelada-le-dan-una-segunda-vida-este-cultivo/ (accessed on 4 March 2021).
2. MINAGRI. Estudio de la Fresa en el Perú y el Mundo. 2008. Available online: https://www.midagri.gob.pe/portal/download/pdf/herramientas/boletines/estudio_fresa.pdf (accessed on 24 October 2022).
3. Olivera, S. *Cultivo de Fresa (Fragaria X ananassa Duch.)*; Ministerio de Agricultura, INIA, Estación experimental Donoso Kiyotada Miyagawa, Programa Nacional de Innovación Agraria en Hortalizas: Huaral, Peru, 2012; pp. 39–49.
4. Xia, S.; Xu, Y.; Hoy, R.; Zhang, J.; Qin, L.; Li, X. The notorious soilborne pathogenic fungus *Sclerotinia sclerotiorum*: An update on genes studied with mutant analysis. *Pathogens* **2020**, *9*, 27. [CrossRef]
5. Marin, M.V.; Peres, N.A. First report of *Sclerotinia sclerotiorum* causing strawberry fruit rot in Florida. *Plant Dis.* **2020**, *104*, 3250. [CrossRef]
6. Ficker, A. Sclerotinia sclerotiorum impacts on host crops. In *Master of Science Monography*; Iowa State University: Ames, IA, USA, 2019.
7. Kamal, M.; Lindbeck, K.; Savocchia, S.; Ash, G. Biological control of *Sclerotinia* stem rot of canola using antagonistic bacteria. *Plant Pathol.* **2015**, *64*, 1375–1384. [CrossRef]
8. Khan, N.; Martínez-Hidalgo, P.; Ice, T.A.; Maymon, M.; Humm, E.A.; Nejat, N.; Sanders, E.R.; Kaplan, D.; Hirsch, A.M. Antifungal activity of *Bacillus* species against *Fusarium* and analysis of the potential mechanisms used in biocontrol. *Front. Microbiol.* **2018**, *9*, 2363. [CrossRef]
9. Memenza, M.; Mostacero, E.; Camarena, F.; Zúñiga, D. Disease control and plant growth promotion (PGP) of selected bacterial strains in *Phaseolus vulgaris*. In *Biological Nitrogen Fixation and Beneficial Plant—Microbe Interactions*; Gonzales-Andres, F., James, E., Eds.; Springer: Cham, Switzerland, 2016; pp. 237–245.
10. Memenza-Zegarra, M.; Zúñiga-Dávila, D. Isolation and characterization of antifungal secondary metabolites produced by rhizobacteria from common bean. In *Microbial Probiotics for Agricultural Systems: Advance in Agronomic Use*; Zúñiga-Dávila, D., Gonzales-Andres, F., Ormeño-Orrillo, E., Eds.; Springer: Cham, Switzerland, 2019; pp. 141–153.
11. Alban, R.; Guerrero, R.; Toro, M. Interactions between a root-knot nematode (*Meloidogyne exigua*) and arbuscular mycorrhizae in coffee plant development (*Coffea arabica*). *Am. J. Plant Sci.* **2013**, *4*, 19–23. [CrossRef]
12. Barea, J.M. Future challenges and perspectives for applying microbial biotechnology in sustainable agriculture based on a better understanding of plant-microbiome interactions. *J. Soil Sci. Plant Nutr.* **2015**, *15*, 261–282. [CrossRef]
13. Jain, P.; Pundir, R.K. Biocontrol of Soil Phytopathogens by ArbuscularMycorrhiza—A Review. In *Mycorrhizosphere and Pedogenesis*; Varma, A., Choudhary, D., Eds.; Springer: Singapore, 2019; pp. 221–237.

14. Dar, M.; Reshi, Z.; Shah, M. Vesicular arbuscular mycorrhizal (VAM) fungi—As a major biocontrol agent in modern sustainable agriculture system. *Russ. Agric. Sci.* **2017**, *43*, 132–137. [CrossRef]

15. Memenza-Zegarra, M.; Zúñiga-Dávila, D. Bioprospection of native antagonistic rhizobacteria from the Peruvian coastal ecosystems associated with *Phaseolus vulgaris. Curr. Microbiol.* **2021**, *78*, 1418–1431. [CrossRef]

16. Gerdemann, J.; Nicolson, T.H. Spores of mycorrhizal Endogone species extracted from soil by wet sieving and decanting. *Trans. Br. Mycol. Soc.* **1963**, *46*, 235–244. [CrossRef]

17. Giovannetti, M.; Mosse, B. An evaluation of techniques for measuring vesicular arbuscular mycorrhizal infection in roots. *New Phytol.* **1980**, *84*, 489–500. [CrossRef]

18. Phillips, J.; Hayman, D. Improved procedures for clearing roots and staining parasitic and vesicular-arbuscular mycorrhizal fungi for rapid assessment of infection. *Trans. Br. Mycol. Soc.* **1970**, *55*, 158–161. [CrossRef]

19. Sieverding, E. *Vesicular-Arbuscular Mycorrhiza Management in Tropical Agrosystems*; GTZ: Eschborn, Germany, 1991.

20. Ouhaibi, N.; Abdeljalil, B.; Vallance, J.; Gerbore, J.; Rey, P.; Daami-Remadi, M. Bio-suppression of *Sclerotinia* stem rot of tomato and biostimulation of plant growth using tomato-associated rhizobacteria. *J. Plant Pathol. Microbiol.* **2016**, *7*, 2.

21. Wilcoxson, R.D.; Skovmand, B.; Atif, A.A. Evaluation of wheat cultivars for the ability to retard development of stem rust. *Ann. Appl. Biol.* **1975**, *80*, 275–287. [CrossRef]

22. Fritze, D. Taxonomy of the genus Bacillus and related genera: The aerobic endospore-forming bacteria. *Phytopathology* **2004**, *94*, 1245–1248. [CrossRef] [PubMed]

23. Shafi, J.; Tian, H.; Ji, M. *Bacillus* species as versatile weapons for plant pathogens: A review. *Biotechnol. Biotechnol. Equip.* **2017**, *31*, 446–459. [CrossRef]

24. Compant, S.; Duffy, B.; Nowak, J.; Clement, C.; Barka, E. Use of plant growth-promoting bacteria for biocontrol of plant diseases: Principles, mechanisms of action, and future prospects. *Appl. Environ. Microbiol.* **2005**, *71*, 4951–4959. [CrossRef]

25. Abdullah, M.T.; Ali, N.Y.; Suleman, P. Biological control of *Sclerotinia sclerotiorum* (Lib.) de Bary with *Trichoderma harzianum* and *Bacillus amyloliquefaciens. Crop Prot.* **2008**, *27*, 1354–1359. [CrossRef]

26. Zhang, J.X.; Xue, A.G. Biocontrol of sclerotinia stem rot (*Sclerotinia sclerotiorum*) of soybean using novel *Bacillus subtilis*strain SB24 under control conditions. *Plant Pathol.* **2010**, *59*, 382–391. [CrossRef]

27. Serret-López, M.; Espinosa-Victoria, D.; Gómez-Rodríguez, O.; Delgadillo-Martínez, J. Tolerance of strawberry (*Fragaria × ananassa* Duch.) plants pre-mycorrhized with *Rhizophagus intraradices* and inoculated with PGPR's to *Phytophthora capsici. Agrociencia* **2016**, *50*, 1107–1121.

28. Colman, V.; Hanif, A.; Farzand, A.; Kibe, M.; Ochieng, S.; Wu, L.; Samad, H.; Gu, Q.; Wu, H.; Gao, X. Volatile Compounds of Endophytic *Bacillus* spp. have Biocontrol Activity Against *Sclerotinia sclerotiorum. Phytopathology* **2018**, *108*, 1373–1385.

29. Richardson, A.E.; Barea, J.-M.; McNeill, A.M.; Prigent-Combaret, C. Acquisition of phosphorus and nitrogen in the rhizosphere and plant growth promotion by microorganisms. *Plant Soil* **2009**, *321*, 305–339. [CrossRef]

30. Idriss, E.E.; Makarewicz, O.; Farouk, A.; Rosner, K.; Greiner, R.; Bochow, H.; Richter, T.; Borriss, R. Extracellular phytase activity of *Bacillus amyloliquefaciens* FZB45 contributes to its plant growth—Promoting effect. *Microbiology* **2002**, *148*, 2097–2109. [CrossRef] [PubMed]

31. Soliman, S.A.; Khaleil, M.M.; Metwally, R.A. Evaluation of the Antifungal Activity of *Bacillus amyloliquefaciens* and *B. velezensis* and Characterization of the Bioactive Secondary Metabolites Produced against Plant Pathogenic Fungi. *Biology* **2022**, *11*, 1390. [CrossRef]

32. Memenza-Zegarra, M.; Zúñiga-Dávila, D. Evaluation of the antagonic activity of metabolites produced by the culture of *Bacillus halotolerans* to control *Rhizoctonia solani.* In *Abstract Book in XVI Symposium on Biological Nitrogen Fixation with Non-Legume*; Embrapa: Foz de Iguazu, Brazil, 2018.

33. Gutierrez-Martínez, P.; Ledezma-Morales, A.; Romero-Islas, L.; Ramos-Guerrero, A.; Romero-Islas, J.; Rodríguez-Pereida, C.; Casas-Junco, P.; Coronado-Partida, L.; González-Estrada, R. Antifungal activity of chitosan against postharvest fungi of tropical and subtropical fruits. In *Chitin-Chitosan—Myriad Functionalities in Science and Technology*; BoD—Books on Deman: Norderstedt, Germany, 2018. [CrossRef]

34. Velázquez-del Valle, M.; Hernández-Lauzardo, A.; Guerra-Sánchez, M.; Mariaca-Gaspar, G. Chitosan as an alternative to control phytopathogenic fungi on fruits and vegetables in Mexico. *Afr. J. Microbiol. Res.* **2012**, *6*, 6606–6611. [CrossRef]

35. Olowe, O.M.; Olawuyi, O.J.; Sobowale, A.A.; Odebode, A.C. Role of arbuscular mycorrhizal fungi as biocontrol agents against *Fusarium verticillioides* causing ear rot of *Zea mays* L. (Maize). *Curr. Plant Biol.* **2018**, *15*, 30–37. [CrossRef]

36. Pérez-de-Luque, A.; Tille, S.; Johnson, I.; Pascual-Pardo, D.; Ton, J.; Cameron, D.D. The interactive effects of arbuscular mycorrhiza and plant growth-promoting rhizobacteria synergistically enhance host plant defenses against pathogens. *Sci. Rep.* **2017**, *7*, 16409. [CrossRef]

37. Pozo, M.J.; Azcón-Aguilar, C. Unravelling mycorrhiza induced resistance. *Curr. Opin. Plant Biol.* **2007**, *10*, 393–398. [CrossRef] [PubMed]

38. Wani, K.A.; Manzoor, J.; Shuab, R.; Lone, R. Arbuscular mycorrhizal fungi as biocontrol agents for parasitic nematodes in plants. *Mycorrhiza-Nutr. Uptake Biocontrol Ecorestor.* **2017**, 195–210. [CrossRef]

39. Weng, W.; Yan, J.; Zhou, M.; Yao, X.; Gao, A.; Ma, C.; Cheng, J.; Ruan, J. Roles of arbuscular mycorrhizal fungi as a biocontrol agent in the control of plant diseases. *Microorganisms* **2022**, *10*, 1266. [CrossRef] [PubMed]

40. Sudhasha, S. Chapter-1 Constructiveness of the biocontrol agents on fusarial wilt of tomato incited by the destructive pathogen *Fusarium oxysporum* f. sp. lycopersici. In *Current Research and Innovations in Plant Pathology*; AkiNik Publications: Delhi, India, 2020.

41. Li, Y.H.; Yanagi, A.; Miyawaki, Y.; Okada, T.; Matsubara, Y. Disease tolerance and changes in antioxidative abilities in mycorrhizal strawberry plants. *J. Jpn. Soc. Hortic. Sci.* **2010**, *79*, 174–178. [CrossRef]

42. Valencia, C.; Zuñiga-Dávila, D. Análisis de la presencia natural de micorrizas en cultivos de algodón (*Gossypium barbadense* L.) inoculados con *Bacillus megaterium* y/o *Bradyrhizobium yuanmingense*. *Ecol. Apl.* **2015**, *14*, 65–69. [CrossRef]

43. Miransari, M. Interactions between arbuscular mycorrhizal fungi and soil bacteria. *Appl. Microbiol. Biotechnol.* **2011**, *89*, 917–930. [CrossRef] [PubMed]

44. Mora, E.; Lopez-Hernández, D.; Toro, M. Arbuscular Mycorrhizae and PGPR applications in tropical savannas. In *Microbial Probiotics for Agricultural Systems: Advance in Agronomic Use*; Zúñiga-Dávila, D., Gonzales-Andres, F., Ormeño-Orrillo, E., Eds.; Springer: Cham, Switzerland, 2019; pp. 169–178.

45. Cai, X.; Zhao, H.; Liang, C.; Li, M.; Liu, R. Effects and mechanisms of symbiotic microbial combination agents to control tomato Fusarium crown and root rot disease. *Front. Microbiol.* **2021**, *12*, 629793. [CrossRef] [PubMed]

46. Hoda, H.; Ghalia, A. Effect of arbuscular mycorrhizal Fungus *Glomus mosseae* and *Glomus subtilis* on the infection of tomato roots by *Fusarium oxysporum*. *Egyp. J. Microbiol.* **2007**, *42*, 29–49.

47. Smolińska, U.; Kowalska, B. Biological control of the soil-borne fungal pathogen *Sclerotinia sclerotiorum*—A review. *J. Plant Pathol.* **2018**, *100*, 1–12. [CrossRef]

sustainability

Review

Foraging Behaviour and Population Dynamics of Asian Weaver Ants: Assessing Its Potential as Biological Control Agent of the Invasive Bagworms *Metisa plana* (Lepidoptera: Psychidae) in Oil Palm Plantations

Moïse Pierre Exélis [1,2,*], **Rosli Ramli** [1], **Rabha W. Ibrahim** [3] **and Azarae Hj Idris** [1]

[1] Institute of Biological Science (Biodiversity and Ecological Research Network), Faculty of Science, Universiti Malaya, Kuala Lumpur 50603, Malaysia

[2] Direction de l'Enseignement Supérieur et de la Recherche Espace Etudiants, Hôtel de la Collectivité Territoriale de la Martinique (CTM), Rue Gaston Defferre, Cluny, CS 30137, Fort-de-France 97201, Martinique

[3] Department of Computer Science and Mathematics, Lebanese American University, Beirut 13-5053, Lebanon

* Correspondence: exelis.pierre@gmail.com

Abstract: The bagworm (*Metisa plana*) is a recurrent indigenous invasive defoliator in oil palm plantations. Moderate foliar injury can cost up to 40% yield loss and more for years. The main objective of this review is to disseminate published research demonstrating the versatile services that would benefit farmers by adopting the Asian weaver ant into their pest management agenda. *Oecophylla smaragdina* is a natural indigenous enemy applied as a successful biological control agent (BCA) and strong component of integrated pest management (IPM) against important damaging pest infestations of commercial crops in the Asia-Pacific region. Farmers facing invasion could benefit by introducing *Oecophylla* ants as a treatment. The foraging behavior and population dynamics of this species are poorly documented, and hence need further evaluation. Ants of the *Oecophylla* genus, while exhibiting an intrinsic obligate arboreal pattern, demonstrate additional lengthy diurnal ground activity. The absolute territorial characteristic via continuous surveillance is significantly valuable to maintain pest balance. The exploratory scheme of major workers over large territories is derived from their inner predation instinct. The insufficient understanding of the population dynamics of this weaver ant species diverges from the knowledge of underground species. However, population density estimations of weaver ants by direct nest visual recordings are practicable and viable. The abundance assessment of individual underground ant species colonies by excavation ends with their extinction, which is not a sustainable model for *O. smaragdina*. Mathematical model estimation by simulation could not resolve this issue, adding inaccuracy to the deficiency of experimental proof. Thus, long-term monitoring of the population dynamics in real time in the field is compulsory to obtain a valid dataset. *Oecophylla* colonies, with the criteria of population stability, individual profusion, and permanent daily patrol services, are eligible as a BCA and alternative IPM treatment. The last decades have witnessed the closing of the scientific applied research gap between Asian and African species in favor of *O. longinoda* with comprehensive novel findings. By introducing *Oecophylla* ants, two main goals are reached: easing the burden of management costs for injurious insects and ending the practice of applying highly toxic pesticides that are harmful to non-target taxa, thus promoting environmental restoration.

Keywords: *Oecophylla* genus; population abundance; territorial foragers; quarantine defoliators; IPM

Citation: Exélis, M.P.; Ramli, R.; Ibrahim, R.W.; Idris, A.H. Foraging Behaviour and Population Dynamics of Asian Weaver Ants: Assessing Its Potential as Biological Control Agent of the Invasive Bagworms *Metisa plana* (Lepidoptera: Psychidae) in Oil Palm Plantations. *Sustainability* **2023**, *15*, 780. https://doi.org/10.3390/su15010780

Academic Editors: Barlin Orlando Olivares Campos, Miguel Araya-Alman and Edgloris E. Marys

Received: 19 November 2022
Revised: 22 December 2022
Accepted: 26 December 2022
Published: 31 December 2022

1. Introduction

The Asian weaver ant (*Oecophylla smaragdina*) is among the ecologically dominant insects in tropical forests, savannas [1], and agricultural landscapes [2]. It is an obligate arboreal, polydomous (multiple nests per occupied tree), absolute territorial species [3].

Few publications [4,5] have exposed the foraging and predation activities of *O. smaragdina* in oil palm plantations in Southeast Asia. The first report focused on the usage of weaver ants as a future potential biological control agent (BCA) for dominant bagworm defoliators (*Pteroma pendula*). Occupied palm trees were protected and demonstrated absence or low level foliar injury with significant higher productivity in comparison to unoccupied trees. Attack chronology patterns in relation to foraging activity were assessed in heavy infested blocks. The second report was a thesis dissertation that discussed the foraging activities of weaver ants in relation to air temperatures and relative humidity. A case study was conducted on a research national station at Teluk Intan, Perak with a preliminary study of population dynamic.

According to foraging activity, there is no major differences between Asian and African weaver ant species [6,7]. Foraging activity is a diurnal task performed exclusively by major workers, continuously patrolling outside the nests for prey along with surveillance duties [6]. Prey transportation by the foragers is performed only during the day period [8]. Infestation outbreak control largely depends on the sustainability of natural enemy populations. Thus, estimating the relative density of individuals to monitor the population dynamics of Asian weaver ants is important for effectively suppressing pests of economic importance to commercial crops [9,10].

The premise of population stability by single or assemblage species with *Oecophylla* ants achieving similar or better protection compared to specialist predators is attractive. Asian weaver ants could become a strong candidate in integrated pest management through direct application on threatened crops [11,12].

The bagworm, *Metisa plana*, an indigenous quarantine pest, is responsible for an average productivity loss of 33–40% in subsequent years of harvesting due to moderate (10–13%) foliar injury [13,14]. However, a more serious infestation can cause up to 43% yield loss over a two-year period [15]. The problems faced by smallholders and large estate plantations due to bagworm are recurrent and affect large, planted areas [16,17]. It is understood that smallholders (comprising many small plantation owners having approximately an average of 4 ha each within the same organization) are unable to properly handle outbreaks due to budget constraints. Further expansion of pest outbreaks is triggered from their small plantations to the neighboring larger cultivated area [18,19].

Plantation owners are very skeptical about using weaver ants to solve the bagworm issue owing to its pugnacious behavior towards humans [20,21]. Previous studies have showed that integrated pest management (IPM) trials [22] for treatment [23] gave conclusive successful outcomes. However, more information on weaver ants (such as mating mechanism, distinctive caste structure, population size-density, and individual behavior as a verified aggressor during foraging activity) is needed before they can be used for IPM [24] or as a BCA [25] in a large-scale management program [26,27].

This review examines other studies in order to understand weaver ant ecology. This understanding can be used to support the novel idea of bagworm control treatment by *Oecophylla* ants as a generalist predator. This review will articulate the information on *O. smaragdina:* (i) foraging behavior, (ii) population dynamics, (iii) the benefits and challengers faced by plantation owners if they adopt weaver ants to mitigate bagworm infestation, and (iv) expose recent research development towards adoption of weaver ants in agriculture and conclude some controversies, rare weaknesses, and strengths.

2. Research Methodology for This Review

2.1. Search and Assessment Inclusion Benchmark

This review was performed to collate the relevant available published academic literature. Only studies that provided information of both *Oecophylla* species were included in a first selective step. The second step of enrichment with broader sources was performed in the absence of enough supportive elements based solely on the first step criteria. Based on this review, title terms such as foraging behavior and population dynamics were the most dominant and relevant attributes to justify *Oecophylla* ants as a biological control agent. This was necessary to extract publications related only by analogy solely within ant taxonomy (http://info-now.org/ants/AntTaxonHierarchy.php, accessed on 5 October 2021) or scientific classification adhering to the Integrated Taxonomic Information System regrouping the Formicidae family. Studies written in "Bahasa Indonesian", French, Spanish, and English were included. We included studies exploring ecology, population modeling, foraging, and predation behavior. To fulfil the main objective of this review (convince farmers of the benefits), topics of the services and disservices of *Oecophylla* ants were given priority in our evaluation. Among them was the potential answer to the looming global food security crisis of including weaver ants in daily diets [28]. Finally, BCA and IPM treatments were the culminating subjects of the research findings. Tables were derived from the most relevant papers describing the associated host plant protection provided by *Oecophylla* ants from pests of economic interest: among them, classified invasive species. *O. smaragdina* was the dominant species.

2.2. Literature Documentation Selection

We started the literature search using the keywords "*Oecophylla* ants", "Asian weaver ants", "*Oecophylla smaragdina*", and "*Oecophylla longinoda*" in the Google search engine. The preliminary relevance of each manuscript was determined from the title based on the content of the abstract. From that initial step, if the content seemed to discuss the content of the review main topic title, we obtained its full reference, including author, year, title, and abstract, for further evaluation. We searched Google Scholar, Web of Science, frequently used databases. Because the two species of *Oecophylla* are rarely evaluated for bagworms in the oil palm plantation industry, we extended the publication date from 1960 to 2022 (articles published in the past sixty-two years), so that the review was constructed based on both older and recent literature. Considering a broader information retrieval and synthesis better demonstrated the hypothesis of *Oecophylla* ants being potent predators for the control of harmful pests. We first applied the Google general search engine to obtain different sources of papers by using keywords "*Oecophylla* foraging activity", "Asian weaver ants population", or added "dynamic", and then copy-pasted it into Google Scholar. The research was fine-tuned by adding "Scholarly articles" before each keyword. Whenever using a less specific term, such as "studies on the predatory activity of Asian weaver ants", the search turnover of 13,600 results was decreased by adding "*P. pendula*" or "*M. plana*". The decrease reached 44 and 25 potentially relevant articles, respectively, of which 15 and 4 articles abided by the intended topic title of this review. For information filtering and final selection of the manuscripts of interest, selection of the quality and eligibility of the published articles was achieved by strongly considering the following authors for most topics of study: Hölldobler & Wilson; Peng & Christian; Peng et al.; Van Mele & Cuc; Offenberg; Dejean; and Way & Khoo. By reading through pre-selected or selected articles, we found more experts doing fundamental and applied research that could significantly contribute to the value of this review as follows: Newey; Robson, Crozier, and Nielsen for the Asian species; and Nene, Vayssieres, and Rwegasira et al. for the African species. The search for keywords "*Oecophylla* foraging activity" and "*Oecophylla* population dynamic" completed the final article selection process. For foraging activity, we obtained approximately 2900 referred articles in Google Scholar, of which only 35 showed a strong relevance to the main title subject. For population dynamics, we obtained a total of 2380 results, of which only 20 were related to the manuscript title with a majority of these articles showing an orientation for applied biological control treatment on various pests of economic interest. After initially screening the titles and reading the abstracts of an average of over 300 related articles, a total of 156 studies were identified as relevant to the title of this review: "Asian weaver ants as potential biological control agents of invasive bagworms *Metisa plana* (Lepidoptera: Psychidae): a review". For each selected article in the review, the "Related articles" option available in the Google Scholar database helped to quickly identify similar studies able to enrich the search for study inclusion in the review. For the final inclusion of identified studies, we scanned through the full-text articles to further evaluate their quality and eligibility by systematically targeting the reputable names of those researchers mentioned above that have a strong record related to the *Oecophylla* ant genus.

3. Foraging Behavior of Weaver Ants

Weaver ants are a well-disciplined and well-organized insect society. Its major workers caste members perform extensive foraging over a large territory to ensure the safety of the entire colony and maintain colony survival [6]. As a diurnal insect, weaver ant foragers are seen patrolling with their special task force of experienced workers to secure the whole perimeter of the colony territory [25]. Although they are strictly arboreal in nature [29,30], weaver ants have been commonly seen actively foraging on the ground [31] and moving by group of foragers [32], even when the canopies are interconnected [5].

During foraging, *Oecophylla* ants use their visual organ to detect encountered items from a distance [33] and olfactory cues to perform daily foraging duties [30]. Various authors [6,34,35] have proposed that the foraging activity of weaver ants can be summarized into five main schemes as follows: (i) the recruitment of ants into a new landscape to fill a gap in their path (i.e., obstacle crossing by bridging with more individuals). Complex chemical compounds are secreted from anus glands coupled with tactile signals. These chemicals form a chain of trails that facilitate the path of recruited nestmates using their antennae to reach the desired destination; (ii) foragers use palpable stimuli by mouth connection, antennae, or feelers, and head shaking to find resources; (iii) to explore new foraging range, fluid droplets from the rectal vesicle are laid to be detected by nestmates; (iv) to resist trespassers, an "alarm" attractant pheromone from the sternal gland is released; (v) defensive long-range recruitment comprising of odor trails, antennae, and thrilling "body jerking". All tasks related to foraging, nest guarding, and repair, along with territorial defense, are carried out by the major workers [6].

Generally, foraging and colony defense is a risky task, substantially impairing survival ability and therefore incurring high mortality rates to ants [36]. This is particularly true for *Oecophylla* ants, where major workers aggressively defend extensive territory against con-specific individuals from different colonies seen as competitors or intruders. Thus, evaluating the general activity of *Oecophylla* ants as a whole colony entity for IPM utilization is well justified. It helps in designing a better method of pest control in the field [37]. The basic main tasks at the colony level comprising the foragers' activity of major workers caste range from foraging to hunting, transporting prey items back to the nest, and surveillance [38–40].

There is still a scarcity in reports concerning the foraging activity of *O. smaragdina* or *O. longinoda* at the colony level based on 24 h monitoring scale [41]. However, another report expressed the importance of defining the appropriate daily time period to perform colony identification, transplantation, and population estimation [37]. The benefits of such manipulations will enhance integrated pest management by defining the multiple duties of weaver ants [37]. Major workers are the sole foragers outside the nest area and responsible for covering extensive grounds for hunting and predation purposes [6]. They also explore more territory to expand the colony boundaries. Figure 1A–D exposes the foraging activity of major workers on canopies, trunks, and ground in Felda oil palm plantations.

Figure 1. (**A–D**). *O. smaragdina* major workers' foraging activity: (**A**) Foragers on palm canopies frond. (**B**) Foragers on palm trunk In Felda Gunung Besout, Perak plantations. (**C**) Nomadic ground foragers around palm trees performing duties of exploring/hunting/surveillance in Felda Keratong Pahang plantations. (**D**) Foragers occupying a different plant species in oil palm plantations in Felda Keratong Pahang. Photo credit: Exélis Moïse Pierre.

4. Population Dynamics of Weaver Ants

After the introduction of any natural enemy, if individual abundance decreases and requires continuous artificial release upon mass-rearing to maintain its stability, this may not be economically feasible [42,43]. Therefore, the basic ecological need is the species status level, which monitors its variations in time and space [44,45]. This concept constitutes one of the main factors for a proper assessment of healthy population dynamics [46,47]. Investigations of the underlying forces (biotic and abiotic elements) responsible for those variations form the other fundamental basic components for checking and estimating PD [46]. Population dynamics are influenced by deterministic (predictable) or stochastic (unpredictable) components operating simultaneously [47]. For instance, in many insect species having short life cycles, predictable seasonal environmental parameters, such as temperature [48], rainfall interception [49], and accessible food web, influence negative or positive fluctuations in population dynamics [47]. Insects are affected by sudden variations in temperature due to their ectothermic nature [48]. The synchrony of Glanville fritillary butterfly (*Melitaea cinxia*) population dynamics during lower summer precipitation is an example of how drought affects the survival of early larvae instar, hence its metapopulation stability in the long term [49]. To successfully use weaver ants in any pest management control, it is fundamental to understand the importance of ecological factors that regulate their population dynamics. In addition, it is also compulsory to evaluate PD in the field for a long period [50]. Manipulation of the *O. smaragdina* population by introducing foreign pupae from different colonies demonstrated a successful boosting with significant worker force increase [51]. Such promotion of incipient colonies to reach growth maturity earlier than usual enables further nest translocation to targeted pest-affected crops [50].

Limited studies have been conducted directly in the field with a large agricultural monoculture over long periods of monitoring (5 to 10 years) that are backed up with empirical database records. This is because most ant colonies are subterranean. An example is the spectacular colossal intricate nest chambers (equal to the size of a house) of the attine leafcutter ant species *Acromyrmex* and *Atta* of tropical America [52]. According to ref. [53], monitoring insect taxa population dynamics by measuring their abundance and biomass based on individual precise count is "historically" an exceptionally rare method. Nest excavation leads to colony habitat destruction [54]. Some researchers answered this hurdle by applying software simulation [55,56]. Ref. [57] gave caution on the adequacy of the ability of such models to predict and explain the overall characteristics of the collective behavior of ants by having scarce quantitative validation and insufficient experimental evidence.

Fortunately, the population dynamics of *O. smaragdina* can be estimated using the direct nest counting method (all individual castes, brood ants, and eggs) (Figure 2A,B). This method is feasible for planters and agricultural officers without the need to consider nest volume and other nest characteristics because none of the parameters are correlated to individuals' distribution in the nest [25]. Nest distribution uniformity within the same habitat or plantation for mature colonies is documented with an average occupancy comprising a range (per tree, per colony) [5,6]. Verifying that the distribution of *O. smaragdina* is not correlated to nest internal and external variables (i.e., volume), this method is acceptable [10].

Figure 2. (**A,B**). Numerous *O. smaragdina* eggs from clusters extracted from a captured nest in Felda Gunung Besout, Perak oil palm plantations and examined by using a stereomicroscope Nikon SMZ800N (**A**). Brood ants, major, and minor workers exposed for direct counting of all individual castes (**B**). Photo Credit (Moïse Pierre Exélis).

As a potential BCA candidate, *O. smaragdina* is viable for practical reasons, such as abundance of individual predators versus that of defoliators [7]. Their surface occupancy is sufficient with fairly large individual numbers, enabling physical counting without needing to destroy the colony for estimation assessment. *O. smaragdina* is never subterranean, but some nests can be found on the ground under heaps of debris or piles of vegetation [31]. In addition, in Peninsular Malaysia, this method produced satisfactory results without complications [5]. This result in oil palm plantations was similar to previous reports on the abundance of individuals per colony for the *Oecophylla* genus [10,58].

In an earlier study, the population size of an *Oecophylla* colony was estimated to be approximately half a million major workers, with more than 1/4 million or more brood ants, without providing data on the total number of minor workers [59]. Similar reports from other studies [60–63] have confirmed the existing range of mature colonies with an average population of several millions of workers. The population abundance and its dynamic may vary with the adopted colony's habitat, such as tropical primary or secondary rain forests, large monoculture, selected preferred variety of fruits trees, including medicinal plants, as well as rural or urban zones [9,58,64].

The widely recorded geographical distribution of *O. smaragdina* in Asia and Oceania [65,66] gives the species some reliable edge as a potential BCA in large agricultural landscapes. To sum up this concept of density interrelated functions for maintaining BCA stability [67,68], the regulation of the population size is dependent on the persistence of positive fluctuations recurrent over generations [69,70], but the lack of data asserting whether any resilient metapopulation is bound by a variable mechanism adds to the ambiguity of this ecological fundamental concept with contemporary challenges [71]. The interdependence between abiotic and biotic factors with coexisting species based on the natural principles of competition is the determinant factor [72–74].

5. Benefits and Challenges

O. smaragdina is an ecologically dominant and aggressive ant species [75]. Asian weaver ants are reputed to be excessively pugnacious generalist predators that prey on a wide range of insects [76]. Their prey comprise eight orders with twenty-six families, for a total of more than one hundred different pests [12]. Being by nature highly predaceous ants, *O. smaragdina* exhibit extensive exploratory behavior [32], with major workers having long, slender, and serrated mandibles exhibiting elongated distal teeth, perfectly adapted to their hunting inner instinct [77]. Records show that in China, *Oecophylla* nests have been introduced in citrus orchards to control pests since 300 A.D [78]. A study [79] in the Solomon Islands showed the smooth dispersal of *O. smaragdina* in coconut trees by having ants naturally infect new plantations, thus establishing new unwavering colonies.

Research has focused mainly on citrus and cashew nut crops [80], but some reports provided solutions for a variety of pests in mango orchards [22]. An integrated pest management model using Asian weaver ants in Australia to control major pests such as leafhoppers, *Idioscopus nitidulus*, red-banded thrips, *Selenothrips rubrocinctus*, the mango tip borer, *Penicillaria jocosatrix*, the fruit spotting bug, *Amblypelta lutescens lutescens*, the kernel weevil, *Sternochetus mangiferae*, the fruit fly, *Bactrocera jarvisi*, numerous leaf waves, and flower caterpillars is well documented [22]. The combined treatment of *Oecophylla* colonies with potassium soap and white oil was performed in comparison with synthetic insecticides, demonstrating a significant increase in yield without affecting pollinators [22].

In their review, ref. [12] listed only seven insect pest families susceptible to weaver ants in tropical crops. Ref. [81] reported that weaver ants significantly reduced the presence of damaging herbivores on *Rhizophora mucronata* in Thailand. Table 1 summarizes the use of Asian weaver ants as a BCA for various insect pests of economic significance affecting major crops in countries in Asia and the Pacific region. The success of using *Oecophylla* spp. as a biological control in fruit orchards (Table 1) has been well documented [81,82]. Initially, weaver ant control of numerous insect pests was associated with their diet orientation. It was suggested that the presence of the African species *O. longinoda* may have impacted the underlying mechanisms of successful pest control. Its ability to initiate the host plant to generate beneficial secondary metabolites in leaves reinforced the plants' defense against insect herbivores [83]. Furthermore, its pheromone density is recognized as a disturbing factor for oviposition by invasive *Ceratitis cosyra* and *Bactrocera invadens* (new invasive species in West-Central Africa) mango fruit flies, capable of achieving noticeable damage reduction [84]. Those pheromones were identified as having a natural fruit fly repellent effect. However, the presence of the synergistic consequence of the weaver ants using the parasitoid *Fopius arisanus* within the same ecosystem may outweigh the subsequent effective suppression on *B. invadens* (foreign invasive species on mango in Africa) [85]. Hence, this factor is important if the two natural enemies are to be adopted in combined efforts against this fruit fly species. Although weaver ants are gaining momentum as a biological control agent in Africa and Asia, there are instances where these ants are a serious hindrance for plantation workers [86]. Their ferociousness is a real nuisance during pruning and harvesting of crops [20]. A protocol helping to alleviate this problem was proposed and offered encouraging measures [80,87]. More study is needed to find a comprehensive, practical, and cheap approach to minimize the painful bites faced by maintenance staff in occupied plant canopies. The following Tables 1–7 present the results of a meta-analysis of *O. smaragdina* functions as a beneficial predator of major agricultural pests from diverse commercial crops (among them, some studies show only potential BCA treatments, see Table 3).

Table 1. Beneficial records of *O. smaragdina* for coconut.

Oecophylla Occupied Plants (Colloquial, Scientific Name)	Control Methods—*O. smaragdina* Presence Effects	Associated Pest Species (Colloquial, Scientific Name)	Damage & Economic Yield Loss/Increase in Presence/Absence of *O. smaragdina* Treatment	Region	Key Reference(s)
Coconut *Cocos nucifera*	Satisfactory: palm base dieldrin spraying prevented *Pheidole megacephala* to induce *Oecophylla*'s population dynamic collapse.	*Amblypelta cocophaga*	Premature bug nut fall, sucks sap of young coconut	Solomon Islands	[89,90]
	High *Oecophylla* predation on *A. cambelli*. Soursop fruit *Annona muricata* buffer zone promoted colony abundance.	*Axiagastus cambelli* *Brontispa longissima* *Promocotheca* spp.	*A. cambelli* causes dry, thin, long nut production (no milk)	NBPNG *	[91–95]
	Oecophylla ants not effective on *O. arenosella. Monomorium floricula* and *Crematogaster* spp. outweighed *Oecophylla* egg predation.	*Opisina arenosella*	Similar	Sri Lanka	[96]
	Iridomyrmex myrmecodiae and *Pheidole*, broke *Oecophylla* population dynamics.	Coconut bug, *A. cocophaga* Hispine beetle *B. longissima.* Palm leaf beetle	Young leaf feeder with seedlings and mature palm damage.	Solomon Islands	[94,97]
	Hispid absence correlated with presence of *Oecophylla*.	*P. papuana* & *P. opacicollis*	Destruction of leaflet distal parts by feeding [88]: 2 years recurrent yield loss before full recovery.	Papua New Guinea	[98]

* New Britain Papua New Guinea.

Table 2. *O. smaragdina* benefits for agarwood, lychee, and cocoa.

Agarwood Gaharu *Aquilaria* spp. *Gyrinops* spp.	Control: 2–4 *Oecophylla* ants per prey	*Heortia vitessoides*	Excessive defoliation	Indonesia	[99]
Lychee *Litchi chinensis*	1 nest managed to prevent foliar injurious insects and pentatomid bug	Lychee stink bug, *Tessaratoma papillosa.*	Fruit: premature fall, external feeding, discoloration. Inflorescence: external feeding, fall of shedding. Stems: external feeding, necrosis.	China	[78]
Cocoa, *Theobroma cacao*	*Oecophylla* abundant population provided complete protection. *P. megacephala* control; *Oecophylla* ants effective protection.	*Helopeltis theobromae* *Amblypelta theobromae* *Peudodoniella laensisPantorhytes* spp.	Mosquito bug nymphs, adults infest cherelles, pods, young shoots. *A. theobromae*: high yield loss. Weevil borer larvae: sapwood of trunks, branches digging 1–3 cm deep burrows causing bark canker water mold disease *Phytophthora palmivora* and termites.	Malaysia PNG Solomon Islands Malaysia	[20,21,101] [102] [93,95,103] [104] [105–107]
	Oecophylla population increased by shrimps, palm sugar pellet feeding: 7.44% and 13.38% less damage, respectively.	*Pantorhytes biplagiatus* *Conopomorpha cramerella* Sn.	Severe pod damage by 21.54%, clumped beans. 64% or more yield loss with significant Average Damage Severity Index (ADSI) of 3.5 (dry season) [100]	Indonesia	
T. cacao	Highly abundant colonies reaching level 5: healthy fruit. Serious fruit damage in control plots. *O. smaragdina* absence/presence > 36–50% pod lesions, respectively. No choice feeding demonstrated highly aggressive predation.	Cocoa pod borer Conopomorph (Acrocerops) cramerella Snellen Cacao mired bug (CMB) *Helopeltis bakeri*	Similar damage as others reports Severe lesions on pods	Indonesia Philippine	[108] [109]

Table 3. *O. smaragdina* benefits for citrus, *Manilkara zapota*.

Citrus *Citrus* spp. C. sisensis C. reticulata C. maxima C. sinensis C. limon Sapodilla- Naseberry @ *M. zapota* Calamansi- Limau kasturi, *Citrofortunella microcarpa*	Buffer conservation zone of associated plants * for weaver ant abundance. Effective with pomelo trees **. IPM by *Oecophylla* replaced WHO classified extremely hazardous insecticides, i.e., methyl parathion. Reduced by 50% pesticide use dependency & 60% vector disease. *Diaphorina citri* reduction but ineffective on mealy bugs. *** IPM: Effective protection on mixed pomelo/orange equal yield, lower cost than chemical treatments. Leafhopper *Idioscopus clypealis* (honeydew) tending = Negativeproductivity in Thailand mangoorchards [23]. Suppressed a wide number of pest species. *O. smaragdina* colonies partially present during the year (5 months). Potential BCA Potential BCA Predation on large number of insect pests [110]	*Tessaratoma papillosa* and other Heteroptera, *Rhynchocoris humeralis* R. serratus *Phyllocnistis citrella* *Toxoptera aurantii—T. citricida* *Diaphorina citriPanonychus citri* *Phyllocoptruta oleivora* *Bactrocera* spp. *Eudocima salaminia* *Ophiusa coronate* *Hypomeces squamosus*. Asiatic citrus psyllid (ACP), *Diaphorina citri*. NA [a] 25 records	*R. humeralis* sucks the juice from fruit, leaves, and branches. *R. serratus* principally seed-feeding and sap feeder *P. citrella* eat leaf tissues *T. aurantii- citricida* responsible for *Citrus tristeza* closterovirus (CTV) a phloem virus *P. oleivora* infests mature branches, green twigs, leaves, and fruit skins causing heavy yield/quality losses. *D. citri* cause greening disease *E. salaminia* green fruit-piercing moth *O. coronate* fruit-piercing moth High density of curculionid beetle *H. squamosus* extensively grazing on young shoots, immature trees, larger trees (no ants or pesticides), See [111] Citrus vein phloem degeneration (CVPD) @ huánglóngbìng. 20% yield increase per year NA [a] NA [a]	China Philippine Vietnam Vietnam Thailand China Indonesia India Malaysia	[78,112,113] [114] [115] [23] [111,116] [117] [118] [110]

* *Eucalyptus tereticorni, Ceiba pentandra, Mangifera indica, Spondias dulcis, Annona glabra, Premna integrifolia* ** Larger and thicker leaves provide better protection from cold temperatures for nest survival during winter periods. *** Mutualistic relation for honey dew energetic source. [a] Not available.

Table 4. *O. smaragdina* benefits for mango.

Eucalyptus *Eucalyptus* spp.	Vegetation clearing around trees in combination with the introduction of *Oecophylla* colonies	*Acria cocophaga*	Adult, nymph causing heavy shoot-tip necrosis	Solomon Islands	[119]
Mango *Mangifera indica*	Potassium soap (1.5%)/white oil (2%) spray reduced scale/mealy bug damage on fruit [26]. Reduction of formic acid damage on fruit by separation of *Oecophylla*. Abundance of weaver ants + soft chemicals < 1% caused much lower downgraded mango than untreated orchards [120]. Control fail: honeydew producer leafhopper *I. clypealis* & *Oecophylla* association = 125% less profit compared to chemical treatments & no fruit setting [23]. No difference in presence/absence of *Oecophylla* on mealy bug *Drosicha mangiferae* & scales *Aulacaspis tubercularis* occurrence [23]	*Cryptorrhynchus gravis* *Idioscopus nitidulus* *Sternochetus mangiferae* *Campylomma austrina* *Selenothrips rubrocinctus* [a] *Bactrocera jarvisi* [23] *I. clypealis* [23] Mango leaf webber larvae *Orthaga euadrusalis*	Hoppers Nymphs & adult suck sap: panicles, tender shoots = withering and dropping of florets, lower photosynthesis. Thrips weaken inflorescence, causing serious bronzing of the fruit surface due to the presence of air in emptied cell cavities; 20% 1st fruit class increase & 80% profit/tree per year [22]. Fruit flies maggots cause rotting	Indonesia Australia Thailand	[121] [26,120] [23]

[a] Extra pests: Mango flower webber *Eublemma versicolor*, Mango shoot webber *Orthaga exvinacea*, Mango leaf hopper *I. nitidulus*, Red bugs *Dysdercus cingulatus*, Leaf twisting weevil *Apoderus tranquebarious* Curculionidae, Brentid beetle *Estenorhinus* spp.

Table 5. *O. smaragdina* benefits for Mahogany, Makassar ebony & teak wood.

African mahogany *Khaya* sp. *K. ivorensis* *Swietenia macrophylla* *K. senegalensis*	Abundant colonies occupation with low cost food supplements, mix cropping, favorite host plant habitats to enhance colonies long term conservation as a maintenance buffer zone support.	*Hypsipyla robusta*	Shoot borer	Malaysia	[58,122,123]
	Comparative study with pesticides demonstrated similar or better protection, yield production	*Amblypelta lutescens*	fruit-spotting bug causing damage tree level 80–100%	Australia	[25]
	Damaged trees mean average was 0–2.6% by weaver ant treatments—14.2–27.0% at Howard Springs, 28.2–48.6% at Berrimah Farm in weaver ants absence. Yearly damage trees: 4.2–32.4% by yellow loopers—0–10.4% by bush crickets	*Gymnoscelis* sp. *Myara yabmanna*	25–70.4% by yellow loopers 25–100% by bush crickets	Australia	[37]

Table 5. *Cont.*

Makassar ebony wood *Diospyros celebica Bakh*	*Oecophylla* colonies presence maintained a low 7.69% rate of attack among 39 trees (others predators available)—important highly commercial luxury wood endemic to Sulawesi	*Arctornis submarginata* *Lymantria marginata*	Leaves are gnawed from the edge to the vein and midrib	Indonesia	[124]
Teak wood *Tectona grandis*	Colonies presence: control pests with an average 80–90% rate of teak trees (most important commercial wood in Indonesia). Absence of weaver ants: 30% level 1, 30% level 2, 25% level 3 and 15% level 4 foliar injury	Undetermined defoliators species *Termites Tectona grandis L.f.*	Teak stands leaves attack in early wet season	Indonesia	[125]
White lead tree, River tamarind *Leucaena leucocephala*	Field surveys observation records: one colony of *Oecophylla* kills an average 2000 psyllid lamtoro jumping plant lice per day. Leucanae trees widely planted offers shadows to crops and livestock feeding for animal production. Agroforestry mix systems field: Coffee—*Coffea Arabica*, Vanilla—*Vanilla planifolia*; Cocoa—*Theobroma cacao*, Oil palms—*Elaeis guineensis* Jacq.	*leucaena psyllid* *Heteropsylla cubana* [a]	Young leaves stems, branches, petioles gregarious feeders. USD 1.5 billion loss from five years of infestation [126,127]	Indonesia	[128]

[a] Invasive species.

Table 6. Beneficial *O. smaragdina* activities in cashew nuts orchards.

Cashew Nuts *Anacardium occidentale*	Low damage, production of high quality nuts and panicles. In absence of insecticides, ant abundance increased from 0 to 80%. First 2 or 3 years, oscillated under 80%. Colony isolation can produce 100% colonization level [129,130]. Monitoring of nest dynamic per naturally occupied cashew trees (11–13 old) [131]. Rearing colonies during 4 years with two blocks (occupied and unoccupied) by nest translocation. *O. smaragdina* nests presence throughout the year. Field, semi-field, and laboratory trials: main method used 3 weaver ant populations, i.e., 0, 5, 10 colonies per 5 plants [132]. *Oecophylla* not affecting *S. indecora* population when combined with *Helopeltis* spp. Plant protection was achieved and nymph predation occurred if cashew shoot hopper infestation occurred in absence of the tea mosquito bug [133].	*Helopeltis pernicialis* *Penicillaria jocosatrix* *Amblypelta lutescens lutescens* *Anigraea ochrobasis* *Helopeltis* spp. *H. antonii* * *Helopeltis* spp. *Sanurus indecora*	Both sap-sucking bugs 75–80% shoot, 98% flower. Positive correlation between yield with levels of ant colonization; the total variation in yield was explained by 83–90% of the results [129,130]. Tea mosquito bug damage significantly reduced with higher fruit yield. Effective protection, preying on adult and nymphs, 13.67% yield increasd [131] Successful control of *Helopeltis* spp. population Lower frequency, number of attacks on flowers of "Jambu Mete"	Australia India Sri Lanka Indonesia	[129,130] [131] [132] [133,134]

* Main and most damaging cashew nuts pest.

Table 7. Medicinal and large monoculture industrial plantations.

Hoop pine *Araucaria cunninghamii*	*O. smaragdina* effective larval predator	*Araucaria* looper or millionair moth *Milionia isodoxa*	Whole day larval feeding causing serious foliar injury	Papua New Guinea	[1] [135]
Climbing vine *Cynanchum pulchellum*	Plant used for medicinal purpose. Predation by *O. smaragdina*.	Common tiger caterpillar, *Danaus genutia genutia*	Larvae feeding on plants	Singapore	[136]
Oil Palm *Elaeis guineensis* Jacq.	<2% foliar injury with Level 0, 1. Higher FFB productivity/quality fruits. Bagworm infestation ≤ 15%. Weevil pollinator *Elaiedobius kamerunicus* safe. Predation trial by non-choice and choice on *S. nitens* 87.50% & 83.33% respectively	*Pteroma pendula—Metisa plana* [a] *Setora nitens Sethosea asigna Thosea sinensis, M. plana, Parasa lapida*	Heavy foliar injury level 4, Less FFB productivity recorded. Severe damage/±44% yield loss/year recurrently. Foliar injury (palm canopies)	Malaysia Indonesia	[4,5] [13,14] [137,138]
Asian weaver ant *O. smaragdina* **beneficial Synthesis**	**Ecosystem services** Productivity & yield enhancer-feces nutrient NPK [3] provider, Regulation of wide variety of pests including invasive species; Supportive of IPM helping reduce harmful pesticides dependency. Weaver ants have longer lifespan, stability factor. High pest predation rate with phytophagous regulation, fruit damage reduction, and pollination neutral effect; herbivory assists nutrient cycle	**Ecosystem disservices** Rare cases of low yield, pollinator abundance and scales insects-mealy bugs foliar damage See refs. [7,80]	**Economic Input-Societal Benefits** High income, rich nutrition, medicinal-antioxidant properties; global food crisis security component: See refs. [28,139–143] *Bacillus thuringiensis, Oecophylla* colony usage in IPM [2] outweigh more expensive treatments.	**Asia**	**Agricultural Systems** Variety of plantations: See refs. [7,80,144] Among all predators, *O. smaragdina* has higher potential for arboreal pest insects (CMB).

[1] available online by 2009. [2] Trials done on same plantations with existing colonies before Bt applications. [3] Nitrogen, Phosphorus, Potassium. [a] Invasive species.

Even though the successful application of Asian weaver ants in cocoa plantations in Malaysia was proven [101], its final adoption was recently abandoned due to the aggressive behavior of *O. smaragdina* [20,21]. This nuisance factor is impossible to avoid since man-made intervention is permanent. During the harvesting process, sometimes highly toxic synthetic chemical poisons were used to eradicate weaver ant colonies. This issue cannot be taken lightly and must be addressed [42,80]. In Africa, the use of ashes in field plantations proved to be very effective against ant bites [27]. Recently, oily repellents proved to be the most effective in Africa, hence giving promising results for eventual adoption and utilization by farmers during work [145]. Plantation staffs could carry out their duties during lowest weaver ant activity periods [37]. By avoiding the active period peaks of Asian weaver ants, pruning and harvesting were safer during early morning hours [5].

Another aspect that needs consideration is the existence of other antagonistic species observed to be a limiting factor to the normal activities of *O. smaragdina*. *Dolichoderus* spp. is suggested to impair *O. smaragdina* activities in cacao plantations in Malaysia, yet *Oecophylla* are surprisingly accepting of the presence of *Dolichoderus* with full passivity [21]. Should *O. smaragdina* to be used as a predator in oil palm plantations, this factor should be seriously considered as to not leave the ground open for any cohabitation shared between *Oecophylla* ants and the black ant *D. thoracicus*. Repeated monitoring during surveys exposed the inability of *Oecophylla* ants to establish their colonies within occupied *D. thoracicus* areas. Another antagonistic effect of *D. thoracicus* was reported in *Citrus sinensis* and *Citrus reticulata* in which colony development of *Oecophylla* was hindered [115]. The possibility of additional existing antagonistic species is real and need further attention.

6. Recent Research Development towards Adoption of Weaver Ants in Agriculture

Over the recent years, a fair investment in research has been engaged in improving the management of *O. smaragdina* by incorporating innovative and effective procedures with new knowledge of the ants' ecological patterns. Such research enabled easily detecting queen nests with the purpose of future transplantation agenda [146,147]. The introduction of weaver ants in cashew nuts trees reduced the menace of the tea mosquito bug, *Helopeltis antonii* (Hemiptera: Miridae) [148]. Gravid queens of *O. smaragdina*, tested for their acceptance of foreign nest pupae, resulted in a drastic increase in the worker population within a short period of time, thus helping colonies mature faster. The proposed study conducted on incipient colonies demonstrated the viability and benefit incurred by the adoption of such new foreign broods as an early colony booster [51]. It is possible that by boosting the population dynamics from the initial stage within an undisturbed environment, those *O. smaragdina* colonies may be introduced later into crop trees [50]. The feasibility of using artificial nests to capture new queens upon nuptial flight has been demonstrated for Asian weaver ants [50]. Incipient colony development to maturity takes as long as three years to potentially produce a new queen brood. Pupae sourced from different colonies added to an incipient colony stimulates early colony growth and gains in significance if the incipient colonies are polygynous [50,51]. Larvae transplantation between different colonies is possible [82] since at this stage the nestmate recognition cues have not yet been formed [149,150]. Pupae adoption by foreign colonies of various ant species is feasible [150]. It is understood that larvae and pupae of *O. smaragdina* do not possess pheromones characteristic to determine recognition cues within each colony, which will only be acquired upon emergence of the adult stage [82]. This is beneficial for the manipulation of colony population growth, knowing that *O. smaragdina* is strictly territorial [151]. In a recent study of Asian weaver ant to promote faster early colony development, the addition of pupae to form a new colony gave promising results without hostile rejection [50]. The combination of polygyny and abundant pupae transfer achieved faster colony growth [50,51]. Comparatively, after 12 days of replacement with 60-pupae transplantation and four queens per colony, it was possible to produce much a higher brood rate than that achieved with two queens and without pupae relocation [50]. Ref. [51] demonstrated that the benefit of having a greater number of gravid queens, resulting in a drastic increase in new workers.

For long-term sustenance of a colony, the availability of food is the major dependent factor in sustainable maintenance. An experiment conducted in cashew nut orchards with *O. longinoda* demonstrated an increase in population for colonies fed additional sugar and fish protein without impeding their predatory activity [152]. Personal observation in cage culture for mass rearing demonstrated that the ants possessed exceptional survival ability when faced with severe food scarcity up to two months (continuous feeding with water, sugar, and protein was provided for the first week only). An average population size of two hundred ants was achieved. With the combination of such advantages and the beneficial factors exhibited by weaver ants, *O. smaragdina* can be a real contender in combatting pests in oil palm plantations [5]. Figure 3A,B illustrates *O. smaragdina* major workers attacking the pupae of *P. pendula* bagworms in an affected plantation [4].

Figure 3. (**A,B**). *O. smaragdina* predation acting on *P. pendula* bagworm pupae in an oil palm plantation. (**A**) Hunting during midday peak foraging period. (**B**) Predation during late evening day period. Photo credit: Exélis Moïse Pierre.

Within the same topic, effective tested queen nurseries were recommended to save time and avoid the hassle of wild capturing ants by providing a continuous direct source of water, sugary solutions, and protein to ensure weaver ants are able to establish a new colony [153]. In the case of failure by *Oecophylla* colony treatment in the face of uncontrollable pest species, such as the mutualistic relationship between the leafhopper *Idioscopus clypealis* and weaver ants to obtain honeydew [23], it is necessary to apply alternative methods. An example of another environmentally safe application includes sex pheromone trapping and Neem (*Azadirachta indica*) application, which demonstrated compatibility with *Oecophylla* control in Ghana cocoa [154].

A recent paleontological study on *Oecophylla* fossils demonstrated an early and middle Eocene appearance from North America, with their chronological distribution related closely to ecology, behavior, and natural competition factors among global ant clusters [155]. Finally, an assessment of *Oecophylla* worker population density and dynamics is feasible using the direct nest counting method, provided that no external nest characteristics are statistically significantly correlated with the number of workers. A simplified multiple linear regression (MLR) model formula demonstrating higher accuracy performance with lower mean squared error (MSE) and root mean squared error (RMSE) has been demonstrated [156].

Potential Controversies Weaknesses and Strengthes

In addition to the mentioned biotic and abiotic potential factors responsible for positively or negatively influencing population dynamics, daily photoperiod cycles have never been reported to harm weaver ant colonies. *O. smaragdina* exhibiting omnivorous diet orientation [6] might invite caution about the possibility that they can prey on both herbivores and other beneficial natural enemies from surrounding crops. In fact, the Asian weaver ant demonstrated clear selective food preferences towards rich protein sources, such as live mealworms over fish, with a balanced lesser attraction for liquid-diluted honey during pilot field trials as a favourable BCA on the shoot borer, *Hypsipyla robusta* [58]. Another report exposed their predilection for chicken meat [157]. However, there is not clear reported evidence of the Asian weaver ant targeting beneficial insects in commercial crops. Few reports emphasize the risk of harming diverse pollinators in agricultural landscapes. It is opportune to revise the possibility of attack causing injuries to pollinators. Ref. [158] surveying pummelo (*C. maxima*) exposed a satisfactory and continuous attendance by diverse pollinators in the presence of *O. smaragdina* [159], which contradicted the findings of repelling pollinators due to Asian weaver ant occupancy in rambutan orchards (*Nephelium lappaceum*). This apparent setback did not disturb the fruit setting mechanism (Kazuki Tsuji, pers. com). The foraging activity of ants on *Polemonium viscosum* was suggested to promote the plant pollination process [160]. The Asian weaver ant, by targeting the less proficient pollinator species, promoted activation of the most efficient pollinators, thus providing a strategic profitable ecological service [161]. Hence, there is no evidence of weaver ants directly physically harming any pollinator. Ref. [159], by narrowing their interpretation of "repelled pollinator", did not explore the benefits of *O. smaragdina* as a more versatile service provider [161]. The mutualistic relationship between weaver ants and trophobiont honeydew producers sheltering and feeding on plants stems damaging host plants is a rare case of disservice [12,58]. The Asian weaver ant obligate territorial stance is not derived as a preventive response to their exposed apparently weak arboreal habitat condition, but rather by natural intrinsic behaviors [6]. The major workers provide excellent protection within and beyond the colony perimeter from three dimensions (canopies, trunks, and ground), denying alien interference to their best ability by sealing all occupied sites [162]. The ground successive defensive layer mechanism, turning colony territory into a fortress of patrolling permanent major workers, is testimony to the intelligence of this species surpassing that of others [162]. The Asian weaver ant is a good natural enemy able to be introduced alone as a biological control agent against the Queensland fruit fly, *Bactrocera tryoni* by 1-octanoil emission [163], the invasive *M. plana* or incorporated as an IPM component in combination with soft chemical support [26,42]. An important slight differentiation in methodology needs to be further explained. Natural enemies are either indigenous or introduced exotic insect species (predators) already existing in various ecosystems function in the ecological balance chain as the dominant regulators of other harmful insects [164], with the second option never to be used. Such predators can be used as biological control agents by performing some artificial manipulations to help them get established and raising their population for massive propagation or for long term field nurseries, thus reaching abundant and stable levels [165]. Even though weaver ants are widely distributed, there are occurrences of poorly occupied territories in need of improvement by forming conservation buffer zones made of favorite hosts [58,78,113] or by massive translocation of their nests [25]. Ant bites are followed by the release of low quantities of formic acid, so the harm incurred by humans is not toxic. Major workers attack all stages of bagworm development, from immature to mature individuals. To conserve energy, foragers first get rid of the immobile pupae, wingless queens, and laid eggs, then all instar larvae stages by conducting a systematic prey hunt [4,5]. Few reports exposed the toxicity of pesticides to *O. smaragdina* [166]. Sometimes in large oil palm plantation monoculture, a campaign of stern elimination is conducted to suppress Asian weaver ant colonies using broad-spectrum, highly toxic, synthetic chemicals. Studies proving the effectiveness of both Asian and African weaver ants as biological control agents and IPM valuable components (combined treatment with other methods) is far beyond the infancy stage and is reaching an advanced level of achievement [7,80,144]. Hence the differentiation of the two methods is fundamental: a biological control agent is used alone for treatment control while IPM is the combination of an array of methods constitutive of biological input with soft chemicals in order to discourage the development of a harmful organism population and guarantee the lowest disruptive impact to the agro-ecosystem's health [167].

7. Conclusions

Weaver ants are reputable natural enemies used as a biological control agent of injurious insects to commercial crops, but a few cases have highlighted their limitation, including rupture of their population dynamics caused by competition with *D. thoracicus* along the promotion of mealy bugs and scale insects in occupied plants for mutualistic benefit. The advantages of *O. smaragdina* as a natural enemy, as a biological control agent or as a component of IPM treatment, are numerous when implemented as a side business of farming entrepreneurship. Among the various ecosystem services is the provision of NPK nutrients sourced from ant feces for absorption through leaves to be assimilated as a productivity-yield enhancer. The regulation of pests of economic interest include invasive species. The method is supportive of IPM in helping to reduce harmful pesticide dependency and weaver ants have a longer lifespan, proving a population stability factor without antagonistic ant interference. The economic input and societal benefits include side income earning, when sold for songbirds or as a nutritious diet delicacy rich in medicinal and anti-oxidants properties. Indeed, weaver ants are suggested as a potential global food crisis security component. To implement the adoption of *O. smaragdina*, understanding its foraging activity and population dynamics is compulsory. Defining the appropriate daily time period to perform colony identification, transplantation, and population estimation will enable avoiding the nuisance of ant bites. Sustained and healthy population dynamics, corresponding to an abundance of major workers, offers more guarantee for effective pest control. It is also necessary to carry out further evaluation to close the knowledge gaps on mating behavior, colony social structure composition, and its functional activity, which are still poorly documented. In view of previous studies, conducting more field practical trials on each targeted potential pest and host plant will be valuable. The phenological differences among diverse plants need to be considered in the study's experimental design to extract more conclusive results. It is also valuable to establish buffer zone small corridors that include favorite *O. smaragdina*-occupied hosts, hence promoting the long-term conservation and population dynamics of colonies. In the last two decades, a great deal of valuable applied research towards the adoption of weaver ants has reinforced the effective biocontrol agent status of *Oecophylla* ants, including IPM applications in large or small commercial orchards. Although some setbacks have occurred due to the nuisance of ant bites in cacao plantations, the interest shown by farmers is gaining momentum. The almost cost-free application would eventually outweigh the tenacious character of ants, especially since the predator is already included by government official agencies in countries such as Australia, Africa, China, and Vietnam.

Author Contributions: M.P.E. proposed and conceptualized the review, analyzed the data, prepared figures and/or tables, authored or reviewed drafts of the paper. All authors contributed to consulting references, writing the paper, and approving the final draft. All authors have read and agreed to the published version of the manuscript.

Funding: This research was funded by the "CollectivitéTerritoriale de la Martinique-CTM" state authority in the French Island of Martinique for providing financial assistance to EMP for his doctoral study under a full scholarship European Union doctoral research grant.

Institutional Review Board Statement: Not applicable.

Informed Consent Statement: Not applicable.

Data Availability Statement: Not applicable.

Conflicts of Interest: The authors declare that they have no competing interests.

References

1. Andersen, A.N. Responses of ant communities to disturbance: Five principles for understanding the disturbance dynamics of a globally dominant faunal group. *J. Anim. Ecol.* **2019**, *88*, 350–362. [CrossRef] [PubMed]
2. Rahim, A.; Ohkawara, K. The Ant Mosaic Distribution of *Oecophylla smaragdina* and Dominant Ant Species: Effects on Ants Communities in Agroforestry in Tarakan Island. *IOP Conf. Ser. Earth Environ. Sci.* **2018**, *197*, 012028. [CrossRef]
3. D'Ettorre, P.; Lenoir, A. Nestmate recognition. In *Ant Ecology*; Oxford Biology: Oxford, UK, 2010; Chapter 1; pp. 194–209.
4. Exélis, M.P.; Idris, A.H. Studies on the predatory activities of *Oecophylla smaragdina* (Hymenoptera: Formicidae) on *Pteroma pendula* (Lepidoptera: Psychidae) in oil palm plantations in Teluk Intan, Perak (Malaysia). *Asian Myrmecol.* **2013**, *5*, 163–176.
5. Exélis, M.P. An Ecological Study of *Pteroma pendula* (Lepidoptera: Psychidae): In Oil Palm Plantation with Emphasis on the Predatory Activities of *Oecophylla smaragdina* (Hymenoptera: Formicidae) on the Bagworm. Master's Thesis, Universiti Malaya, Kuala Lumpur, Malaysia, 2014.
6. Hölldobler, B.; Wilson, E.O. *The Ants*; Harvard University Press: Cambridge, MA, USA, 1990.
7. Offenberg, J. Ants as tools in sustainable agriculture. *J. Appl. Ecol.* **2015**, *52*, 1197–1205. [CrossRef]
8. Crozier, R.H.; Schlüns, E.A.; Robson, S.K.A.; Newey, P.S. A masterpiece of evolution *Oecophylla* weaver ants (Hymenoptera: Formicidae). *Myrmecol. News* **2010**, *13*, 57–71.
9. Verghese, A.; Kamala Jayanthi, P.D.; Sreedevi, K.; Sudha Devi, K.; Viyolla Pinto, V. A quick and non-destructive population estimate for the weaver ant *Oecophylla smaragdina* Fab. (Hymenoptera: Formicidae). *Curr. Sci.* **2013**, *104*, 641–646.
10. Pinkalski, C.; Damgaard, C.; Jensen, K.-M.; Gislum, R.; Peng, R.; Offenberg, J. Non-destructive biomass estimation of *Oecophylla smaragdina* colonies: A model species for the ecological impact of ants. *Insect Conserv. Divers.* **2015**, *8*, 464–473.
11. Symondson, W.O.C.; Sunderland, K.D.; Greenstone, M.H. Can generalist predators be effective biocontrol agents? *Annu. Rev. Entomol.* **2002**, *47*, 561–594.

12. Thurman, J.H.; Northfield, T.D.; Snyder, W.E. Weaver ants provide ecosystem services to tropical tree crops. *Front. Ecol. Evol.* **2019**, *7*, 120. [CrossRef]
13. Basri, M.W. Life History, Ecology and Economic Impact of the Bagworm, *Metisa plana* Walker (Lepidoptera: Psychidae) on the Oil Palm, *Elaeis guineensis* Jacquin (Palmae) in Malaysia. Ph.D. Thesis, University of Guelph, Guelph, ON, Canada, 1993.
14. Basri, M.W.; Norman, K.; Hamdan, A.B. Natural enemies of the bagworm, *Metisa plana* Walker (Lepidoptera: Psychidae) and their impact on host population regulation. *Crop Prot.* **1995**, *14*, 637–645. [CrossRef]
15. Norman, K.; Basri, M.W. Interactions of the bagworm, *Pteroma pendula* (Lepidoptera: Psychidae), and its natural enemies in an oil palm plantation in Perak. *J. Oil Palm Res.* **2010**, *22*, 758–764.
16. Norman, K.; Basri, M.W. Status of common oil palm insect pest in relation to technology adoption. *Planter* **2007**, *83*, 371–385.
17. Ho, T.C.; Ibrahim, Y.; Chong, K.K. Infestations by the bagworms *Metisa plana* and *Pteroma pendula* for the period 1986–2000 in major oil palm estates managed by Golden Hope Plantation Berhad in Peninsular Malaysia. *J. Oil Palm Res.* **2011**, *23*, 1040–1050.
18. Wood, B.J.; Norman, K. A review of developments in integrated pest management (IPM) of bagworm (Lepidoptera: Psychidae) infestation in oil palms in Malaysia. *J. Oil Palm Res.* **2019**, *31*, 529–539. [CrossRef]
19. Wood, B.J.; Norman, K. Bagworm (Lepidoptera: Psychidae) Infestation in the centennial of the Malaysian oil palm industry—A review of causes and control. *J. Oil Palm Res.* **2019**, *31*, 364–380. [CrossRef]
20. Way, M.J.; Khoo, K.C. Colony dispersion and nesting habits of the ants, *Dolichoderus thoracicus* and *Oecophylla smaragdina* (Hymenoptera: Formicidae), in relation to their success as biological control agents on cocoa. *Bull. Entomol. Res.* **1991**, *81*, 341–350. [CrossRef]
21. Way, M.J.; Khoo, K.C. Role of ants in pest management. *Annu. Rev. Entomol.* **1992**, *37*, 479–503. [CrossRef]
22. Peng, R.K.; Christian, K. The control efficacy of the weaver ant, *Oecophylla smaragdina* (Hymenoptera: Formicidae), on the mango leafhopper, *Idioscopus nitidulus* (Hemiptera: Cicadellidea) in mango orchards in the Northern Territory. *Int. J. Pest Manag.* **2005**, *51*, 297–304. [CrossRef]
23. Offenberg, J.; Cuc, N.T.T.; Wiwatwitaya, D. The effectiveness of weaver ant (*Oecophylla smaragdina*) biocontrol in Southeast Asian citrus and mango. *Asian Myrmecol.* **2013**, *5*, 139–149.
24. Peng, R.; La, P.L.; Christian, K. Weaver Ant Role in Cashew Orchards in Vietnam. *J. Econ. Entomol.* **2014**, *107*, 1330–1338. [CrossRef]
25. Peng, R.; Christian, K.; Gibb, K. The best time of day to monitor and manipulate weaver ant colonies in biological control. *J. Appl. Entomol.* **2012**, *136*, 155–160. [CrossRef]
26. Peng, R.; Christian, K. The effect of the weaver ant, *Oecophylla smaragdina* (Hymenoptera: Formicidae), on the mango seed weevil, *Sternochetus mangiferae* (Coleoptera: Curculionidae), in mango orchards in the Northern Territory of Australia. *Int. J. Pest Manag.* **2007**, *53*, 15–24. [CrossRef]
27. Van Mele, P.; Cuc, N.T.T. *Ants as Friends: Improving your Tree Crops with Weaver Ants*, 2nd ed.; Africa Rice Center (WARDA): Cotonou, Benin, 2007; 72p.
28. Van Itterbeeck, J.; Sivongxay, N.; Praxaysombath, B.; Van Huis, A. Indigenous Knowledge of the Edible Weaver Ant *Oecophylla smaragdina* Fabricius Hymenoptera: Formicidae from the Vientiane Plain, Lao PDR. *Ethnol. Lett.* **2014**, *5*, 4–12. [CrossRef]
29. Dejean, A. Orcadian rhythm of *Oecophylla longinoda* in relation to territoriality and predatory behaviour. *Physiol. Entomol.* **1990**, *15*, 393–403. [CrossRef]
30. Dejean, A.; Corbara, B.; Orivel, J.; Leponce, M. Rainforest canopy ants: The implications of territoriality and predatory behavior. *Funct. Ecosyst. Communities* **2007**, *1*, 105–120.
31. Offenberg, J. The use of artificial nests by weaver ants: A preliminary field observation. *Asian Myrmecol.* **2014**, *6*, 119–128.
32. Rastogi, N. Prey concealment and spatiotemporal patrolling behaviour of the Indian tree ant *Oecophylla smaragdina* (Fabricius). *Insectes Sociaux* **2000**, *47*, 92–93. [CrossRef]
33. Mishra, M.; Bhadani, S. Daily activity and visual discrimination reflects the eye organization of weaver ant *Oecophylla smaragdina* (Insecta: Hymenoptera: Formicidae). *bioRxiv* **2017**. [CrossRef]
34. Bradshaw, J.W.S.; Howse, P.E. Sociochemicals in ants. In *Chemical Ecology of Insects*; Bell, W.J., Cardé, R.T., Eds.; Chapman and Hall: London, UK, 1984; pp. 429–473.
35. Roux, O.; Billen, J.; Orivel, J.; Dejean, A. An Overlooked Mandibular-Rubbing Behavior Used during Recruitment by the African Weaver Ant, *Oecophylla longinoda*. *PLoS ONE* **2010**, *5*, e8957. [CrossRef]
36. Chapuisat, M.; Keller, L. Division of labour influences the rate of ageing in weaver ant workers. *Proc. R. Soc. London B* **2002**, *1494*, 909–914. [CrossRef]
37. Peng, R.; Christian, K.; Reilly, D. Biological control of the fruit-spotting bug *Amblypelta lutescens* using weaver ants *Oecophylla smaragdina* on African mahoganies in Australia. *Agric. For. Entomol.* **2012**, *14*, 428–433. [CrossRef]
38. Hölldobler, B.; Wilson, E.O. Journey to the Ants: A Story of Scientific Exploration. *Wilson Q.* **1976**, *19*, 119.
39. Hölldobler, B. The Chemistry of Social Regulation: Multicomponent Signals in Ant Societies. *Proc. Natl. Acad. Sci. USA* **1995**, *92*, 19.
40. Holldobler, B.; Wilson, E. *Journey to the Ants: A Story of Scientific Exploration*; The Belknap Press of Harvard University: London, UK, 1994.

41. Vayssieres, J.F.; Sinzogan, A.; Korie, S.; Adandonon, A. Field observational studies on circadian activity pattern of *Oecophylla longinoda* (Latreille) (Hymenoptera: Formicidae) in relation to abiotic factors and mango cultivars. *Int. J. Biol. Chem. Sci.* **2011**, *5*, 790–802.

42. Peng, R.; Christian, K.; Gibb, K. *Implementing Ant Technology in Commercial Cashew Plantations and Continuation of Transplanted Green Ant Colony Monitoring*; Report for the Rural Industries Research and Development Corporation, Publication No. W04/088 Project No. UNT-5A; Australian Government, Rural Industries Research and Development Corporation: Kingston, ACT, Australia, 2004; p. 72. Available online: https://agrifutures.com.au/wp-content/uploads/publications/W04-088.pdf (accessed on 9 September 2022).

43. Advento, A.D.; Yusah, K.M.; Naim, M.; Caliman, J.P.; Fayle, T.M. Which Protein Source is Best for Mass-Rearing of Asian Weaver Ants? *J. Trop. Biol. Conserv.* **2022**, *19*, 93–107. [CrossRef]

44. Waite, S. *Population Ecology: A Unified Study of Animals and Plants*, 3rd ed.; Begon, M., Mortimer, M., Thompson, D.J., Eds.; John Wiley & Sons: Hoboken, NJ, USA, 2009.

45. Turchin, P. Population Dynamics from First Principles. In *Complex Population Dynamics: A Theoretical/Empirical Synthesis*; Monographs in Population Biology Series; Princeton University Press: Princeton, NJ, USA, 2003; Volume 35, pp. 17–46.

46. Sibly, R.M.; Hone, J. Population growth rate and its determinants: An overview. *Philos. Trans. R. Soc. Lond. Ser. B: Biol. Sci.* **2002**, *357*, 1153–1170. [CrossRef]

47. Lande, R.; Engen, S.; Saether, B.E. *Stochastic Population Dynamics in Ecology and Conservation*; Oxford Series in Ecology and Evolution; Oxford University Press Inc.: New York, NY, USA, 2003.

48. Seidl, R.; Fernandes, P.M.; Fonseca, T.F.; Gillet, F.; Jönsson, A.M.; Merganičová, K.; Mohren, F. Modelling natural disturbances in forest ecosystems: A review. *Ecol. Model.* **2011**, *222*, 903–924. [CrossRef]

49. Tack, A.J.; Mononen, T.; Hanski, I. Increasing frequency of low summer precipitation synchronizes dynamics and compromises metapopulation stability in the Glanville fritillary butterfly. *Proc. R. Soc. B Biol. Sci.* **2015**, *282*, 20150173. [CrossRef]

50. Peng, R.; Christian, K.; Reilly, D. Utilisation of multiple queens and pupae transplantation to boost early colony growth of weaver ants *Oecophylla smaragdina*. *Asian Myrmecol.* **2013**, *5*, 177–184.

51. Offenberg, J.; Peng, R.; Nielsen, M.G.; Birkmose, D. The effect of queen and worker adoption on weaver ant (*Oecophylla smaragdina* F.) queen fecundity. *J. Insect Behav.* **2012**, *25*, 478–485.

52. Hölldobler, B.; Wilson, E.O. *The Leafcutter Ants: Civilization by Instinct*; WW Norton & Company Inc.: New York, NY, USA, 2010; Available online: http://www.dictionnaireamoureuxdesfourmis.fr/L/Livres/Livres%20scientifiques/Incroyable-instinctfour.mis/The%20leafcutter%20ants-Lenoir.pdf (accessed on 17 March 2019).

53. Montgomery, G.A.; Belitz, M.W.; Guralnick, R.P.; Tingley, M.W. Standards and Best Practices for Monitoring and Benchmarking Insects. *Front. Ecol. Evol.* **2021**, *513*, 579193. Available online: https://www.frontiersin.org/article/10.3389/fevo.2020.579193 (accessed on 21 June 2021). [CrossRef]

54. Tschinkel, W.R. *Ant Architecture: The Wonder Beauty and Science of Underground Nests*; Princeton University Press: Princeton, NJ, USA, 2021.

55. Britton, N.F.; Partridge, L.W.; Franks, N.R. A mathematical model for the population dynamics of army ants. *Bull. Math. Biol.* **1996**, *58*, 471. [CrossRef]

56. Baker, C.M.; Hodgson, J.C.; Tartaglia, E.; Clarke, R.H. Modelling tropical fire ant (*Solenopsis geminata*) dynamics and detection to inform an eradication project. *Biol. Inv.* **2017**, *19*, 2959–2970. [CrossRef]

57. Yamanaka, O.; Shiraishi, M.; Awazu, A.; Nishimori, H. Verification of mathematical models of response threshold through statistical characterisation of the foraging activity in ant societies. *Sci. Rep.* **2019**, *9*, 8845. [CrossRef] [PubMed]

58. Lim, G.T. Enhancing the Weaver Ant, *Oecophylla smaragdina* (Hymenoptera: Formicidae), for Biological Control of a Shoot Borer, *Hypsipyla robusta* (Lepidoptera: Pyralidae), in Malaysian Mahogany Plantations. Ph.D. Thesis, Virginia Tech University, Blacksburg, VA, USA, 2007. Available online: https://vtechworks.lib.vt.edu/handle/10919/26850 (accessed on 18 January 2019).

59. Way, M. Studies of the life history and ecology of the ant *Oecophylla longinoda* Latreille. *Bull. Entomol. Res.* **1954**, *45*, 93–112. [CrossRef]

60. Vanderplank, F.L. The Bionomics and Ecology of the Red Tree Ant, *Oecophylla* sp.; and its Relationship to the Coconut Bug *Pseudotheraptus wayi* Brown (Coreidae). *J. Anim. Ecol.* **1960**, *29*, 15–33. [CrossRef]

61. Lokkers, C. The Distribution of the Weaver Ant, *Oecophylla smaragdina* (Fabricius) (Hymenoptera, Formicidae) in Northern Australia. *Aust. J. Zool.* **1986**, *34*, 683–687. [CrossRef]

62. Lokkers, C. Colony Dynamics of the Green Tree Ant (*Oecophylla smaragdina* Fab.) in a Seasonal Tropical Climate. Ph.D. Thesis, James Cook University, Douglas, QLD, Australia, 1990; 301p.

63. Hölldobler, B.; Wilson, E.O. The multiple recruitment systems of the African weaver ant *Oecophylla longinoda* (Latreille), (Hymenoptera: Formicidae). *Behav. Ecol. Sociobiol.* **1978**, *3*, 19–60. [CrossRef]

64. Offenberg, J.; Wiwatwitaya, D. Sustainable weaver ant (*Oecophylla smaragdina*) farming: Harvest yields and effects on workers ant density. *Asian Myrmecol.* **2010**, *3*, 55–62.

65. Dlussky, G.M.; Wappler, T.; Wedmann, S. New Middle Eocene Formicid Species from Germany and the Evolution of Weaver Ants. *Acta Pal. Pol.* **2008**, *4*, 615. [CrossRef]

66. Wetterer, J.K. Geographic distribution of the weaver ant *Oecophylla smaragdina*. *Asian Myrmecol.* **2017**, *9*, e009004.

67. Turchin, P. Population dynamics: New Approaches and Synthesis. In *Population Regulation: Old Argument and a New Synthesis*; Academic Press Inc.: Cambridge, MA, USA, 1995; pp. 19–40.

68. Turchin, P. Population Regulation: A Synthetic View. *Oikos* **1999**, *84*, 153. [CrossRef]
69. Nicholson, A.J. The balance of animal populations. *J. Anim. Ecol.* **1993**, *2*, 132–178.
70. Elton, C. Population interspersion: An essay on animal community patterns. *J. Ecol.* **1949**, *37*, 1–23. [CrossRef]
71. Hixon, M.A.; Pacala, S.W.; Sandin, S.A. Population regulation: Historical context and contemporary challenges of open vs. closed systems. *Ecology* **2002**, *83*, 1490–1508. [CrossRef]
72. Martin, S.J. The role of Varroa and viral pathogens in the collapse of honeybee colonies: A modelling approach. *J. Appl. Ecol.* **2001**, *38*, 1082–1093. [CrossRef]
73. El Keroumi, A.; Naamani, K.; Soummane, H.; Dahbi, A. Seasonal dynamics of ant community structure in the Moroccan Argan Forest. *J. Insect Sci.* **2012**, *12*, 94. [CrossRef]
74. Cardoso, D.C.; Schoereder, J.H. Biotic and abiotic factors shaping ant (Hymenoptera: Formicidae) assemblage in Brazilian coastal sand dunes: The case of Restinga in Santa Catarina. *Fla. Entomol.* **2014**, *97*, 1443–1450. [CrossRef]
75. Rastogi, N. Seasonal pattern in the territorial dynamics of the arboreal ant (Hymenoptera: Formicidae). *J. Bombay Nat. Hist. Soc.* **2007**, *104*, 14–17.
76. Peng, R.K.; Christian, K. Effective control of Jarvis's fruit fly, *Bactrocera jarvisi* (Diptera: Tephritidae), by the weaver ant, *Oecophylla smaragdina* (Hymenoptera: Formicidae), in mango orchards in the Northern Territory of Australia. *Int. J. Pest Manag.* **2006**, *52*, 275–282. [CrossRef]
77. Chowdhury, R.; Rastogi, N. Comparative analysis of mandible morphology in four ant species with different foraging and nesting habits. *bioRxiv.* 2021. [CrossRef]
78. Huang, H.T.; Yang, P. The Ancient Cultured Citrus Ant. *Bio Sci.* **1987**, *37*, 665–671. [CrossRef]
79. Greenslade, P.J.M. Interspecific competition and frequency changes among ants in Solomon Islands coconut plantations. *J. Appl. Ecol.* **1971**, *8*, 323–349. [CrossRef]
80. Van Mele, P. A historical review of research on the weaver ant *Oecophylla* in biological control. *Agric. For. Entomol.* **2008**, *10*, 13–22. [CrossRef]
81. Offenberg, J.; Havanon, S.; Aksornkoae, S.; MacIntosh, D.J.; Nielsen, M.G. Observations on the Ecology of Weaver Ants (*Oecophylla smaragdina* Fabricius) in a Thai Mangrove Ecosystem and Their Effect on Herbivory of *Rhizophora mucronata* Lam. *Biotropica* **2004**, *36*, 344.
82. Krag, K.; Lundegaard, R.; Offenberg, J.; Nielsen, M.G.; Wiwatwittaya, D. Intercolony Transplantation of *Oecophylla smaragdina* (Hymenoptera: Formicidae) larvae. *J. Asia-Pac. Entomol.* **2010**, *13*, 97–100. [CrossRef]
83. Vidkjær, N.H.; Wollenweber, B.; Gislum, R.; Jensen, V.K.M.; Fomsgaard, I.S. Are ant feces nutrients for plants? A metabolomics approach to elucidate the nutritional effects on plants hosting weaver ants. *Metabolomics* **2015**, *11*, 1013–1028. [CrossRef]
84. Adandonon, A.; Vayssières, J.-F.; Sinzogan, A.; Van Mele, P. Density of pheromone sources of the weaver ant *Oecophylla longinoda* affects oviposition behaviour and damage by mango fruit flies (Diptera: Tephritidae). *Int. J. Pest Manag.* **2009**, *55*, 285–292. [CrossRef]
85. Appiah, E.F.; Ekesi, S.; Afreh-Nuamah, K.; Obeng-Ofori, D.; Mohamed, S.A. African weaver ant-produced semiochemicals impact on foraging behaviour and parasitism by the Opiine parasitoid, *Fopius arisanus* on *Bactrocera invadens* (Diptera: Tephritidae). *Biol. Cont.* **2014**, *79*, 49–57. [CrossRef]
86. Sinzogan, A.A.C.; Van Mele, P.; Vayssieres, J.F. Implications of on-farm research for local knowledge related to fruit flies and the weaver ant *Oecophylla longinoda* in mango production. *Int. J. Pest Manag.* **2008**, *54*, 241–246. [CrossRef]
87. Van Mele, P.; Nguyen Thi Thu, C.; Seguni, Z.; Camara, K.; Offenberg, J. Multiple sources of local knowledge: A global review of ways to reduce nuisance from the beneficial weaver ant *Oecophylla*. *Int. J. Agric. Res. Gov. Ecol.* **2009**, *8*, 484–504.
88. Kalshoven, L.G.E. *Pest of Crops in Indonesia*; van der Lann, P.A., Translator; Ichtiar Baru-Van Houve PT: Jakarta, Indonesia, 1981; 70p.
89. Brown, E.S. Immature nutfall of coconuts in the Solomon Islands. I. Distribution of nutfall in relation to that of *Amblypelta* and of certain species of ants. *Bull. Entomol. Res.* **1959**, *50*, 97–113. [CrossRef]
90. Phillips, J.S. Immature nutfall of coconuts in the Solomon Islands. *Bull. Entomol. Res.* **1940**, *31*, 295–316. [CrossRef]
91. Baloch, G.M. Natural enemies of *Axiagastus cambelli* Distant (Hemiptera: Pentatomidae) on the Gazelle Peninsula, New Britain. *Papua New Guin. Agric. J.* **1973**, *24*, 41–45.
92. O'Sullivan, D.F. Observations on the coconut spathe bug *Axiagastus cambelli* Distant (Hemiptera: Pentatomidae) and its parasites and predators in Papua New Guinea. *Papua New Guin. Agric. J.* **1973**, *24*, 79–86.
93. Stapley, J.H. Insect pests of coconuts in the Pacific Region. *Outlook Agric.* **1973**, *7*, 211–217. [CrossRef]
94. Stapley, J.H. Coconut leaf beetle (*Brontispa*) in the Solomons. *Alafua Agric. Bull. West. Samoa* **1980**, *5*, 17–22.
95. Stapley, J.H. Using the predatory ant, *Oecophylla smaragdina*, to control insect pests of coconuts and cocoa. *Inf. Circular. South Pac. Comm. CAB Direct* **1980**, *85*, 7.
96. Way, M.J.; Cammell, M.E.; Bolton, B.; Kanagaratnam, P. Ants (Hymenoptera: Formicidae) as egg predators of coconut pests, especially in relation to biological control of the coconut caterpillar, *Opisina arenosella* Walker (Lepidoptera: Xyloryctidae), in Sri Lanka. *Bull. Entomol. Res.* **1989**, *79*, 219–234. [CrossRef]
97. O'connor, B.A. Premature nutfall of coconuts in the British Solomon Islands Protectorate. *Agric. J. Fiji* **1950**, *21*, 21–42.
98. Gressitt, J.L. The coconut leaf-mining beetle *Promecotheca papuana*. *Papua New Guin. Agric. J.* **1959**, *12*, 119–148.

99. Rahmanto, B.; Lestari, F. *Predator Ulat Heortia Vitessoides Pada Tanaman Penghasil Gaharu*; Balai Penelitian Kehutanan: Banjarbaru, Indonesia, 2013; Volume 6, pp. 1–5.
100. Saripah, B.; Azhar, I. Five years of using cocoa black ants, to control cocoa pod borer at farmer plot—An epilogue. *Malays. Cocoa J.* **2012**, *7*, 8–14.
101. Way, M.J.; Khoo, K.C. Relationships between *Helopeltis theobromae* damage and ants with special reference to Malaysian cocoa smallholdings. *J. Plant Prot. Trop.* **1989**, *6*, 1–11.
102. Szent-lvany, J.J.H. Insect pests of *Theobroma cacao* in the territory of Papua New Guinea. *Papua New Guin. Agric. J.* **1961**, *13*, 127–147.
103. Room, P.M.; Smith, E.S.C. Relative abundance and distribution of insect pests, ants and other components of the cocoa ecosystem in Papua New Guinea. *J. Appl. Ecol.* **1975**, *12*, 31–46. [CrossRef]
104. Lim, G.T. Biology, ecology and control of Cocoa Pod Borer, *Conopomorpha cramerella* (Snellen). In *Cocoa Pest and Diseases Management in Southeast Asia and Australia*; Keane, P.J., Putter, C.A.J., Eds.; FAO: Rome, Italy, 1992.
105. Daha, L.; Gassa, A. Weaver Ant, Oecophylla smaragdina as a Potential Biological Control of CPB in Sulawesi Cocoa Plantation. In *Technical Brain-Storming Meeting on Bio-Control Technologies for Integrated Pest Management (IPM) of Cocoa Makassar–Indonesia. Effem Prima Project*; Effem Prima Project; Acdi-Voca: Washington, DC, USA; USAID: Washington, DC, USA; Hasanuddin University: Makassar, Indonesia, 2003.
106. Gassa, A.; Abdullah, T.; Junaid, M. Formulation of Artificial Diet to Increase Population Distribution and Aggressive Behavior of Weaver Ant (*Oecophylla Smaragdina* F.) For Controling Cocoa Pod Borer (*Conopomorpha Cramerella* Sn.). *Acad. Res. Int.* **2014**, *5*, 1.
107. Gassa, A.; Abdullah, T.; Junaid, M. The use of several types of artificial diet to increase population and aggressive behavior of weaver ants (*Oecophylla Smaragdina* F.) in reducing cocoa pod borer infestation (*Conopomorpha Cramerella* Sn.). *Acad. Res. Int.* **2015**, *6*, 63.
108. Fatahuddin, F.; Gassa, A.; Junaidi, J. Population Development of Several Species of Ants on the Cocoa Trees in South Sulawesi. *Pelita Perkebunan Coffee Cocoa Res. J.* **2010**, *26*, 101–110. [CrossRef]
109. Pag-Ong, A.I. Taxonomy and Diversity of Ants Associated with Cacao and Evaluation of *Oecophylla smaragdina Fabricius* (Hymenoptera: Formicidae) for Biological Control of *Helopeltis bakeri Poppius* (Hemiptera: Miridae) in Quezon and Laguna, Philippines. Ph.D. Thesis, De La Salle University, Manila, Philippines, 2021.
110. Nur Adila, K.; Zolkepli, N.; Kamarudin, N.S.W.; Basari, N. A predatory activity of *Oecophylla smaragdina* (Hymenoptera: Formicidae) on Citrus pests. *Serangga* **2022**, *27*, 83–93.
111. Niu, J.Z.; Hull-Sanders, H.; Zhang, Y.X.; Lin, J.Z.; Dou, W.; Wang, J.J. Biological control of arthropod pests in citrus orchards in China. *Bio Control* **2014**, *68*, 15–22. [CrossRef]
112. Needham, J. Science and Civilisation in China. In *Biology and Biological Technology Part 1: Botany*; Cambridge University Press: Cambridge, MA, USA, 1986; Volume VI, p. 109.
113. Yang, P. Biology of the yellow citrus ant, *Oecophylla smaragdina* and its utilisation against citrus insect pests. *Acta Sci. Nat. Univ. Sunyatseni* **1982**, *3*, 102–105.
114. Garcia, C.E. A field study of the citrus green bug, *Rhynchocoris serratus* Donovan. *Philipp. J. Agric.* **1935**, *6*, 311–325.
115. Van Mele, P.; Cuc, N.T.T. Evolution and status of *Oecophylla smaragdina* (Fabricius) as a pest control agent in citrus in the Mekong Delta. Vietnam. *Int. J. Pest Manag.* **2000**, *46*, 295–301. [CrossRef]
116. Yang, P. Historical perspective of the red tree ant, *Oecophylla smaragdina* and its utilization against citrus insect pests. *Chin. J. Biol. Con.* **2002**, *18*, 28.
117. Beattie, G.A.C.; Holford, P. Can *Oecophylla smaragdina* be used to suppress incidence of CVPD in citrus orchards in Indonesia? *IOP Conf. Ser. Earth Environ. Sci.* **2022**, *1018*, 012030. [CrossRef]
118. Ambika, S.; Nalini, T. Nest composition, inward and outward flows of *Oecophylla smaragdina* Fabricius (Hymenoptera: Formicidae) in selected fruit crops. *Plant Arch.* **2019**, *19*, 2781–2784.
119. MacFarlane, R.; Jackson, G.V.H.; Marten, K.D. Die-back of Eucalyptus in the Solomon Islands. *Commonw. For. Rev.* **1976**, *55*, 133–139.
120. Peng, R.K.; Christian, K. The dimpling bug, *Campylomma austrina* Malipatil (Hemiptera: Miridae): The damage and its relationship with ants in mango orchards in the Northern Territory of Australia. *Int. J. Pest Manag.* **2008**, *54*, 173–179. [CrossRef]
121. Voûte, A.D. *Cryptorrhynchus gravis* and the causes of its multiplication in Java (*Cryptorrhynehus gravis* F. and the causes of its mass increased in Java). *Ar. Néer. Zool.* **1935**, *2*, 112–142. [CrossRef]
122. Lim, G.T.; Kirton, L.G. A preliminary study on the prospects for biological control of the mahogany shoot borer, *Hypsipyla robusta* (Lepidoptera: Pyralidae), by ants (Hymenoptera: Formicidae). In *Tropical Forestry Research in the New Millennium: Meeting Demands and Challenges, Proceedings of The International Conference on Forestry and Forest Products Research (CFFPR 2001), Kuala Lumpar, Malaysia, 1–3 October 2001*; Forest Research Institute Malaysia (FRIM): Kuala Lumpur, Malaysia, 2001; pp. 240–244.
123. Lim, G.T.; Kirton, L.G.; Salom, S.M.; Kok, L.T.; Fell, R.D.; Pfeiffer, D.G. Mahogany Shoot Borer Control in Malaysia and Prospects for Biocontrol Using Weaver Ants. *J. Trop. For. Sci.* **2008**, *20*, 147–155.
124. Mokodompit, H.S.; Pollo, H.N.; Lasut, M.T. Identifikasi Jenis Serangga Hama Dan Tingkat Kerusakan Pada *Diospyros celebica* Bakh. *Eugenia* **2019**, *24*, 64–75. [CrossRef]
125. Musyafa, H.; Bahri, S.; Supriyo, H. Potential of Weaver Ant (*Oecophylla smaragdina* Fabricius, 1775) as Biocontrol Agent for Pest of Teak Stand in Wanagama Forest, Gunungkidul, Yogyakarta, Indonesia in The UGM Annual Scientific Conference Life Sciences 2016. *KnE Life Sci.* **2019**, *4*, 239–244.

126. CABI Digital Library-CABI Compendium. *Economic Review of Psyllid Damage on Leucaena in Southeast Asia and Australia*; Australian International Development Assistance Bureau: Canberra, Australia, 2019; Available online: https://www.cabidigitallibrary.org/doi/10.1079/cabicompendium.27919 (accessed on 11 February 2020).

127. Oka, I.N. Progress and future activities of the *leucaena psyllid* research programmes in Indonesia. In *Leucaena psyllid: Problems and Management*; Napompeth, B., MacDicken, K.G., Eds.; Winrock International/IDRC/NFTA/F-FRED: Bangkok, Thailand, 1990; pp. 25–27.

128. Mangoendihardjo, S.; Mahrub, E.; Warrow, J. *Endemic Natural Enemies of the Leucaena Psyllid in Indonesia: In LEUCAENA PSYLLID: Problems and Management, Proceedings of the International Workshop, Bogor, Indonesia, 16–21 January 1989*; Napompeth, B., MacDicken, K.G., Eds.; Winrock International Institute for Agricultural Development: Nairobi, Kenya, 1990; pp. 159–161.

129. Peng, R.K.; Christian, K.; Gibb, K. The effect of colony isolation of the predacious ant, *Oecophylla smaragdina* (F.) (Hymenoptera: Formicidae), on protection of cashew plantations from insect pests. *Int. J. Pest Manag.* **1999**, *45*, 189–194. [CrossRef]

130. Peng, R.K.; Christian, K.; Gibb, K. The effect of levels of green ant, *Oecophylla smaragdina* (F.), colonisation on cashew yield in northern Australia. Biological control in the tropics: Towards efficient biodiversity and bioresource management for effective biological control. In Proceedings of the Symposium on Biological Control in the Tropics held at MARDI Training Centre, Serdang, Malaysia, 18–19 March 1999; pp. 24–28.

131. Mahapatro, G.K.; Jose, M. Role of red-ant, *Oecophylla smaragdina* Fabricius (Formicidae: Hymenoptera) in managing tea mosquito bug, *Helopeltis* species (Miridae: Hemiptera) in cashew. *Proc. Natl. Acad. Sci. India Sect. B Biol. Sci.* **2016**, *86*, 497–504. [CrossRef]

132. Wijetunge, P.M.A.P.K.; Ranaweera, B. Rearing of red weaver ant (*Oecophylla smaragdina* F.) for the management of *Helopeltis antonii* Sign. (Hemiptera: Miridae) in cashew (*Anacardium occidentale* L.). *Acta Hortic.* **2015**, *1080*, 401–408. [CrossRef]

133. Karmawati, E. Pengendalian hama *Helopeltis* spp. pada jambu mete berdasarkan ekologi: Strategi dan implementasi. *Pengemb. Inov. Pertan.* **2010**, *3*, 102–119.

134. Karmawati, E.; Wikardi, E.A. Peranan Semut (*Oecophylla Smaragdina* Dan *Dolichoderus* SP.) Dalam Pengendalian *Helopeltis* Spp.; Dan *Sanurus indecora* Pada Jambu Mete. *Ind. Crops Res. J.* **2004**, *10*, 1–7. [CrossRef]

135. Wylie, F.R. The distribution and life-history of *Milionia isodoxa* Prout (Lepidoptera, Geometridae), a pest of planted hoop pine in Papua New Guinea. *Bull. Entomol. Res.* **1974**, *63*, 649–659. [CrossRef]

136. Lim, I.H.P.; Jain, A. Biodiversity Record: Predation of a common tiger caterpillar by Asian weaver ants. *Nat. Singap.* **2022**, *15*, e2022040. [CrossRef]

137. Falahudin, I. Peranan Semut Rangrang (*Oecophylla smaradigna*) dalam Pengendalian Biologis pada Perkebunan Kelapa Sawit. In Proceedings of the Annual International Conference on Islamic Studies (AICIS) XII, Surabaya, Indonesia, 5–8 November 2012; Universitas Islam Negeri Sunan Ampel Surabaya: Jawa Timur, Indonesia, 2013. AICIS XII.2604-2618.

138. Falahudin, I. Diversitas Semut Arboreal (Hymenoptera: Formicidae) Dan Potensinya Sebagai Pengendali Ulat Api (Lepidoptera: Limacodidae) Pada Tanaman Kelapa Sawit. Ph.D. Thesis, Universitas Andalas, Padang City, Indonesia, 2016.

139. Césard, N. Le kroto (*Oecophylla smaragdina*) dans la région de Malingping, Java Ouest, Indonésie: Collecte et commercialisation d'une ressource animale non négligeable. *Anthropozoologica* **2004**, *39*, 15–31.

140. Payne, C.L.; Van Itterbeeck, J. Ecosystem services from edible insects in agricultural systems: A review. *Insects* **2017**, *8*, 24. [CrossRef] [PubMed]

141. Alagappan, S.; Chaliha, M.; Sultanbawa, Y.; Fuller, S.; Hoffman, L.C.; Netzel, G.; Mantilla, S.M.O. Nutritional analysis, volatile composition, antimicrobial and antioxidant properties of Australian green ants (*Oecophylla smaragdina*). *Futur. Foods.* **2021**, *3*, 100007. [CrossRef]

142. Van Itterbeeck, J.; Pelozuelo, L. How Many Edible Insect Species Are There? A Not So Simple Question. *Diversity* **2022**, *14*, 143. [CrossRef]

143. Jena, S.; Das, S.S.; Sahu, H.K. Traditional value of Red weaver ant (*Oecophylla smaragdina*) as food and medicine in Mayurbhanj district of Odisha, India. *Int. J. Res. Appl. Sci. Eng. Technol.* **2020**, *8*, 936–946. [CrossRef]

144. Diame, L.; Rey, J.Y.; Vayssieres, J.F.; Grechi, I.; Chailleux, A.; Diarra, K. Ants: Major Functional Elements in Fruit Agro-Ecosystems and Biological Control Agents. *Sustainability* **2018**, *10*, 23. [CrossRef]

145. Chailleux, A.; Stirnemann, A.; Leyes, J.; Deletre, E. Manipulating natural enemy behavior to improve biological control: Attractants and repellents of a weaver ant. *Entomol. Gen.* **2019**, *38*, 191. [CrossRef]

146. Peng, R.; Christian, K.; Gibb, K. How many queens are there in mature colonies of the green ant, *Oecophylla smaragdina* (Fabricius)? *Aust. J. Entomol.* **1998**, *37*, 249–253. [CrossRef]

147. Nene, W.A.; Rwegasira, G.M.; Mwatawala, M. Optimizing a method for locating queen nests of the weaver ant *Oecophylla longinoda* Latreille (Hymenoptera: Formicidae) in cashew, *Anacardium occidentale* L. plantations in Tanzania. *Crop Prot.* **2017**, *102*, 81–87. [CrossRef]

148. Ambethgar, V. Recognition of Red weaver ants in integrated control of tea mosquito bug in cashew plantations in India. *Acta Hortic.* **2015**, *1080*, 393–400. [CrossRef]

149. Lenoir, A.; Fresneau, D.; Errard, C.; Hefetz, A. Individuality and colonial identity in ants: The emergence of the social representation concept. In *Information Processing in Social Insects*; Birkhauser Verlag: Basel, Switzerland, 1999; pp. 219–237.

150. Lenoir, A.; d'Ettorre, P.; Errard, C.; Hefetz, A. Chemical ecology and social parasitism in ants. *Annu. Rev. Entomol.* **2001**, *46*, 573–599. [CrossRef]

151. Newey, P.S.; Robson, S.K.A.; Crozier, R.H. Weaver ants *Oecophylla smaragdina* encounter nasty neighbors rather than dear enemies. *Ecology* **2010**, *91*, 2366–2372. [CrossRef] [PubMed]
152. Abdulla, N.R.; Rwegasira, G.M.; Mwatawala, M.W.; Jensen, K.-M.V.; Offenberg, J. Effect of supplementary feeding of *Oecophylla longinoda* on their abundance and predatory activities against cashew insect pests. *Biocontrol Sci. Technol.* **2015**, *25*, 1333–1345. [CrossRef]
153. Rwegasira, R.G.; Mwatawala, M.W.; Rwegasira, G.M.; Axelsen, J. Optimizing methods for rearing mated queens and establishing new colony of *Oecophylla longinoda* (Hymenoptera: Formicidae). *Int. J. Trop. Insect Sci.* **2017**, *37*, 217. [CrossRef]
154. Ayenor, G.K.; Van Huis, A.; Obeng-Ofori, D.; Padi, B.; Röling, N.G. Facilitating the use of alternative capsid control methods towards sustainable production of organic cocoa in Ghana. *Int. J. Trop. Insect Sci.* **2007**, *57*, 85–94. [CrossRef]
155. Perfilieva, K.S. Distribution and differentiation of fossil *Oecophylla* (Hymenoptera: Formicidae) species by wing imprints. *Paleon. J.* **2021**, *55*, 76–89. [CrossRef]
156. Exélis, M.P.; Ramli, R.; Azarae, I.H.; Sani, M.S.A. Estimation of worker population size-density by nest counting in the Asian weaver ant, *Oecophylla smaragdina* (Formicidae: Formicinae) and its dynamic in oil palm plantations industry. *Preprints* **2022**. [CrossRef]
157. Hasan, M.U.; Su Santi, F.S.; Raffiudin, R. Perilaku Pemilihan Makanan dan Pengenalan Anggota Koloni pada Semut Rangrang *Oecophylla smaragdina*. *Jur. Sumberdaya Haya* **2021**, *7*, 41–48. [CrossRef]
158. Cholis, M.N.; Atmowidi, T.; Kahono, S. The diversity and abundance of visitor insects on pummelo (*Citrus maxima* (burm) Merr) cv. nambangan. *J. Entol. Zool. Stud.* **2020**, *8*, 344–351.
159. Tsuji, K.; Hasyim, A.; Nakamura, K. Asian weaver ants, *Oecophylla smaragdina*, and their repelling of pollinators. *Ecol. Res.* **2004**, *19*, 669–673. [CrossRef]
160. Galen, C.; Butchart, B. Ants in your plants: Effects of nectar-thieves on pollen fertility and seed-siring capacity in the alpine wildflower, *Polemonium viscosum*. *Oikos* **2003**, *101*, 521–528. [CrossRef]
161. Gonzálvez, F.G.; Santamaría, L.; Corlett, R.T.; Rodríguez-Gironés, M.A. Flowers attract weaver ants that deter less effective pollinators. *J. Ecol.* **2013**, *101*, 78–85. [CrossRef]
162. Exélis, M.P.; Rosli, R.; Idris, A.H.; Latif, S.A.A.; Yaakop, S.; Ibrahim, R.W. Relationships between weather parameters and foraging daily activity of *Oecophylla smaragdina* (Hymenoptera: Formicidae) in oil palm plantations: Case Study. *Preprints* **2022**. [CrossRef]
163. Kempraj, V.; Park, S.J.; Cameron, D.N.; Taylor, P.W. 1-Octanol emitted by *Oecophylla smaragdina* weaver ants repels and deters oviposition in Queensland fruit fly. *Sci. Rep.* **2022**, *12*, 15768. [CrossRef] [PubMed]
164. DeBach, P.; Rosen, D. *Biological Control by Natural Enemies*, 2nd ed.; CUP Archive; Cambridge University Press: London, UK, 1991.
165. Hajek, A.E.; Eilenberg, J. *Natural Enemies: An Introduction to Biological Control*, 2nd ed.; Cambridge University Press: London, UK, 2018.
166. Selvam, K.; Nalini, T. Toxicity of selected insecticides to the weaver ant, *Oecophylla smaragdina* fabricius (Hymenoptera: Formicidae). *Int. J. Entomol. Res.* **2021**, *6*, 1–5.
167. Barzman, M.; Bàrberi, P.; Birch, A.N.E.; Boonekamp, P.; Dachbrodt-Saaydeh, S.; Graf, B.; Sattin, M. Eight principles of integrated pest management. *Agron. Sustain. Dev.* **2015**, *35*, 1199–1215. [CrossRef]

 *sustainability*

Article

Prediction of Banana Production Using Epidemiological Parameters of Black Sigatoka: An Application with Random Forest

Barlin O. Olivares [1,*], **Andrés Vega** [2], **María A. Rueda Calderón** [3], **Edilberto Montenegro-Gracia** [4], **Miguel Araya-Almán** [5] and **Edgloris Marys** [6]

1   Programa de Doctorado en Ingeniería Agraria, Alimentaria, Forestal y del Desarrollo Rural Sostenible, Universidad de Córdoba, Carretera Nacional IV, km 396, 14014 Córdoba, Spain

2   Facultad de Ciencias Agropecuarias, Universidad Nacional de Córdoba, Av. Haya de la Torre s/n, Córdoba X5000HUA, Argentina

3   Laboratorio de Genética y Genómica Aplicada, Escuela de Ciencias del Mar, Pontificia Universidad Católica de Valparaíso, Chile. S/Brasil, Valparaíso 2950, Chile

4   Facultad de Ciencias Agropecuarias, Universidad de Panamá, CRUBO Bocas del Toro, Finca 15, Changuinola 01001, Panama

5   Departamento de Ciencias Agrarias, Universidad Católica del Maule, km 6 Camino Los Niches, Curicó 3466706, Chile

6   Laboratorio de Biotecnología y Virología Vegetal, Centro de Microbiología y Biología Celular, Instituto Venezolano de Investigaciones Científicas (IVIC), Caracas 1204, Venezuela

*   Correspondence: barlinolivares@gmail.com

Abstract: Accurate predictions of crop production are critical to developing effective strategies at the farm level. Knowing banana production is due to the need to maximize the investment–profit ratio, and the availability of this information in advance allows decisions to be made about the management of important diseases. The objective of this study was to predict the number of banana bunches from epidemiological parameters of Black Sigatoka (BS), using random forests (RF) for its ability to predict crop production responses to epidemiological variables. Weekly production data (number of banana bunches) and epidemiological parameters of BS from three adjacent banana sites in Panama during 2015–2018 were used. RF was found to be very capable of predicting the number of banana bunches, with variance explained as 70.0% and root mean square error (RMSE) of 1107.93 ± 22 of the mean banana bunches observed in the test case. The site, week, youngest leaf spotted and youngest leaf with symptoms in plants with 10 weeks of physiological age were found to be the best predictor group. Our results show that RF is an efficient and versatile machine learning method for banana production predictions based on epidemiological parameters of BS due to its high accuracy and precision, ease of use, and usefulness in data analysis.

Keywords: Black Sigatoka; Musa; production; banana disease; random forest; machine learning

Citation: Olivares, B.O.; Vega, A.; Rueda Calderón, M.A.; Montenegro-Gracia, E.; Araya-Almán, M.; Marys, E. Prediction of Banana Production Using Epidemiological Parameters of Black Sigatoka: An Application with Random Forest. *Sustainability* **2022**, *14*, 14123. https://doi.org/10.3390/su142114123

Academic Editor: Mariolina Gullì

Received: 2 October 2022
Accepted: 27 October 2022
Published: 29 October 2022

1. Introduction

The banana (*Musa* spp.) economically and globally represents the most important fruit species in production and trade [1,2]. In the case of Panama, it is one of the largest producers in Latin America, ranking tenth in the FAO ranking [1,3] with a production of 308,835 tons, a cultivated area of 8449 ha, and a yield of 36,553 Kg/ha, which makes it an essential crop both for its economic importance and for its deep cultural roots [4,5].

However, banana production has been seriously harmed by the emergence of Black Sigatoka (BS), or black streak disease, of banana leaves, caused by the pathogenic fungus *Pseudocercospora fijiensis* (asexual phase, [Morelet] Deighton; the sexual phase of *P. fijiensis* is *Mycosphaerella fijiensis* Morelet) [6–8]. The control of this disease in Central and South America represents 27% of production costs, this being approximately USD 10 million annually. It is for this reason that BS is considered the most severe and destructive banana disease in the world [9,10], causing the largest harvest losses in banana plantations and outweighing all other banana fungal diseases and the high costs for their control [11–13].

These financial limitations, coupled with the decrease in production in many local farmers in developing countries, are the basis for generating studies that seek to predict banana production based on BS epidemiological parameters [7,14]. Therefore, predicting the next harvests through rates of BS infections in banana plants would allow farmers to take appropriate preventive measures and mitigate disease management costs. Although there are studies linking BS forecasts to meteorological variables [15–17], powerful models to forecast BS infections have not yet been developed (so far as we know), much less to predict banana production.

Advances in machine learning algorithms in bananas, such as neural networks (NN) [18–20], random forest (RF) [11,20–23], support vector machines (SVM) [21,24,25], and orthogonal partial least-squares-discriminant analysis (OPLS-DA) [11], have addressed time series models. However, these models have difficulty modeling irregularity over time. Recently, a study by Olivares et al. [11] established through supervised methods such as RF and OPLS-DA relationships between incidences of banana wilt and soil properties, demonstrating the great classifying and predictive power that RF has in this type of situation. Today, RF regression applications in crop science remain lacking, with few exceptions. Numerous studies have pointed out various promising advantages of RF as a regression tool compared to traditional regression models [21–23]; therefore the initiative to use the RF algorithm arises in this study focused on its usefulness as a prediction tool in banana production.

For the prediction of yield (number of bunches), the farmer's experience is important when there are no methods for estimation and this becomes the only resource; however, these approximations may be insufficient, and what is needed is systematically stored information that an example includes, first of all, average historical production records; and, secondly, yield variations due to agricultural management, climatic factors, epidemiological parameters or disease management, among others, to reduce any bias or error. This situation causes new methodologies to be developed and other data to be considered, for example: physical, environmental, epidemiological, or climatic, that can improve the quality of predictions.

Although various scientific studies have produced mathematical models of infectious diseases, it is difficult to adapt these models to incorporate the production approach in terms of the number of banana bunches or the number of exportable boxes, for example. This study aimed to validate the hypothesis that BS epidemiological parameters can predict banana bunch numbers in large areas of Bocas del Toro, Panama using an RF prediction model. In this case, the quantitatively estimated parameters of this disease could be used to improve the evaluation of banana production. Analysis used with highly predictive RF can provide a working model for bananas in an area where there is little history of such problems.

2. Materials and Methods

2.1. Study Area

The study refers to three commercial banana farms belonging to the same cooperative, with banana plants (AAA) Cavendish cv. Williams and Gran Enano, located in the Changuinola District of the Bocas del Toro province in Panama (Table 1). The climate is tropical and rainy, and the amount of precipitation fluctuates between 2500 mm annually and has a non-seasonal rainfall regime, where the lowest rainfall is in February (143 mm), and most of the precipitation falls in December (284 mm). The absolute maximum temperature is 36 °C and the minimum is 15 °C, with an annual mean temperature between 25–26 °C [26].

Table 1. The geographical location of cultivated area (ha) until December 2018 and the number of banana lots of the three banana farms in Bocas del Toro, Panama.

Site	Geographic Coordinates		Cultivated Area (ha)	Banana Lots (n)
1	9°25′43.0″ N	82°32′57.1″ W	160.75	56
2	9°28′09.2″ N	82°31′27.9″ W	212.91	77
3	9°30′05.5″ N	82°35′42.2″ W	184.51	47

2.2. Epidemiological Evaluation of Black Sigatoka in Host Plants

The epidemiological parameters to evaluate the incidence and severity of BS were carried out in three ages of plants: plants in the acorn stage, and plants seven weeks (49 days) and ten weeks (70 days) after the appearance of the acorn, in reports of identical format. The only exception is in the parameter called symptom status through the gross sum, which was evaluated only in plants in the acorn stage.

Sampling consisted of weekly monitoring of 10 representative production areas per site. Within each area, 15 plants obtained randomly at the three physiological ages were sampled: five young plants in the acorn stage (1 to 4 days), five plants at seven weeks of age (49 days), and five plants at ten weeks of age (70 days) during the period 2015–2018. The evaluation method based on the severity scale developed by Stover [27], and later modified by Gauhl [28,29], was used. It consisted of visually estimating the affected leaf area, showing necrotic lesions of the stalks go from 4 to 6 on the Fouré scale [30]. The variable related to production was the total number of bunches harvested weekly (52 weeks per year) during the 2015–2018 period for the three sites evaluated (Site 1, Site 2, and Site 3). The epidemiological parameters evaluated were:

a. The infection index (INC): Expresses the magnitude of the damage caused by the disease on a quantitative scale, whose ideal limit is 0.10. It was calculated using Equation (1). This index is a weighted average considering the severity of the BS attack in all the evaluated leaves. The INC gives us a single value indicating the magnitude of foliage damage, on a scale of 0 to 1. In plants in the acorn stage, a value of <0.10 is considered low, $\geq$ 0.10–0.15 as intermediate, and >0.15 as high [28,29].

$$\text{INC} = \sum \frac{nb}{(N-1)\text{T} \times 100} \tag{1}$$

where INC = infection index, n = number of leaves in each grade, b = grade, N = number of grades used in the scale, and T = total number of leaves evaluated.

b. State of the symptom (gross sum) (SIN): comes from the observations of the most developed state of the present symptom and the density of these symptoms in sheet No. 3 plants in acorn state. Through a spreadsheet, a numerical value (weight) was obtained for each state of the symptom to subsequently obtain the gross sum. The gross sum reflects the development of BS at an early stage and allows the development of new BS infections to be determined several weeks earlier than with all other parameters. It also allows us to evaluate if the last applied fungicide treatment had a significant effect on the control of BS, or if it is not controlling the disease. The gross sum can range between 0 and 140. A value below 50 is considered good disease control [28,29].

c. The number of functional leaves (LEAF): represents the average of the total number of leaves per plant; only those leaves that, visually, presented 80% of their surface free of necrosis by BS [28,29].

d. Youngest leaf with symptoms (YLWS): number of the leaf-bearing symptoms. Leaves were numbered from the first (topmost) open leaf downward, according to the methodology proposed by Hernández et al. [31]. Youngest leaf with symptoms visible from the ground. These are usually stage 2 and 3 striae that can be identified from the ground.

e. Youngest leaf spotted (YLS): number of the leaf-bearing spots with dry centers. A higher value of YLS indicates more healthy leaves on the plant. Leaves were numbered as for YLWS. It is the average of the youngest leaf with 10 or more spots with a grayish and dry center, which is the last stage of the development of BS. In the spot state with a grayish and dry center, the fungus produces ascospores [28,29].

These last two variables, YLWS and YLS, represent two concepts of epidemiological importance, such as the incubation period and the latency period. The fluctuations in these two variables are associated with the behavior of the disease in the field, and, therefore, are used as elements of judgment to make decisions in the selection of the fungicide and the days between applications.

2.3. Data Analysis

2.3.1. Exploratory Analysis

The data matrix X was constituted by the set of vectors of the observations $X_{[ij]}$, $j = 1$, ..., p, and where each vector $X_{[ij]}$ presented the jth variable for all the observations and where X the data matrix was formed by "$n = 623$" observations with "$p = 13$" variables. A total of 13 predictor variables were considered in this study, being the following: infection index (INC) in plants in the acorn stage; infection index in plants at 7 weeks of age (INC_7) and 10 weeks of age (INC_10); youngest leaf with symptoms on plants at the acorn stage (YLWS) (n°); youngest leaf with symptoms on plants at 7 weeks of age (YLWS_7), and on plants at 10 weeks of age (YLWS_10); youngest leaf spotted in plants in the acorn stage (YLS) (n°), at 7 weeks of age (YLS_7) and 10 weeks of age (YLS_10); more developed state of the BS symptoms in plants in the acorn stage (SIN); and the number of functional leaves in plants in the acorn stage (LEAF) (n°), at 7 weeks of age (LEAF_7) and 10 weeks of age (LEAF_10).

Before data analysis, we checked the data integrity. A principal component analysis (PCA) was performed as an exploratory analysis to check the presence of outliers and identify patterns in the data using the statistical package in R software version 4.0.2 (R Core Team, Vienna, Austria) and the function "PCA" [32]. The objective of this application was to summarize the information of many variables (13 in our study) in a few latent variables, trying to avoid overfitting as new components are added. Pearson's correlation coefficient was also used as a method of parametric statistics, which is not only used to determine the relationship between two quantitative variables, but also to predict a variable.

2.3.2. Random Forest Prediction

The RF algorithm was applied, as a machine learning method based on binary trees, to predict the number of banana bunches. In the scope of our study, RF was used as a regression tool. Predictor variables are evaluated by how much they decreased in node impurity when selected for splits. The impurity of the node is defined as the root mean square error (RMSE) of the node in the RF regression [33].

The statistical program R [32] was used together with the 'randomForest' package with the following configuration (number of variables randomly sampled as candidates at each split (mtry) = 9, number of the trees (ntree) = 500, node size = 5) [34]. Three analysis measures in the package were used: (1) mean decrease accuracy, (2) mean decrease Gini, which are the measures of the performance of the model without each variable, and (3) partial dependence plots.

First, the mean decrease accuracy plot expresses how much accuracy the model loses by excluding each variable. The more precision suffers, the more important the variable is for successful classification. The variables are presented in decreasing importance. The second measure is the mean decrease in the Gini coefficient, which is a measure of how each variable contributes to the homogeneity of nodes and leaves in the resulting random forest. The higher the mean decline precision value or the mean decline Gini score, the greater the importance of the variable in the model [34–36].

The third analysis measure is the partial dependence plots; this shows how each predictor influences the RF model predictions when all other model predictors are controlled. The value of the Y-axis of a partial dependence plot is determined by the average of all possible predictions of the model with the data set when the value of the target predictor is X [37].

2.3.3. Model Performance Evaluation

The k-fold cross-validation procedure was performed, with k = 3 becoming 3-fold cross-validation. Cross-validation was used primarily in this case to estimate the skill of the machine learning model on unseen data; that is, using a limited sample to estimate how the model is expected to perform overall when used to make predictions on data that was not used during model training.

Four methods were used to evaluate model performance: (1) root mean square error (RMSE), (2) the explanation of variance (%), (3) an observed versus predicted plot to visualize model performance, and the coefficient of determination (R^2). A simple linear regression line was drawn on the plot to compare the accuracy of the model's predictions. These are commonly used measures for farming systems and models of different crops [22,38].

3. Results

3.1. Descriptive Analysis

Table 2 shows the descriptive statistics of the epidemiological parameters and the number of clusters per site evaluated. Site 2 presents the highest average values of the number of banana bunches with 10,855.82 ± 1575.60, and the lowest CV value of 14.5%. On the other hand, site 1 presented the lowest average cluster number value with 7611.77 ± 1648.12 and the highest CV with 21.65%. The asymmetry measures presented served to provide an idea about the shape of the frequency distribution through the coefficient shown in Table 2. The coefficient measures the degree of asymmetry of the distribution concerning the mean. A positive value for this indicator means that the distribution is skewed to the left (positive orientation), as is the case for the banana bunch variables at site 1, YLWS_10, and SIN at all sites. A negative result means that the distribution is skewed to the right, as in the case of the banana bunch variables at sites 2 and 3, and YLS_10 in all the sites evaluated.

Table 2. Summary of descriptive statistics for the three sites evaluated: means values, standard deviations (S.D), standard error (S.E), coefficient of variation, (CV) minimum, maximum, skewness (S), kurtosis (K) and percentiles 25, 50, 75 for the production variable and epidemiological variables.

Site	Variable	Mean	S. D	S. E	CV	Min	Max	S	K	P (25)	P (50)	P (75)
1 (n = 208)	Bunch (n°)	7611.77	1648.12	114.28	21.65	4152.0	12,921.0	0.59	0.25	6404.00	7360.00	8620.00
	INC	0.06	0.03	0.00	41.00	0.02	0.17	1.35	3.09	0.04	0.06	0.08
	YLS (n°)	14.08	1.42	0.10	10.12	9.35	16.60	−1.08	1.09	13.25	14.40	15.16
	YLWS (n°)	11.06	1.41	0.10	12.71	7.45	14.28	−0.08	−0.43	10.10	11.00	11.96
	SIN	52.36	9.91	0.69	18.93	40.00	99.25	1.45	2.62	44.80	49.60	57.60
	LEAF (n°)	13.67	0.65	0.05	4.77	11.50	15.40	−0.47	0.60	13.25	13.70	14.16
	LEAF_7(n°)	10.17	0.89	0.06	8.79	7.25	11.80	−0.62	0.15	9.60	10.28	10.80
	LEAF_10(n°)	9.30	1.20	0.08	12.86	5.15	11.12	−0.76	0.06	8.48	9.48	10.28
	INC_7	0.13	0.05	0.00	42.37	0.03	0.31	1.06	2.02	0.09	0.12	0.16
	YLS_7 (n°)	10.08	1.93	0.13	19.12	4.75	13.20	−0.71	−0.07	8.88	10.48	11.48
	YLWS_7 (n°)	6.43	2.04	0.14	31.73	1.20	10.72	−0.03	−0.22	5.15	6.50	7.56
	INC_10	0.17	0.06	0.00	37.41	0.05	0.39	0.72	0.74	0.13	0.17	0.21
	YLS_10 (n°)	8.66	2.24	0.16	25.86	2.55	12.32	−0.65	−0.27	7.25	9.04	10.24
	YLWS_10 (n°)	4.58	2.08	0.14	45.42	1.00	9.32	0.28	−0.65	3.10	4.32	6.04
2 (n = 208)	Bunch	10,855.82	1575.60	109.25	14.51	6297.00	15,843.00	0.30	0.42	9899.00	10,640.00	11,775.00
	INC	0.08	0.03	0.00	35.06	0.03	0.17	0.85	0.62	0.06	0.07	0.09
	YLS (n°)	13.24	1.93	0.13	14.60	7.95	16.04	−0.78	−0.40	11.96	13.84	14.56
	YLWS (n°)	10.13	1.40	0.10	13.84	6.50	13.36	−0.05	−0.36	9.00	10.28	11.00
	SIN	54.49	10.47	0.73	19.22	40.00	103.00	1.33	2.41	47.20	52.00	59.20
	LEAF (n°)	13.19	1.03	0.07	7.81	10.45	15.12	−0.61	−0.30	12.48	13.36	13.96
	LEAF_7 (n°)	9.47	1.30	0.09	13.73	5.85	11.56	−0.67	−0.53	8.48	9.84	10.52
	LEAF_10(n°)	8.42	1.60	0.11	18.99	4.73	10.92	−0.55	−0.97	7.08	9.00	9.80
	INC_7	0.15	0.06	0.00	40.05	0.06	0.35	1.05	0.85	0.11	0.14	0.18
	YLS_7 (n°)	8.89	2.27	0.16	25.57	3.10	12.56	−0.66	−0.53	7.40	9.40	10.68
	YLWS_7(n°)	5.36	1.94	0.13	36.20	1.00	9.56	−0.19	−0.67	3.80	5.56	6.84
	INC_10	0.21	0.08	0.01	38.67	0.09	0.45	0.90	0.20	0.14	0.19	0.26
	YLS_10 (n°)	7.31	2.51	0.17	34.34	1.00	11.68	−0.75	−0.42	5.85	7.92	9.28
	YLWS_10(n°)	3.53	1.88	0.13	53.35	1.00	7.40	0.22	−1.20	1.70	3.50	5.32

Table 2. *Cont.*

Site	Variable	Mean	S. D	S. E	CV	Min	Max	S	K	P (25)	P (50)	P (75)
3 ($n = 207$)	Bunch (n°)	8984.02	1479.62	102.84	16.47	5543.00	12,517.00	−0.01	−0.05	8137.00	8969.00	9915.00
	INC	0.07	0.03	0.00	41.15	0.03	0.17	1.17	1.59	0.05	0.06	0.08
	YLS (n°)	13.04	1.49	0.10	11.45	8.45	15.85	−0.94	0.63	12.30	13.36	14.04
	YLWS (n°)	10.53	1.22	0.08	11.59	7.45	13.30	−0.09	−0.40	9.60	10.56	11.35
	SIN	56.76	12.60	0.88	22.19	40.00	106.00	1.17	1.37	48.00	52.80	64.80
	LEAF (n°)	12.89	0.72	0.05	5.61	11.05	14.85	−0.19	−0.30	12.43	12.96	13.40
	LEAF_7 (n°)	9.62	0.75	0.05	7.84	7.45	11.45	−0.66	0.77	9.28	9.70	10.12
	LEAF_10 (n°)	8.73	0.93	0.06	10.70	5.95	10.25	−0.90	0.42	8.30	8.88	9.44
	INC_7	0.14	0.05	0.00	40.45	0.05	0.32	1.17	1.65	0.09	0.13	0.16
	YLS_7 (n°)	9.27	1.78	0.12	19.21	3.78	12.65	−0.81	0.13	8.18	9.72	10.52
	YLWS_7 (n°)	6.07	1.66	0.12	27.36	1.55	9.45	−0.26	−0.28	4.88	6.08	7.28
	INC_10	0.19	0.17	0.01	88.06	0.05	2.43	10.90	137.99	0.13	0.17	0.22
	YLS_10 (n°)	7.85	2.00	0.14	25.48	2.85	11.15	−0.84	−0.23	6.30	8.40	9.36
	YLWS_10 (n°)	4.31	1.74	0.12	40.42	1.00	7.75	0.04	−0.82	3.20	4.12	5.80

Note: Bunch: number of harvested bunches; infection index (INC) in plants in the acorn stage. infection index in plants at 7 weeks of age (INC_7) and 10 weeks of age (INC_10); youngest leaf with symptoms on plants at the acorn stage (YLWS) (n°), youngest leaf with symptoms on plants at 7 weeks of age (YLWS_7) and on plants at 10 weeks of age (YLWS_10); youngest leaf spotted in plants in the acorn stage (YLS) (n°) at 7 weeks of age (YLS_7) and 10 weeks of age (YLS_10); state of the symptom (gross sum) in the acorn stage plants (SIN); the number of functional leaves in plants in the acorn stage (LEAF) (n°) at 7 weeks of age (LEAF_7) and 10 weeks of age (LEAF_10).

On the other hand, some variables present a higher degree of concentration (less dispersion) of the values around their mean, such as the case of LE-AF_10 in sites 1 and 3; and other variables, on the contrary, present a lower degree of concentration (greater dispersion) of their values around their central value, such as the variables YLWS_10 and YLS_10 in all the sites. Therefore, kurtosis informs the pointed (higher concentration) or flattened (lower concentration) of these distributions. Regarding the percentile as a measure of non-central position, for site 2, the P75 of the bunch variable is 9915.0 banana bunches. This means that 75% of the weeks show a bunch production equal to or less than 9915.0, followed by P75 = 9865.0 and 8620.0 banana bunches for sites 3 and 1, respectively.

The correlation analysis is shown as a heat map (Figure 1), useful to try to understand the relationships between the multiple variables studied. The lower triangle of the matrix shows the heatmap of correlations between pairs of variables. The bunch variable has an extremely low correlation coefficient concerning the epidemiological variables; high correlations ($p \leq 0.05$), whose Pearson's r was greater than | 0.85 | are also shown between the variables derived from the YLS, YLWS, LEAF, and INC.

The PCA was carried out to evaluate the changes in the pattern of the epidemiological variables by evaluated sites. PCA was unable to distinguish the sites evaluated in this study (Figure 2a). The first two principal components (PC) explained 81.6% of the variables; however, no trends were detected in the differences (Figure 2b). According to the value of the contribution in the biplot, we obtained four epidemiological parameters, including YLS_10, INC_7, YLWS_10, and LEAF_7 (Figure 2c). The goodness-of-fit of the PCA model was $R^2X = 0.70$. By analyzing the projections perpendicular to PC1 of the points that represent the cases, certain weeks were identified during the evaluated years with greater inertia, i.e., the points that are at a greater distance from zero (Figure 2c), either moving away to the right or the left. Likewise, the analysis of the projections of the points that represent the variables on the PC1 allowed us to identify the variables with the greatest inertia and with the greatest collinearity, e.g., (YLS10 and YLS7); (LEAF, LEAF 7, and LEAF10); (YLWS, YLWS7, and YLWS10).

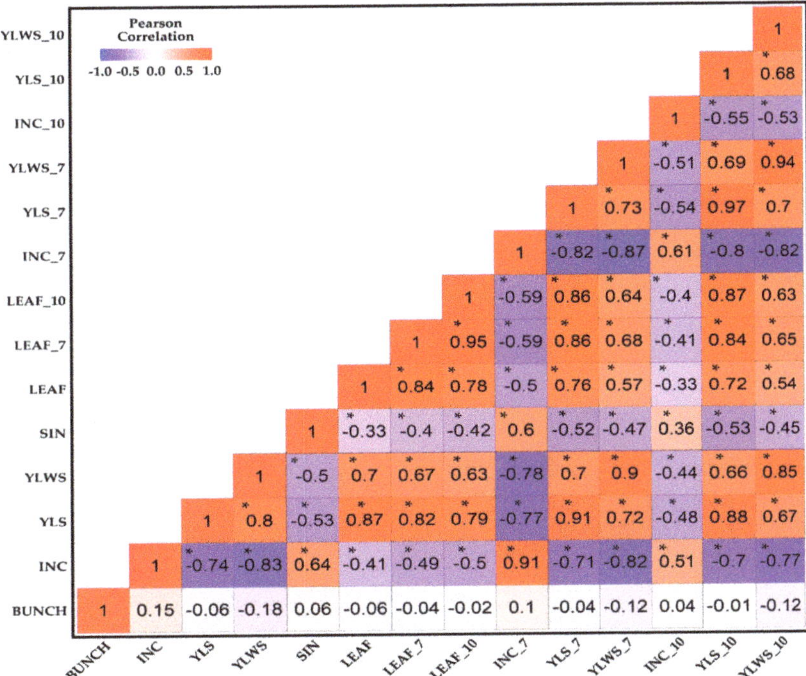

Figure 1. Heatmap of correlation coefficient among epidemiological parameters and numbers of banana bunches. (*) Pearson correlation ($p \leq 0.05$) is shown between variables described in this study. Note: Bunch: number of harvested bunches; infection index (INC) in plants in the acorn stage, infection index in plants at 7 weeks of age (INC_7) and 10 weeks of age (INC_10); youngest leaf with symptoms on plants at the acorn stage (YLWS) (n°); youngest leaf with symptoms on plants at 7 weeks of age (YLWS_7) and on plants at 10 weeks of age (YLWS_10); youngest leaf spotted in plants in the acorn stage (YLS) (n°) at 7 weeks of age (YLS_7) and 10 weeks of age (YLS_10); state of the symptom (gross sum) (SIN); the number of functional leaves in plants in the acorn stage (LEAF) (n°) at 7 weeks of age (LEAF_7) and 10 weeks of age (LEAF_10).

In the interpretation of "correlations" between variables according to the angles of the vectors that represent them, it was found that acute angles indicated positive correlations (all variables derived from YLS, YLWS, and LEAF), obtuse angles corresponded to negative correlations (SIN and INC) and right angles indicated that there was no correlation between the variables. The length of the vectors corresponding to the variables is not of interest when the data has been previously standardized.

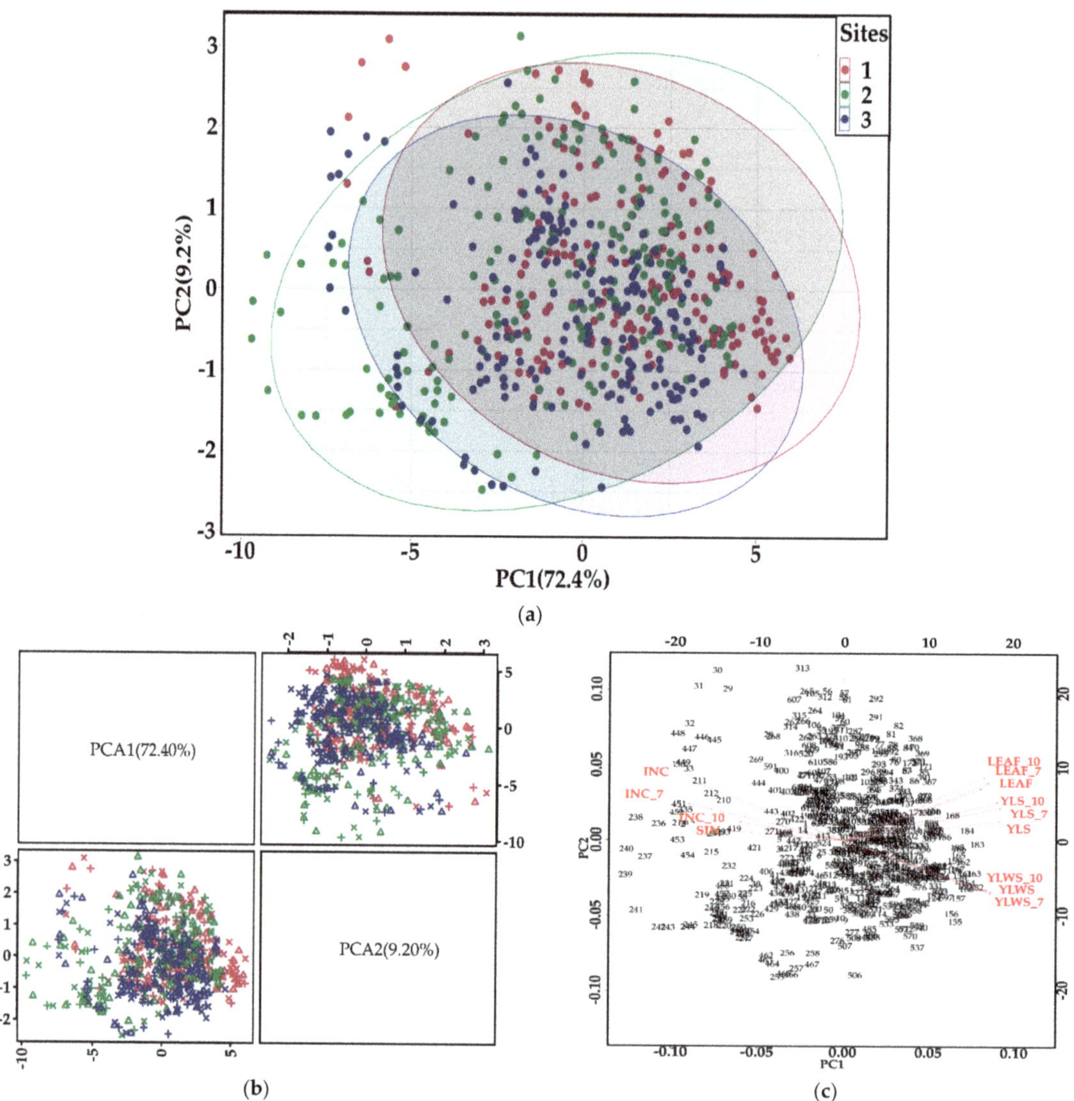

(**a**)

(**b**)

(**c**)

Figure 2. Classifying PC from site based on the epidemiological parameters; (**a**) PCA based on the first two principal components; (**b**) sample scatterplot displays the first two components in each data set in PCA; (**c**) contribution of each feature selected on the first component in PCA using the biplot. Note: Infection index (INC) in plants in the acorn stage, infection index in plants at 7 weeks of age (INC_7) and 10 weeks of age (INC_10); youngest leaf with symptoms on plants at the acorn stage (YLWS) (n°); youngest leaf with symptoms on plants at 7 weeks of age (YLWS_7) and on plants at 10 weeks of age (YLWS_10); youngest leaf spotted in plants in the acorn stage (YLS) (n°) at 7 weeks of age (YLS_7) and 10 weeks of age (YLS_10); state of the symptom (gross sum) (SIN); the number of functional leaves in plants in the acorn stage (LEAF) (n°) at 7 weeks of age (LEAF_7) and 10 weeks of age (LEAF_10).

3.2. Global Banana Bunches Predictions

RF successfully predicted the number of banana bunches on a global scale when compared to test data that had not been included in model training. Figure 1 shows the observed versus predicted plot for the number of banana bunches at all sites tested. The RF model explained 70.6% of the mean yield variation with good agreement between predicted and observed values in the test data (Table 3; Figure 3).

Table 3. Random forest model performance evaluation statistics.

K-Folds	Training Data			Testing Data		
	RMSE (n° Bunches)	Variance Explained (%)	R^2	RMSE (n° Bunches)	Variance Explained (%)	R^2
1	1112.93	69.1	0.69	1132.31	72.0	0.72
2	1112.33	71.0	0.71	1101.23	71.0	0.71
3	1145.67	70.0	0.70	1090.26	69.0	0.69
k-folds CI 95% *	1076–1171	68.0–72.0	0.68–0.72	1054–1162	67.0–74.0	0.67–0.74

* 95% confidence interval (CI).

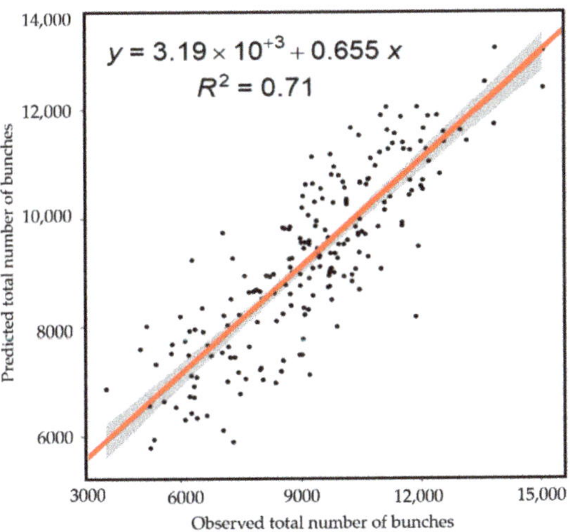

Figure 3. Random Forest model performance for test data sets. The dashed lines indicate a 1:1 relationship and the solid line represents the linear regression between the observations and the predictions made for the test data sets.

RMSE values were within the 95% confidence interval of 1054–1162 banana bunches, which is 11.51% of the observed average yield. The comparison between the number of banana bunches observed and predicted (Figure 3) indicates that the predictions of the RF model are in good agreement with the observations with a slope of 0.65 and the R^2 of 0.71.

The RF method was performed with Mtry = 9, ntree = 500; the OOB (out-of-bag) error was found to be minimal, as shown in Figure 4a. About two-thirds of the sample, called the in-bag sample, was used to train the model, and the remaining third of the sample, called the out-of-bag sample, was used for internal cross-validation on the RF model.

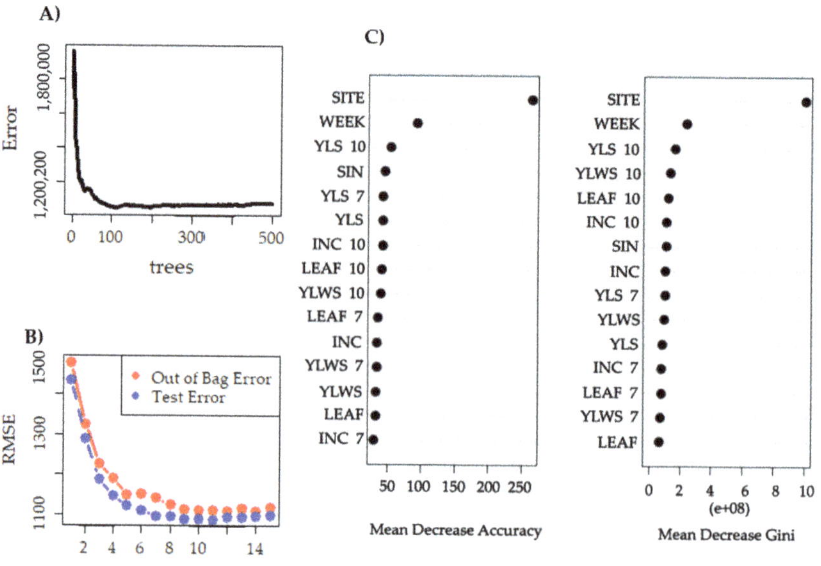

Figure 4. RF estimation for the number of banana bunches: (**A**) OOB error, (**B**) predictor variable importance graph, and (**C**) selection of a few variables using 3-fold cross-validation based on least RMSE.

The prediction of the number of banana bunches was based on the 13 independent variables on a global scale over four years. The main predictor variables in the graph of the importance of the variable obtained from the RF are shown in Figure 4b. The influence of a set of six variables for the prediction of the number of banana bunches is observed through the recursive elimination of characteristics with the cross-validation technique, as shown in Figure 4c.

Additionally, RF can also provide useful information about the importance and dependence of the variable on its predictive capacity. The range of importance of the epidemiological variables and the partial impact of the variable on the response represented by the number of banana bunches can be evaluated for systems analysis purposes. The use of the measure of variable importance, represented by mean decrease accuracy and mean decrease Gini, allowed us to identify the most influential variables that determine banana production in the evaluated sites that we tested (Figure 4b).

Partial dependence plots were useful to assess the relationship between each predictor and the response variable (Figure 5), measure the effect of explanatory variables, such as site (Figure 5a), week (Figure 5b), SIN (Figure 5c), YLS_10 (Figure 5d) and YLWS_10 (Figure 5e), on the variable number of banana bunches. Variable importance measures and partial dependence plots revealed unique responses concerning the evaluated variables. The RF model identified the trend of the YLS parameter in plants with 10 weeks of physiological age, being one of the most influential variables responsible for the increase in the number of bunches (>1000) during the four years; the ideal values of this parameter are between blade position 10 and 12 (Figure 5d). The site was ranked as the first and most important variable for estimating banana yield (Figure 5a). Likewise, the week was classified as the second most important variable for the number of banana bunches, since the yield or the banana bunches fell in weeks 18 to 22, which indicates that the week of the year can be a key factor to achieve a high yield of bananas in this site, based on the evaluation of the BS parameters (Figure 5b). For example, the partial dependence plot shows that the epidemiological parameters such as SIN were one of the main predictors of the number of banana bunches, with a saturation response at SIN values less than 40 (Figure 5c). The partial dependence graphs for other variables studied presented minor effects with very complex patterns that are difficult to explain through certain physiological or agronomic responses of bananas.

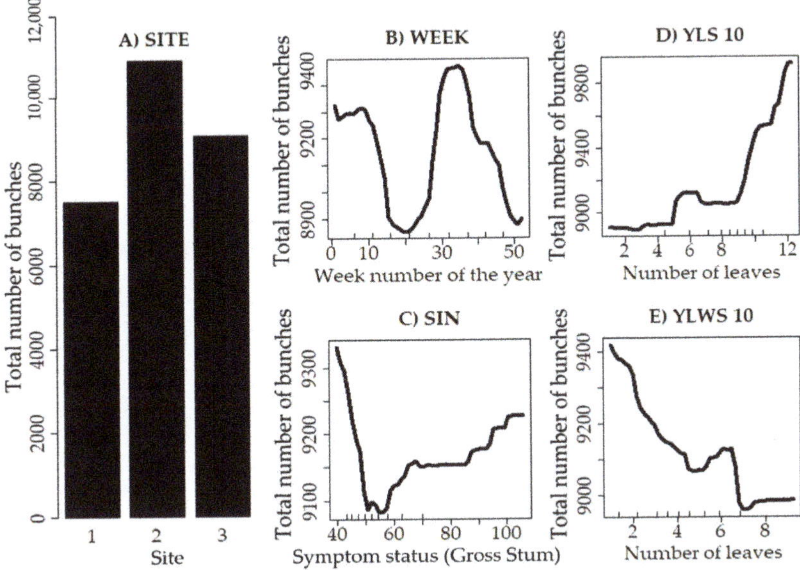

Figure 5. Partial dependence plots for the top-ranked predictor variable from variable importance measures of random forest models: (**A**) site, (**B**) week, (**C**) SIN, (**D**) YLS_10, and (**E**) YLWS_10. The Y-axis of each plot indicates the average of all the possible model predictions for the X predictor value. The X-axis hash marks indicate deciles.

3.3. Identification of the Main Variables

The behavior of the disease was subject to a dynamic that depended on environmental conditions during the 2015–2018 period, which showed different phases of development, indicating that the disease is cyclical, showing variable fluctuations during the year and between years. In terms of evaluation, sharp exponential growth was seen in the month of August weeks 32–35 according to the SIN, with 2015 being the year with the highest reported values of SIN (99.25–106.0) for all sites (Appendix A Figure A1(1–3)). Then, a phase of growth slowdown manifested itself in August 2016.

For the evaluated sites, a generalized trend of the maximum number of bunches harvested globally was observed, located in weeks 32 to 36 with values higher than 11,700 bunches for the period 2017–2018 in site 1; for site 2, it was evidenced that higher values of the number of clusters were obtained in weeks 32 to 35, being higher than 15,000; and finally in site 3, the maximum number of clusters harvested was during weeks 30–33, with values greater than 12,000 bunches. This establishes a clear pattern that the number of the week, together with the evaluation of epidemiological parameters, have an important weight for the management of BS and production. The YLWS_10 presented a curve with maximums in three seasons during the years 2017 and 2018, with week 4/2018 corresponding to 22–28 January, being the highest value (9.32) for site 1, followed by week 26/2018, corresponding to the 25 June to 1 July, with 9.16 for site 1 and 7.75 for site 3, which indicates that the higher the value of YLWS_10, the slower the development of the disease. On the contrary, low values of YLWS_10, such as those recorded during weeks 31–36/2015 correspond to the end of July and the month of August, which indicate that the development of the disease is faster during that time.

Regarding the YLS_10 in general, it was evidenced that the maximum values of this parameter were in weeks 5/2018 and 7/2018, corresponding to the first fortnight of February with values of 11.68 and 12.30, respectively (Appendix A Figure A1(1–3)); likewise, all the sites presented maximum values of YLS_10 during the weeks 30/2018 (10.64) and 33/2018 (11.8), corresponding to the days of 23 to 27 July and August 13 to 19. The lowest values of this parameter were in weeks 32, 33, and 35 of 2015, corresponding to the month of August, whose values fluctuated between 1.0–2.95.

In the case of the variable LEAF_10, all the sites showed a tendency to a considerable reduction in the number of functional leaves during weeks 34–38/2015, corresponding to the month of August and mid-September, with values that ranged between 5 and 6 functional leaves. Site 2 presented the highest concentration of functional leaves below 8 leaves in 2018, with values of 5 functional leaves (Appendix A Figure A1(1–3)).

The SIN variable presented its maximum points mostly during week 32/2015, corresponding to 3–9 August, for all sites evaluated with values of 103.0–106.0. On the other hand, the lowest values were located mostly in week 17/2017, corresponding to 24–30 April, whose values were 40.0. However, the trend of the curve is more important than the absolute value of the SIN. In all cases, the SIN continued to rise three weeks in a row, entering critical levels (>50) (Appendix A Figure A1(1–3)); therefore, a careful review of the fungicide application program is necessary for the weeks before and after these levels, since it is an indicator that there is a high potential for a strong infection in the plantation.

The time when the symptoms of the disease manifested with greater severity was in the months of July, August, and part of September in most years; after this time, it can be said that a self-destruction phase of the disease was defined. Fungus, in the months of December–January, and May–July; since a recovery period of the plant could be observed, corroborated by the increase in the values of the YLWS, YLS, and LEAF (Appendix A Figure A1(1–3)).

Once the dynamics of the disease are known, an integrated management program for BS can be established, according to the results obtained, the phytosanitary control practices in the evaluated area and areas with similar climatic conditions, in the months of growth or exponential phase of the disease. In this area, it is appropriate to carry out preventive control in April, June–July, and November.

4. Discussion

In the process of filling the fruits of banana cultivation, the progressive development of the leaf area and its physiological activity are determining factors in the productive performance, for which it is essential that, during its growth, the plant has enough functional leaves, which guarantee that photosynthesis is carried out optimally [39]. In this sense, to obtain a good production of banana bunches, a minimum of eight functional leaves must be maintained at the time of flowering [40]. The foregoing indicates the importance of agronomic management to reach flowering, at least, with eight healthy leaves free of BS [41,42].

The mean lowest number of functional leaves was found in site 2 with 4.73, indicating that the management conditions and even the environmental conditions in the plantation of this site could provide the crop with a negative effect on the number of healthy functional leaves. In this regard, a study showed that greater health in the plantation is determined by a greater number of functional leaves when the crop is grown in an environment of 20% shade [42]. These results coincide with the investigations of Belalcázar et al. [43] which indicate that it is necessary to establish a simultaneous agroforestry system with banana cultivation at distances that do not interfere directly with the plants, such as, for example, 4 m between rows of plants and 2 m between plants in the row, since it helps to reduce the severity of BS disease.

The research reported by Barrera et al. [41] established that banana plants that kept between eight and twelve leaves during flowering had larger bunches, so to guarantee a good filling the plant at harvest must end with at least six functional leaves. Additionally, the results of studies, such as those by Cayón [44] and Siles et al. [45], describe that the highest rates of photosynthesis occur in the youngest leaves of the plant (two, three, four, and five), drastically reducing in leaves six, seven, eight and nine; these are older leaves, which shows the possibility that certain microclimatic conditions of shade in early stages of plant development may favor the crop, for its general vigor and better response to the appearance of BS.

Bananas shed between 0.8 and 1.2 leaves each week. From this, it is inferred that an important part of the symptoms that are observed weekly in the fourth leaf come from inoculations that occurred during the different stages of opening of the leaves or when they are just opened [46]. Likewise, in Costa Rica with a climate like that of our study, the symptoms took 9.6 days more to develop after flowering of the False Horn banana cv. 'Currare' than before flowering [47]. These results suggest that the host's response to BS depends on the host's metabolism. The difference in host response observed between the vegetative and generative stages of plantain could be due to changes in the hormonal system of the central meristem that starts to flower and simultaneously stops the formation of new leaves [48].

In leaves affected by BS, photosynthesis and respiration processes are affected, causing irreparable morphological and physiological damage, which prevents the proper functioning of the crop, directly damaging its development and production [7,49], which generates, therefore, low productivity, premature ripening of fruits and a decrease in quality [50].

Banana plants were more sensitive to BS, presenting the YLWS between leaves 1 and 4 (Table 2), which are below the number of leaves necessary to develop a good bunch, affecting yield and the quality of the fruit. The YLWS provides information on the quality of control in young leaves and the BS progress in the plantation [49].

YLWS is a widely used variable, captured weekly by inspectors from the plant health departments of aerial spraying companies. This parameter presented an important influence in the RF model and in a certain way it has a good relationship with the production of bunches. That is, it is possible to predict the future level of the number of banana bunches knowing the status of this epidemiological parameter, requiring in general terms a greater number of variables and especially the evaluation of their interaction.

The study by Cedeño et al. [51] established that, with defoliation and surgery at weekly intervals, the severity of the disease was reduced. Other practices in addition to sanitary defoliation could also help in its management, such as piling or cordoning off diseased tissue on the ground and the application of 10% urea as a sporulation inhibitor [52].

The rank of the youngest spotted leaf (counted from the last emerging leaf), called YLS, is an especially important index that provides information on the growth rate of necrotic lesions responsible for the destruction of large portions of leaf tissue. The lower the YLS, the faster the development of the disease. The YLS_10 was on average between leaf number seven and eight; ideally, it would be between leaf number ten and eleven, indicating that the banana plant can reach the end of its production cycle, which would allow a good filling of the fruits and, therefore, high production of bunches.

When a younger diseased leaf is detected in the field among the first leaves of the plant, obtaining good yields and bunch production is aggravated, since within the agronomic management of BS, if it is decided to carry out defoliation operations and pruning to reduce the more rapid ascent of the disease towards the upper leaves, all old and infected leaves should be carefully cut before flowering, which would result in several leaves below six, the minimum quantity necessary to avoid a decrease bunch production [53].

In the reproductive cycle of the banana plant (*Musa* AAB cv Harton) from the exit of the bunch to its harvest, 70 to 98 days (12 $\pm$ 2 weeks) pass under the conditions of the banana zone of the South of Lake Maracaibo, Venezuela [49]. The reports of this last study indicate that the minimum number of leaves present at the beginning of the cycle should be 12 to 13 in bananas. It is understood that the three upper leaves (the youngest) supply the needs of the plant, and older leaves help bunch growth.

In this regard, in the climate of the South Pacific of Mexico, Black Sigatoka showed that the greatest damage (12 to 25% severity) occurs during the months of June to December, which coincides with the season of greatest rainfall [54]. In this period, symptoms appear in a state of spot between leaf nos. 4 to 6 and 25 to 58% of diseased leaves. The lower severity of the disease (January to May) in the South Pacific is related to the period of lower rainfall [55], where spots appear between leaves No. 7 to 9 and 7 to 25% diseased leaves [56].

Unfortunately, this disease reduces leaf longevity and threatens the plant's ability to achieve full fruit fill. In practical terms, it is recommended to use the variables YLWS, YLS, and LEAF to evaluate the seasonal behavior of the disease at the farm level. Although the analysis methods may be complex for technicians in the region, at a later stage, it is recommended to implement, through an algorithm such as the one developed in this study, all the analysis procedures required for modeling the disease, in such a way that a technician only has to enter the data from the monitoring of BS to make adjustments to the models and, through a platform, predict the future evolution of bunch production.

The results generated in this study emphasize the obtaining of scientific information from commercial plantations, which can be complemented by the addition of data and information on cultural practices and chemical control carried out, applications of biological inputs, such as promoters of growth and resistance to disease, or biological regulators of pathogen populations, among other practices, which could increase the prediction accuracy of the RF model by reducing the mean square error of prediction, which would finally translate in timely applications of fungicides (protectants) and not calendar applications as is currently done, thus generating a reduction in the number of cycles of systemic fungicides by being used only when there is an increase in the evolution of the disease that justifies it.

The models developed, such as the RF model, can become a planning tool for production and disease management in terms of cycles, product rotation, and times to use the different types of products. On the other hand, as more farms adopt the methodology, the spraying frequencies will probably not be the same for all sites or farms.

The decision to use RF for this type of problem was associated with the main advantage of RF regression when the explanatory variables are highly correlated (e.g., the number of functional sheets, YLS, and YLWS at different stages, and physiological characteristics of the plant) [57]. Many predictors of crop production, such as these types of variables that are repeated in the evaluation, are often highly correlated with each other and can have multicollinearity. RF uses the best single variable when splitting the responses at each node of the decision trees and averaging the predictions from the trees in the forest to make a multidimensional step function [36]. This means that even if several variables are correlated and drive the response similarly, only one of them can affect the RF regression model at a time [38].

The results demonstrate that RF regression is highly effective for farm-scale crop yield predictions. Although RF has been widely used as a classification algorithm for various applications recently [11,21], nowadays, few studies have analyzed its regression capabilities for banana production in Latin American territories. Our results are shown here to establish that the RF regression has significant merits that are highly suitable for predicting the production of crops, such as bananas with SB epidemiological parameters.

Finally, a wide range of factors that affect Sigatoka has been reported in the literature; for example, shade, soil type, drainage, use of fertilizers, plant density, and irrigation [8,14], as well as the mineral nutrition of plants, considered as an exogenous factor, which can be modified, and constitutes an additional fundamental point to combat diseases [11,13]. Any reduction in the number of bunches harvested or in the weight of the bunch in Musaceae could be the effect of the low initial number of leaves and/or their deterioration due to the action of pathogens or other biotic agents. In the case of BS attacks, the economic critical point must be established to proceed with the control of the disease, and this value is related to the effective leaf area over time.

5. Conclusions

This study evaluated the effectiveness of banana bunch prediction of the RF algorithm, which has certain advantages for the regression of crop systems such as bananas but is not yet widely used in this field. We show that RF provides a production based on epidemiological parameters of black sigatoka. The result of this study shows great potential for the use of the RF algorithm with the accuracy of the model's predictions of 0.71, so RF is an alternative statistical modeling method for production predictions in bananas.

The information on the prediction of the production of banana bunches can help the producer to estimate the profitability of the harvest and encourage the generation and monitoring of studies to verify which epidemiological parameters of black sigatoka most influence the final production. In addition, one of the biggest problems in the management of this disease is the time or incubation period and the latency period of black sigatoka, represented by the youngest leaf with symptoms and youngest leaf spotted parameters, being important elements of judgment in decisions related to the selection of the fungicide and the days between applications. Therefore, any tool that reduces this period will be of great help.

In summary, the results support that RF regression may be a suitable algorithm to predict the number of banana bunches at such sites with careful selection of a training dataset that includes a diverse range of predictors, with the site, week, youngest leaf spotted, and youngest leaf with symptoms in plants with 10 weeks of physiological age as the best predictor groups.

Author Contributions: Conceptualization, B.O.O., M.A.R.C. and A.V.; methodology, M.A.R.C. and A.V.; software, B.O.O., M.A.R.C., M.A.-A. and A.V.; validation, B.O.O., M.A.R.C. and A.V.; formal analysis, M.A.R.C. and A.V.; investigation, B.O.O., M.A.-A., E.M. and E.M.-G.; resources, B.O.O., M.A.-A., E.M. and E.M.-G.; data curation, B.O.O. and E.M.-G.; writing—original draft preparation, B.O.O.; writing—review and editing, M.A.-A., E.M. and E.M.-G.; visualization, M.A.-A., M.A.R.C. and A.V.; supervision, M.A.-A., E.M.-G. and E.M.; project administration, E.M.-G.; funding acquisition, B.O.O. All authors have read and agreed to the published version of the manuscript.

Funding: This research received no external funding.

Institutional Review Board Statement: Not applicable.

Informed Consent Statement: Not applicable.

Data Availability Statement: Not applicable.

Acknowledgments: The authors are grateful for the international mobility scholarship from the Campus of International Excellence for the Environment, Biodiversity and Global Change (CEI-Cambio) based in Seville, Spain; for promoting outreach activities in indigenous territories of Panama. Additionally, to the Board of Directors of the Cooperativa Bananera del Atlántico (COOBANA RL), the Management, collaborators, and technical personnel, especially Ing. Ernesto Ortiz (Phytosanitary Technical Advisor), Luis Escobar (Sigatoka Technical Supervisor), Juan Corella, and Diomedes Rodríguez, for his valuable support in obtaining epidemiological information, and that will serve as a guide for decision-making in the sustainable production of bananas for export by COOBANA, RL.

Conflicts of Interest: The authors declare no conflict of interest.

Appendix A

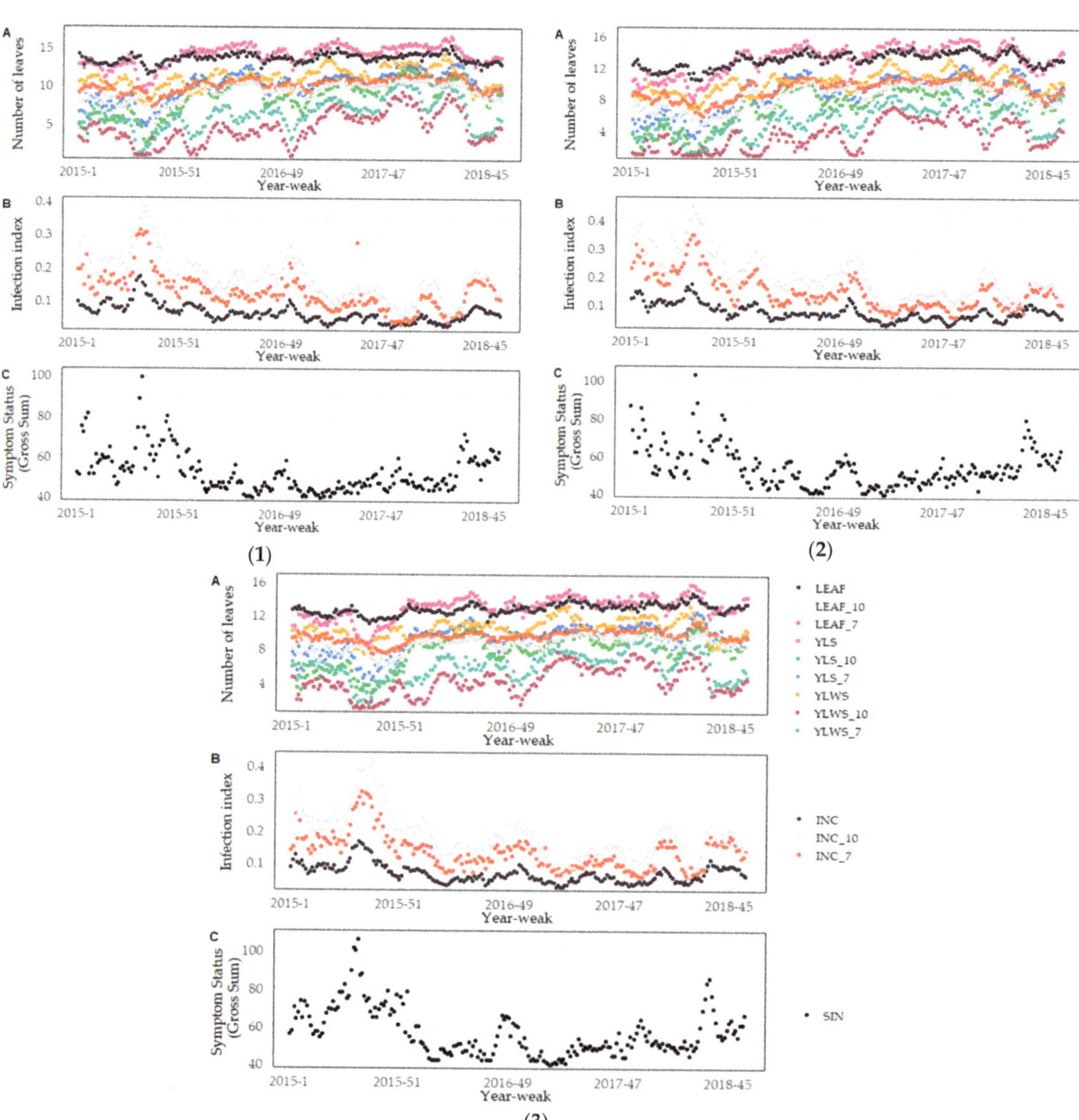

Figure A1. Weekly distribution of epidemiological parameters (2015–2018) at site 1 (**1**), site 2 (**2**), and site 3 (**3**). (**A**) Youngest leaf with symptoms on plants at the acorn stage (YLWS) (n°), youngest leaf with symptoms on plants at 7 weeks of age (YLWS_7) and on plants at 10 weeks of age (YLWS_10); youngest leaf spotted in plants in the acorn stage (YLS) (n°) at 7 weeks of age (YLS_7) and 10 weeks of age (YLS_10); the number of functional leaves in plants in the acorn stage (LEAF) (n°) at 7 weeks of age (LEAF_7) and 10 weeks of age (LEAF_10). (**B**) Infection index (INC) in plants in the acorn stage, infection index in plants at 7 weeks of age (INC_7) and 10 weeks of age (INC_10). (**C**) State of the symptom (gross sum) in the acorn stage plants (SIN).

References

1. Food and Agriculture Organization of the United Nation. Banana Market Review: Preliminary Results 2020. Available online: https://www.fao.org/3/cb5150en/cb5150en.pdf (accessed on 16 September 2022).
2. Evans, E.A.; Ballen, F.H.; Siddiq, M. Banana production, global trade, consumption trends, postharvest handling and processing. In *Handbook of Banana Production, Postharvest Science, Processing Technology and Nutrition*; Siddiq, M., Ahmed, J., Lobo, M.G., Eds.; Wiley Online Library: New York, NY, USA, 2020; pp. 1–18. [CrossRef]
3. Pitti, J.; Olivares, B.O.; Montenegro, E.; Miller, L.; Ñango, Y. The role of agriculture in the Changuinola district: A case of applied economics in Panama. *Trop. Subtrop. Agroecosyst.* **2021**, *25*, 017. [CrossRef]
4. Martínez-Solórzano, G.E.; Rey-Brina, J.C. Bananos (Musa AAA): Importance, production, and trade in COVID-19 times. *Agron. Mesoam.* **2021**, *32*, 1034–1046. [CrossRef]
5. Montenegro, E.J.; Pitti-Rodríguez, J.E.; Olivares-Campos, B.O. Identification of the main subsistence crops of Teribe: A case study based on multivariate techniques. *Idesia* **2021**, *39*, 83–94. [CrossRef]
6. Rhodes, P.L. A new Banana disease in Fiji. *Commonw. Phytopathol. News* **1964**, *10*, 38–41.
7. Marin, D.H.; Romero, R.A.; Guzman, M.; Sutton, T.B. Black Sigatoka: An increasing threat to banana cultivation. *Plant Dis.* **2003**, *87*, 208–222. [CrossRef] [PubMed]
8. Fullerton, R.A. Sigatoka Leaf Diseases. In *Compendium of Tropical Fruit Diseases*; Ploetz, R.C., Zentmyer, G.A., Nishijinia, W.T., Rohrbach, K.G., Ohr, H.D., Eds.; American Phytopathological Society: St. Paul, MN, USA, 1994; pp. 12–14.
9. Jacome, L.H.; Schuh, W.; Stevenson, R.E. Effect of temperature and relative humidity on germination and germ tube development of *Mycosphaerella fijiensis* var. *difformis*. *Phytopathology* **1991**, *81*, 1480–1485. [CrossRef]
10. Fullerton, R.A.; Casonato, S.G. The infection of the fruit of 'Cavendish' banana by *Pseudocercospora fijiensis*. cause of black leaf streak (black Sigatoka). *Eur. J. Plant Pathol.* **2019**, *155*, 779–787. [CrossRef]
11. Olivares, B.O.; Vega, A.; Calderón, M.A.R.; Rey, J.C.; Lobo, D.; Gómez, J.A.; Landa, B.B. Identification of Soil Properties Associated with the Incidence of Banana Wilt Using Supervised Methods. *Plants* **2022**, *11*, 2070. [CrossRef] [PubMed]
12. Dita, M.; Barquero, M.; Heck, D.; Mizubuti, E.S.; Staver, C.P. Fusarium wilt of banana: Current knowledge on epidemiology and research needs toward sustainable disease management. *Front. Plant Sci.* **2018**, *9*, 1468. [CrossRef]
13. Olivares, B.O.; Paredes, F.; Rey, J.C.; Lobo, D.; Galvis-Causil, S. The relationship between the normalized difference vegetation index. rainfall. and potential evapotranspiration in a banana plantation of Venezuela. *ST-JSSA* **2021**, *18*, 58–64. [CrossRef]
14. Churchill, A.C. *Mycosphaerella fijiensis*. the black leaf streak pathogen of banana: Progress towards understanding pathogen biology and detection. disease development. and the challenges of control. *Mol. Plant Pathol.* **2011**, *12*, 307–328. [CrossRef]
15. Wang, Y.; Chee, M.C.; Edher, Z.; Hoang, M.D.; Fujimori, S.; Kathirgamanathan, S.; Bettencourt, J. Forecasting black sigatoka infection risks with latent neural odes. *arXiv* **2020**, arXiv:2012.00752. [CrossRef]
16. Ochoa, A.; Abaunza, F.; Rey, V. Forecasting black sigatoka in banana crops with stochastic models. In Proceedings of the VI International Banana Congress CORBANA, Miami, FL, USA, 19–20 April 2016.
17. Ganry, J.; de Bellaire, L.D.L.; Mourichon, X. A biological forecasting system to control Sigatoka disease of bananas and plantains. *Fruits* **2008**, *63*, 381–387. [CrossRef]
18. de Hernández, R.M.A.G.; Rodríguez, V.; Caraballo, E.A.H. Predicción del rendimiento de un cultivo de plátano mediante redes neuronales artificiales de regresión generalizada. *Publ. Cienc. Tecnol.* **2012**, *6*, 31–40.
19. Soares, J.D.R.; Pasqual, M.; Lacerda, W.S.; Silva, S.O.; Donato, S.L.R. Utilization of artificial neural networks in the prediction of the bunches' weight in banana plants. *Sci. Hortic.* **2013**, *155*, 24–29. [CrossRef]
20. Ye, H.C.; Huang, W.J.; Huang, S.Y.; Cui, B.; Dong, Y.Y.; Guo, A.T.; Ren, Y.; Jin, Y. Identification of banana fusarium wilt using supervised classification algorithms with UAV-based multi-spectral imagery. *Int. J. Agric. Biol.* **2020**, *13*, 136–142. [CrossRef]
21. Selvaraj, M.G.; Vergara, A.; Montenegro, F.; Ruiz, H.A.; Safari, N.; Raymaekers, D.; Ocimati, W.; Ntamwira, J.; Tits, L.; Omondi, A.B.; et al. Detection of banana plants and their major diseases through aerial images and machine learning methods: A case study in DR Congo and Republic of Benin. *ISPRS* **2020**, *169*, 110–124. [CrossRef]
22. Olivares, B.O.; Araya-Alman, M.; Acevedo-Opazo, C.; Rey, J.C.; Cañete-Salinas, P.; Kurina, F.G.; Balzarini, M.; Lobo, D.; Navas-Cortés, J.A.; Landa, B.B.; et al. Relationship between soil properties and banana productivity in the two main cultivation areas in Venezuela. *J. Soil Sci. Plant Nutr.* **2020**, *20*, 2512–2524. [CrossRef]
23. Sangeetha, T.; Lavanya, G.; Jeyabharathi, D.; Rajesh Kumar, T.; Mythili, K. Detection of pest and disease in banana leaf using convolution Random Forest. *Test Eng. Manag.* **2020**, *83*, 3727–3735.
24. Hou, J.C.; Hu, Y.H.; Hou, L.X.; Guo, K.Q.; Satake, T. Classification of ripening stages of bananas based on support vector machine. *Int. J. Agric. Biol. Eng.* **2015**, *8*, 99–103. [CrossRef]
25. Sabilla, I.; Wahyuni, C.S.; Fatichah, C.; Herumurti, D. Determining banana types and ripeness from image using machine learning methods. In Proceedings of the 2019 International Conference of Artificial Intelligence and Information Technology (ICAIIT), Ouargla, Algeria, 13–15 March 2019; IEEE: New York, NY, USA, 2019; pp. 407–412.
26. Montenegro, E.; Pitti, J. Analysis of climate risks in the indigenous food system The Teribe, Panama. *Acta Nova* **2020**, *9*, 713–736.
27. Stover, R.H. A proposed international scale for estimating intensity of Banana leaf spot (*Mycosphaerella musicola* Leach). *Trop. Agric. Trinidad Tobago* **1971**, *48*, 185–196.
28. Gauhl, F. *Epidemiología y Ecología de la Sigatoka Negra (Mycosphaerella fijiensis, Morelet) en Plátano (Musa sp.), en Costa Rica*; Trad; de Alemán, J.E., Ed.; UPEB: Panama City, Panama, 1990; p. 126.

29. Gauhl, F.; Pasberg-Gauhl, C.; Jones, D.R. Disease cycle and epidemiology. In *Diseases of Banana. Abacá and Enset*; Jones, D.R., Ed.; CAB International: Wallingford, UK, 2000; pp. 56–62.

30. Fouré, E. *Black Leaf Streak Disease of Bananas and Plantains (Mycosphaerella fijiensis Morelet). Study of the symptoms and Stages of the Disease in Gabon*; IRFA-CIRAD: Paris, France, 1985; p. 20.

31. Hernández, L.; Hidalgo, W.; Linares, B.; Hernández, J.; Romero, N.; Fernández, S. Surveillance and forecast preliminary study for black sigatoka (*Mycosphaerella fijiensis* Morelet) disease in *Musa* AAB cv Hartón plantain crop in "Macagua-Jurimiquire" Yaracuy state. *Rev. Fac. Agron.* **2005**, *22*, 325–329.

32. R Core Team. *R: A Language and Environment for Statistical Computing*; R Foundation for Statistical Computing: Vienna, Austria, 2020.

33. Breiman, L.; Last, M.; Rice, J. Random forests: Finding quasars. In *Statistical Challenges in Astronomy*; Feigelson, E., Jogesh, G., Eds.; Springer: New York, NY, USA, 2003; pp. 243–254. [CrossRef]

34. Breiman, L. Random forests. *Mach. Learn.* **2001**, *45*, 5–32. [CrossRef]

35. Strobl, C.; Boulesteix, A.L.; Zeileis, A.; Hothorn, T. Bias in random forest variable importance measures: Illustrations, sources, and a solution. *BMC Bioinform.* **2007**, *8*, 1–25. [CrossRef] [PubMed]

36. Han, H.; Guo, X.; Yu, H. Variable selection using mean decrease accuracy and mean decrease gini based on random forest. In Proceedings of the 2016 7th IEEE International Conference on Software Engineering and Service Science (ICSESS), Beijing, China, 26–28 August 2016; pp. 219–224. [CrossRef]

37. Liaw, A.; Wiener, M. Classification, and regression by random Forest. *R News* **2002**, *2*, 18–22.

38. Jeong, J.H.; Resop, J.P.; Mueller, N.D.; Fleisher, D.H.; Yun, K.; Butler, E.E.; Timlin, D.J.; Shim, K.M.; Gerber, J.S.; Reddy, V.R.; et al. Random forests for global and regional crop yield predictions. *PLoS ONE* **2016**, *11*, e0156571. [CrossRef] [PubMed]

39. Arcila, M.; Giraldo, G.; Duarte, J. *Influencia de Las Condiciones Ambientales Sobre Las Propiedades Físicas y Químicas Durante La Maduración del Fruto de Plátano Dominico-Hartón (Musa AAB Simonds) en la Zona Cafetera Central*; Cayón, G., Ed.; Poscosecha y Agroindustria del Plátano en el eje Cafetero de Colombia; Universidad del Quindío: Quindío, Colombia, 2000; pp. 101–124.

40. Torres, N.; Hernández, J. Efecto del número de hojas en el desarrollo del racimo de plátano Hartón *Musa* AAB. *Agroalim. Des. Sost.* **2004**, *5*, 17–22.

41. Barrera, J.; Cayón, G.; Robles, J. Influence of leaf and fruit epicarp exposition on development and quality of 'Hartón' plantain (*Musa* AAB Simmonds) bunch. *Agron. Colomb.* **2009**, *27*, 73–79.

42. Barrera, V.J.; Barraza, A.F.; Campo, A. Shadow effect on black sigatoka (*Mycosphaerella fijiensis* Morelet) in plantain cultivation cv harton (*Musa* AAB Simmonds). *Rev. UDCA Actual. Divulg. Cient.* **2016**, *19*, 317–323. [CrossRef]

43. Belalcázar, S.; Merchán, V.; Mayorga, M. *Control de Enfermedades. El Cultivo Del Plátano en el Trópico*; Belalcázar, S., Ed.; Feriva: Bogota, Colombia, 1991; pp. 241–297.

44. Cayón, G. Evolución de la fotosíntesis, transpiración y clorofila durante el desarrollo de la hoja de plátano (*Musa* AAB Simmonds). *Infomusa* **2001**, *10*, 12–15.

45. Siles, P.; Bustamante, O.; Valdivia, E.; Burkhardt, J.; Staver, C. Photosynthetic performance of banana (Gross Michel, AAA) under a natural shade gradient. *Acta Hortic.* **2013**, *986*, 71–77. [CrossRef]

46. Porras, A.; Hernández, A.; Pérez, L. Epidemiología de la Sigatoka Negra (*Mycosphaerella fijiensis* Morelet) en Cuba. II. Pronóstico Bio-Climático de los Tratamientos contra la Enfermedad en Plátanos (*Musa* spp. AAB). *Rev. Mex. Fitopatol.* **2000**, *18*, 27–35.

47. Gauhl, F.; Pasberg-Gauhl, C. Epidemiology of black sigatoka disease on plantain in Nigeria. *Phytopathology* **1994**, *84*, 1080.

48. Mobambo, K.N.; Gauhl, F.; Pasberg-Gauhl, C.; Zuofa, K. Season and plant age effect evaluation of plantain for response to black sigatoka disease. *Crop Prot.* **1996**, *15*, 609–614. [CrossRef]

49. Nava, C.; Vera, J. Relation of leaves numbers a florewing time and wasted at reproductive cycle with bunch weight in plantain bunch plants under black Sigatoka attack. *Rev. Fac. Agron.* **2004**, *21*, 335–342.

50. Rodríguez, P.; Cayón, G. Efecto de *Mycosphaerella fijiensis* sobre la fisiología de la hoja de banano. *Agron. Colomb.* **2008**, *26*, 256–265.

51. Cedeño-Zambrano, J.R.; Díaz-Barrios, E.J.; de Jesús Conde-López, E.; Cervantes-Álava, A.R.; Avellán-Vásquez, L.E.; Zambrano-Mendoza, M.E.; Sánchez-Urdaneta, A.B. Evaluación de la severidad de Sigatoka negra (*Mycosphaerella fijiensis* Morelet) en platano "Barraganete" bajo fertilización con magnesio. *Rev. Tec.* **2021**, *44*, 4–12. [CrossRef]

52. Villalta, R.; Guzmán, M. Capacidad de esporulación de *Mycosphaerella fijiensis* en tejido foliar de banano depositado en el suelo y efecto antiesporulante de la urea. *Corbana* **2005**, *31*, 41–43.

53. Etebu, E.; Young, W. Control of black sigatoka disease: Challenges and prospects. *Afr. J. Agric. Res.* **2011**, *6*, 508–514. [CrossRef]

54. Vázquez-Euán, R.; Chi-Manzanero, B.; Hernández-Velázquez, I.; Tzec-Simá, M.; Islas-Flores, I.; Martínez-Bolaños, L.; Garrido-Ramírez, E.R.; Canto-Canché, B. Identification of new hosts of *Pseudocercospora fijiensis* suggests innovative pest management programs for black sigatoka disease in banana plantations. *Agronomy* **2019**, *9*, 666. [CrossRef]

55. Gómez-Correa, J.C.; Torres-Aponte, W.S.; Cayón-Salinas, D.G.; Hoyos-Carvajal, L.M.; Castañeda-Sánchez, D.A. Modelación espacial de la Sigatoka negra (*Mycosphaerella fijiensis* M. Morelet) en banano cv. Gran Enano. *Rev. Ceres* **2017**, *64*, 47–54. [CrossRef]

56. Bebber, D.P. Climate change effects on Black Sigatoka disease of banana. *Philos. Trans. R. Soc. B* **2019**, *374*, 20180269. [CrossRef]

57. Grömping, U. Variable importance assessment in regression: Linear regression versus random forest. *Am. Stat.* **2009**, *63*, 308–319. [CrossRef]

 *sustainability*

Article

Assessing the Plant Health System of Burundi: What It Is, Who Matters and Why

Willis Ndeda Ochilo [1,*], Stefan Toepfer [2], Privat Ndayihanzamaso [3], Idah Mugambi [1], Janny Vos [4] and Celestin Niyongere [3]

1   CABI, Canary Bird, 673 Limuru Road, Muthaiga, Nairobi P.O. Box 633-00621, Kenya
2   CABI, Rue des Grillons 1, CH-2800 Delemont, Switzerland
3   Institut des Sciences Agronomiques du Burundi (ISABU), Avenue de la Cathedrale Regina Mundi, Bujumbura P.O. Box 795, Burundi
4   CABI, Egham TW20 9TY, Surrey, UK
*   Correspondence: w.ochilo@cabi.org

Abstract: The concept of a plant health system (PHS) is mainly anchored on experiences from human health where varied sources of knowledge, expertise, and technology are combined to provide healthcare. While diverse human health systems have been proven, little is known about PHS and what is needed to base effective plant healthcare services. A stakeholder analysis was carried out in Burundi. The aim is to understand the system as it is presently and to identify constraints and opportunities. This paper reports on the process and results of this assessment. The initial step in this process was to define PHS and its functions and to evaluate stakeholders' interests and influence. The first step was followed by examining stakeholders' perceptions concerning the sustainability of interventions geared at strengthening PHS functions. The process included a document review and stakeholder workshops. After the stakeholders defined the PHS functions, they proceeded to identify valuable actors. The assessment process highlighted several key challenges, including inadequate skills to serve farmers and insufficient capacity to diagnose pests, as significant impediments to effective PHS performance. Based on the information marshalled here, seven broad interventions are proposed for practitioners to strengthen Burundi's PHS rapidly.

Keywords: plant health system; stakeholder analysis; plant health problems

Citation: Ochilo, W.N.; Toepfer, S.; Ndayihanzamaso, P.; Mugambi, I.; Vos, J.; Niyongere, C. Assessing the Plant Health System of Burundi: What It Is, Who Matters and Why. *Sustainability* **2022**, *14*, 14293. https://doi.org/10.3390/su142114293

Academic Editors: Barlin Orlando Olivares Campos, Edgloris E. Marys and Miguel Araya-Alman

Received: 2 July 2022
Accepted: 26 October 2022
Published: 1 November 2022

1. Introduction

Burundi, a landlocked country in the Great Lakes Region of central-eastern Africa, occupies an area of approximately 27,834 km^2 [1]. The total population was estimated at 11.89 million in 2020 (463 inhabitants/km^2), making it the fourth most densely populated country in Africa [2]. The economy of Burundi is driven mainly by agriculture, with the sector contributing close to 40% of the gross domestic product (GDP), employing 84% of the working population, and providing 95% of the food supply [3–5]. Furthermore, the sector accounts for more than 90% of foreign exchange earnings and is the leading supplier of raw materials for the agro industry. The portion of the country's land area appropriated for agriculture is 73.3%, of which 38.9% is arable land (though prone to erosion and low soil fertility). Average land ownership is 0.27 ha per household, which is well below the 0.90 ha thought as the minimum for economic viability [6].

Agricultural productivity in Burundi is impeded by several constraints, including outbreaks of crop pests and diseases [7], occasional droughts and floods [8,9], limited cash for inputs (fertilizers, plant protection products, productive seeds, and planting materials), land fragmentation [10], inefficient use of the existing water resources [11], lack of cash and credit facilities among smallholder farmers [12], and limited access to research and extension services [8,13].

Despite the challenges mentioned above, there are considerable opportunities to increase farm production and food security. The climatic conditions in most parts of the country favor agriculture [14]. There is usually enough rain, and the countryside is diverse to sustain different crops. The country is endowed with abundant water resources [15] and potential irrigable lands, which are yet underdeveloped. The emergence of farmers' organizations (e.g., the National Confederation of Coffee Growers' association, cooperatives at the colline level, etc.), although still developing, is demonstrating the potential role of farmers' organizations [16]. The government of Burundi has also developed policies in support of agriculture [17,18].

To respond to the challenges impeding agricultural productivity in Burundi, CABI, with funding support from the Embassy of the Netherlands in Bujumbura and Nuffic, is introducing its award-winning Plantwise program in Burundi [19]. Plantwise is a global program led by CABI, which supports farmers to lose less of what they grow to plant health problems. The program has helped millions of farmers in various countries to identify and manage threats posed by pests at the farm level and to minimize crop losses by strengthening in-country plant health systems (PHS) [20].

Plantwise in Burundi will address some of the main challenges in the country's agricultural sector (mainly pests and diseases). The strategies proposed for managing plant health problems include setting up an early-warning system for communicating pest threats and strengthening the capacity of farmers' organizations in pest management [19].

A stakeholder analysis was carried out in 2021 to investigate whether the proposed Plantwise project can achieve its aims of strengthening the PHS of Burundi. The objective is to understand the system as it is presently and to identify challenges and opportunities from the perspective of various actors. Stakeholder analyses are crucial for identifying significant stakeholders and their particular sets of interests, power/policy influence, and roles [21]. In addition, stakeholder analyses allow the exploration of the individual, group, and institutional landscape amongst pertinent stakeholders, their associations, what concerns them the most, and how these could affect the project [22].

The specific objectives of the Plantwise Burundi stakeholder analysis included: (1) to examine the PHS of Burundi; (2) to evaluate and map stakeholders' interests and influence concerning the functions of the PHS; (3) to assess stakeholders' perceptions concerning the success and sustainability of the proposed project's broad work areas. This paper reports on the process and results of this stakeholder analysis.

2. Methods

The analysis of the PHS of Burundi and its key stakeholders, undertaken from March to December 2021, was structured in four major phases: (i) a document review, (ii) a stakeholder workshop, (iii) a needs assessment workshop, and (iv) a stakeholder forum meeting.

2.1. Document Review

The assessment partially understood Burundi's plant health system and functioning through a literature review. Information on contextual influences is often available in grey literature [23,24]. Where the assessors could not obtain information from the CAB Abstracts database, a search of the grey literature was carried out in March 2021 within the open-access collections on Google™ (Menlo Park, CA, USA). The target of the search was contemporary reports on Burundi focusing on institutional structures, politics and organizational culture, development partners' influences, and local and international strategies. Additionally, the investigators obtained key statistics relating to Burundi within the databases FAOSTAT, a widely used database in peer-reviewed literature as the base for many global agriculture perspectives analyses [25], and the World Bank Open Data, an open-access collection on global development data. The investigators used universal terms associated with agriculture to search for country-specific reports. These terms were: food security, plant health, crop protection, crop production, agricultural extension, farmer advisory services, and agricultural policy. From the documents retrieved, the investigators searched reference lists for other sources. A total of 15 documents were included: (1) original research and research reviews (*n* = 10); (2) government policy (*n* = 3); and (3) international organizations strategic plans (*n* = 2). The assessors present these documents in Table 1.

Table 1. Key documents reviewed on the contextual influences of the plant health system of Burundi.

Documents Reviewed
Government policy
Republique du Burundi Portant Organisation Du Ministere De L'environment, De L'agrculture et de L'elevage; Burundi, 2020
Republique du Burundi Plan National de Developpement Du Burundi—PND Burundi 2018–2027; 2018
Republique du Burundi Vision Burundi 2025; 2011
Original research and reviews
Bamber, P.; Gereffi, G. Burundi in the Agribusiness Global Value Chain: Skills for Private Sector Development; 2014
IFPRI Burundi: Recent Developments in Public Agricultural Research; 2011
Niragira, S. Feasibility Study for the Development of Public-Private Seed Delivery Systems in Burundi; 2020;
Chauvin, N.D.; Mulangu, F.; Porto, G. Food Production and Consumption Trends in Sub-Saharan Africa: Prospects for the Transformation of the Agricultural Sector; 2012
Curtis, M. Improving African Agriculture Spending: Budget Analysis of Burundi, Ghana, Zambia, Kenya and Sierra Leone; 2013
Ludgate, N.S.; Tata, J.S. Integrating Gender and Nutrition within Agricultural Extension Services: Burundi Landscape Analysis; 2015
Niragira, S.; Ndikumana, D.; Muvunyi, R.; Nzigamasabo, D.; Nduwimana, A. Situational Analysis of Food Safety Control Systems in Burundi; 2020
Kinuthia, R.; Kiptot, E.; Nkurunzinza, C. The Extension System in Burundi: Kayanza Province, Muruta Commune; 2016
FAO FAOSTAT Database on Agriculture. Available online: https://www.fao.org/faostat/en/#data
The World Bank Burundi Available online: https://data.worldbank.org/country/BI
International organizations strategic plans
IFAD Republic of Burundi: Country Strategic Opportunities Programme (2022–2027); 2022
WFP Burundi Interim Country Strategic Plan (2018–2020); 2018

However, policy documents and institutional frameworks do not always mirror the reality on the ground. Consequently, the document review only served as a basis for understanding the system. At the same time, the workshops were essential for gaining a deeper understanding of Burundi's overall plant health system.

2.2. Stakeholder Workshop

The investigators convened stakeholders in March for a 2-day stakeholder workshop. The participants numbering 25 were identified based on document review, snowball technique where stakeholders identified additional stakeholders, and the result of brainstorming amongst the project team members to determine: (i) groups of stakeholders and how they could be affected by the project's interventions, (ii) who are the individuals expected to be the direct beneficiaries of the project interventions, and (iii) the potential impact of the project interventions on stakeholders, including their numbers. The investigators then prioritized the following groups of stakeholders: The Ministry of Environment, Agriculture, and Livestock (MINEAGRIE), institutions of higher learning, provincial leadership, development partners, non-governmental organization (NGOs), international organizations, farmer associations and cooperatives, agricultural input supply, and other private sector players.

The workshop focused on: defining the PHS of Burundi; stakeholder roles, responsibilities, and mandates in delivering PHS functions; nature of interactions between stakeholders; and policies, politics, and institutions that influence those interactions.

Based on the functions described in the literature [26], the PHS functions of Burundi were examined and defined through quick brainstorming and group discussions. After the investigators delimited the functions, they identified the activities/roles involved per each function and the stakeholders carrying out those activities/roles. After identifying stakeholders and their actions, the investigators assessed how much interest and influence each stakeholder has regarding PHS functions. The investigators requested the participants to draw a box/circle in flipcharts in the middle with the name of the PHS function [26]. Investigators asked participants around the PHS function to draw a circle for each stakeholder involved, with the explanation that the length from the central function ought to show the interest that the stakeholder has in the process, and the size of the circle ought to show the influence that a stakeholder has on the PHS function. The closer the stakeholder circle is to the central function, the higher their interest level in the process. The larger the stakeholder circle, the more influence the stakeholder wields.

The assessors discussed the findings in plenary to highlight points of departure, the things that stand out, associations, and stakeholders who appeared isolated. Afterward, critical stakeholders—stakeholders with high interest and a significant influence (larger circles close to the center)—were identified for each PHS function. Identification of essential stakeholders was made by reviewing the diagrams drawn by participants for each function and deciding where the stakeholders fall within the following categories [26]:

- High interest, significant influence—critical
- High interest, low influence—important
- Low interest, significant influence—optional
- Low interest, low influence—irrelevant

Finally, the participants focused on how the different stakeholders identified above interact with each other within the PHS as a whole and explored the strengths and weaknesses of the critical stakeholder interactions.

2.3. Needs Assessment Workshop

The investigators undertook in June a 1-day needs assessment workshop. The aim of the workshop, which was an extension of the stakeholder workshop conducted in March, was to examine the strengths, weaknesses, and needs of the diverse functions of the PHS of Burundi and identify opportunities for interventions. The needs assessment workshop involved 15 in-country experts drawn from the same organizations, earlier engaged in the stakeholder workshop. As a kick-off to the meeting, the investigators presented an overview of the PHS functions. Subsequently, for each PHS function, the participants discussed the country's strengths and weaknesses regarding the same. Afterward, the researchers engaged the participants to discuss and agree by consensus on fundamental needs (gaps) that the project can realistically address within the available time and resources.

2.4. Stakeholder Forum Meeting

The researchers assembled influential stakeholders of the PHS of Burundi in a national forum conducted in November and attended by 28 representatives from the same organizations involved in the stakeholder and needs assessment workshops. In the course of the meeting, findings from the stakeholder and needs assessment workshops were presented by researchers and validated by participants—meaning the participants compared results, identified areas of commonality and divergence, and determined how best to move forward. The workshop included presentations, work within small groups, a section on questions and answers, and a plenary discussion.

3. Results

3.1. Policy and Institutional Framework of the Plant Health System of Burundi

In tandem with the country's Vision 2025, the ambition of the government of Burundi is to transform the country into an emerging country by 2040 through, among others, reduction in poverty, reduction of population pressure and improvement of agricultural development, and increase in the country's GDP. Coupled with Vision 2025, the country's National Development Plan (NDP) (2018–2027) aims to fundamentally change the country's economy, with the modernization of agriculture and attainment of food self-sufficiency at the heart of this objective. Within the framework of NDP, the government of Burundi in April 2021 adopted the National Program for the Capitalization of Peace, Social Stability, and the Promotion of Economic Growth (NPCP-SS-PEG) [6]. In addition, the MINEAGRIE produced an environmental, agricultural, and livestock policy (DOPEAE) (2020–2027). Other relevant national policies/strategies include: (i) multisectoral strategic plan for food security and nutrition (2019–2023); (ii) national action plan (2017–2021) for implementation of United Nations Security Council resolution 1325 for women, peace, and security; (iii) national gender policy (2012–2025); (iv) action plan for youth employment (2021–2024); (v) national health development plan (2021–2025); (vi) investment code (2008); (vii) national road map for strengthening food systems; (viii) national employment policy (2016–2025); and (ix) national strategy for financial inclusion [6].

In terms of institutional arrangement, the Ministry of Environment, Agriculture, and Livestock (MINEAGRIE) essentially coordinates agriculture in Burundi by overseeing agricultural development programs in the country and developing agricultural policies [6,18]. The ministry comprises seven general directorates (Figure 1): (i) the General Directorate of Agricultural and Livestock Planning—which supports the development of projects and programs within the sector; (ii) the General Directorate of Mobilization for Self-Development and Agricultural Extension (DGMAVAE)—develops methods and approaches to extension and designs and organizes training for various officials; (iii) the General Directorate of Agriculture—promotes the development of the agricultural sector, including seed, and policy formulation; (iv) the General Directorate of Livestock—promotes the development of the livestock sector; (v) the General Directorate of Land use planning, irrigation, and protection of land assets; (vi) the General Directorate of Environment, Water Resources and Sanitation; and (vii) the General Directorate of Human Resources. Additionally, MINEA-GRIE oversees various national institutions, including Institut des Sciences Agronomiques du Burundi (ISABU)—responsible for agricultural research; Centre National de Technologie Alimentaire (CNTA)—supports food processing technical innovations; and Autorité de Régulation de la Filière Café au Burundi (ARFIC)—regulatory authority for coffee [18].

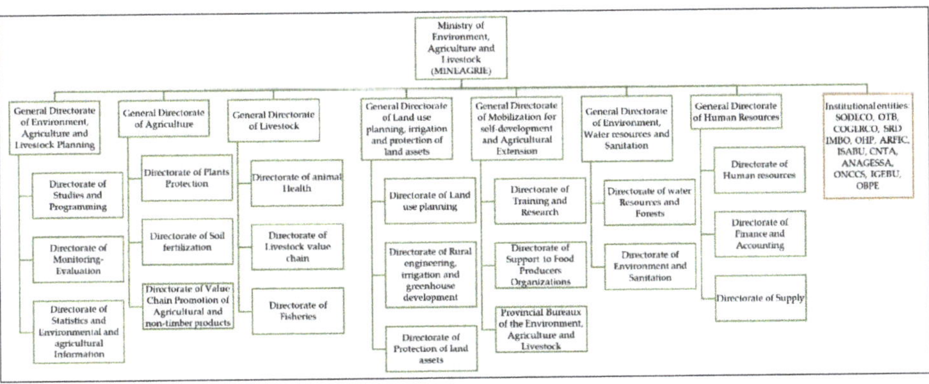

Figure 1. Structure of the Ministry of Environment, Agriculture, and Livestock of Burundi (MINEAGRIE).

3.2. Functions of the Plant Health System of Burundi

The concept of PHS is based on lessons and experiences from human health, where varied sources of knowledge, expertise, and technology are combined to provide healthcare [27]. In human health, in an attempt to advance a universal understanding of what a health system is and the basis for health system strengthening, the World Health Organization (WHO) defined, based on functions, a discrete number of "building blocks" making up the system [28].

While diverse human health systems have been advanced, proven, and studied over time, little is known regarding PHS and what is needed to base effective plant healthcare services. A step towards addressing this gap is a need to identify and define the structure and functioning of a PHS. However, such functions would not represent a 'universal truth' as a country's perspective on things sways them.

During the discussions at the stakeholder workshop, the participants confirmed that the PHS functions pre-defined in [26] (namely advisory services, information management, diagnostic services, policy and regulation, input supply, and research and technology development) were accurate and appropriate for Burundi. In addition, the participants highlighted two more PHS functions (crop production and agricultural training), bringing the total number of identified PHS functions for Burundi to eight (8):

1. Farmer advisory (extension) services. A complete set of institutions facilitating and supporting individuals engaged in agricultural production to solve problems and access skills, information, and technologies to enhance their livelihoods.
2. Plant health information management. Structures and approaches for availing plant health information to diverse actors at various levels.
3. Diagnostic services. Expertise and facilities required for identifying plant health problems, including pests, diseases, nutrient deficiencies, and soil health problems
4. Research and technology development. Process of invention and innovation to establish better practices and to implement technology as an added value
5. Agricultural input supply. Provision of agricultural inputs (seeds, fertilizer, pest control products, etc.)
6. Policy, regulation, and control. The mechanism through which government can manage the sector to achieve sustainable outcomes
7. Crop production. The process of growing crops for subsistence and commercial purposes
8. Agricultural training. Developing skills and knowledge relating to agriculture.

While looking at the PHS, one should consider the system in terms of its functions (e.g., service delivery, training, research, etc.) and, more crucially, its interrelations.

3.3. Stakeholders Involved in the Functions of the Plant Health System of Burundi

Stakeholders can be defined as individuals and organizations concerned with a particular activity owing to their participation in generating, consuming, regulating, managing, or evaluating the same [29]. While looking at the PHS of Burundi, one should include the institutional or supply side of the PHS and the individuals making up the system (population). In a progressive view, the population is not an extraneous recipient of the system; it is an integral part of it. This understanding of the population is because, when it comes to plant health, persons play at least two roles: (i) as farmers with specific needs requiring care; and (ii) as consumers with expectations.

After the investigators defined the PHS functions, they identified twenty (20) stakeholders to be helpful to engage when it comes to various PHS functions (Table 2). Most of the stakeholders were from the government sector (65%), followed by NGOs/International agencies/development partners/civil society (15%), farmers/farmers associations/farmers cooperatives, academia, consumers, and the private sector—all at 5% each. The researchers provide the interactions between the stakeholders in Figure 2.

The government of Burundi plays a significant role in the PHS of the country (Table 3). It sets policies and standards, supports plant health services research, passes laws and regulations, and provides technical assistance and resources to local PHS. In addition, the government helps to finance plant health services, provides protection against international plant health threats, undertakes surveillance to establish status and plant health needs, and supports international efforts towards better plant health (Table 3).

Table 2. Importance of stakeholders in each function of the plant health system of Burundi according to 68 participants involved in three workshops.

Critical to Engage	Important to Engage	Optional to Engage
Farmer advisory (extension) services		
- General Directorate of Mobilization for Self-Development and Agricultural Extension (DGMAVAE) - Bureau Provinciale de L'Environnement, de l'Agriculture et de l'Elevage (BPEAE) - Directorate of Plant Protection (DPV) - Institut des Sciences Agronomiques du Burundi (ISABU)	- Agricultural input suppliers - Farmers, farmer associations & cooperatives - NGOs - Institutions of higher learning - Ministry of Finance - Office National de Contrôle et de Certification des Semences (ONCCS)	- Consumers
Plant health information management		
- Communication unit of Ministry of Environment, Agriculture and Livestock (MINEAGRIE) - Department of Statistical and Agricultural Information (DSIA)	- Institut de Statistiques et d'etudes Economiques du Burundi (ISTEEBU) - NGOs, international organizations, development partners (e.g., FAO) - ISABU - Institutions of higher learning - BPEAE - Agricultural input suppliers - Ministry of Finance	- Farmers, farmer associations, and cooperatives
Diagnostic services		
- ISABU - DPV - Farmers, farmer associations, and cooperatives - BPEAE - DGMAVAE	- Institutions of higher learning - Agricultural input suppliers - ONCCS - Ministry of Finance	- NGOs, international organizations, development partners
Research and technology development		
- ISABU - DPV	- NGOs, International Organization, Development Partners - Farmers, farmer associations & cooperatives - Institutions of higher learning - BPEAE - Ministry of Finance	- Agricultural input suppliers - Consumers
Agricultural input supply		
- Agricultural input suppliers - Collectif des Compagnies et Coopératives de Production des Semences du Burundi (COPROSEBU) - DPV	- Farmers, farmer associations, and cooperatives - BPEAE - ONCCS - Ministry of Finance	- ISABU - NGOs, international organizations, development partners (e.g., FAO) - Institutions of higher learning
Policy, regulation and control		
- Cabinet of MINEAGRIE - DPV	- National assembly - NGOs, international organizations, development partners (e.g., FAO, IPPC) - BPEAE - Agricultural input suppliers - Farmers, farmer associations, and cooperatives - ISABU	- Ministry of Finance - Institutions of higher learning

Table 2. *Cont.*

Critical to Engage	Important to Engage	Optional to Engage
Crop production - Farmers, farmer associations, and cooperatives - General Directorate of Agriculture	- Agricultural input suppliers - Institutions of higher learning - Consumers - NGOs, international organizations, development partners - ISABU	
Agricultural training - Institutions of higher learning - Ministry of Education	- BPEAE - Farmers, farmers association, and cooperatives - DPV - Agricultural input suppliers	- NGOs, International organization, development partners (e.g., IFDC, INADES, IFAD) - Ministry of Finance - ISABU

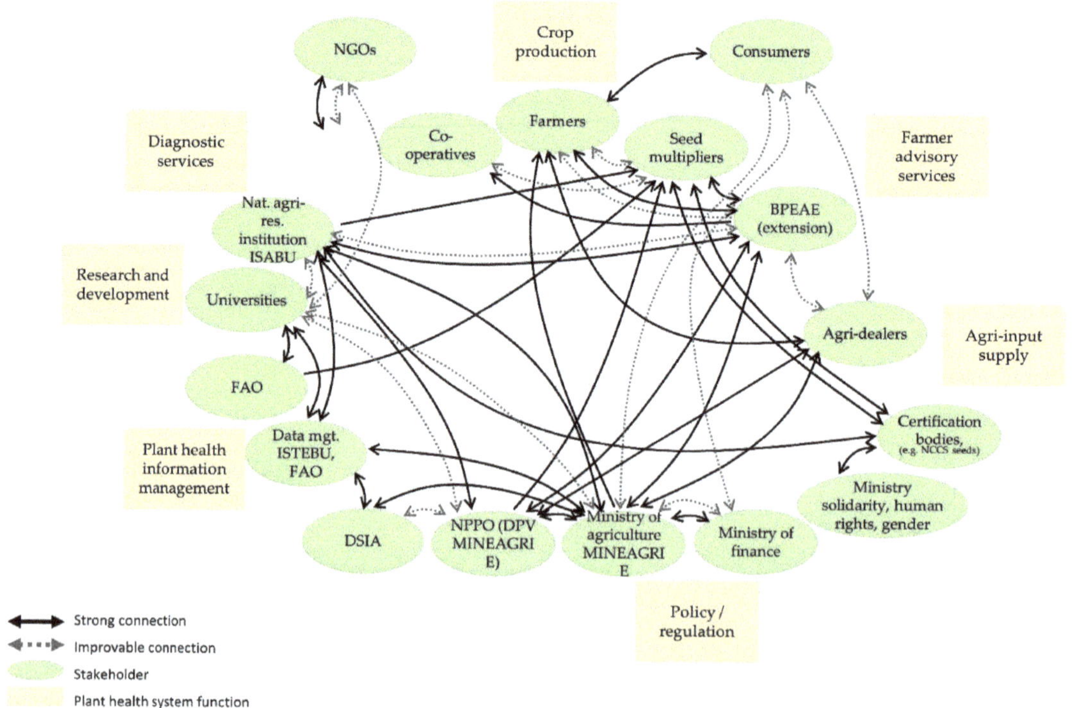

Figure 2. Interactions among stakeholders in the plant health system of Burundi identified by 68 participants involved in three workshops.

Table 3. Role of plant health system stakeholders in each function of the plant health system of Burundi according to 68 participants involved in three workshops.

Key Actors	Role
Farmer advisory (extension) services	
(a) General Directorate of Mobilization for Self-Development and Agricultural Extension (DGMAVAE)	- Promotion of extension service delivery - Coordination of extension services nationally
(b) Bureau Provinciale de L'Environnement, de l'Agriculture et de l'Elevage (BPEAE)	- Coordination of extension activities in the provinces
(c) Institut des Sciences Agronomiques du Burundi (ISABU)	- Provision of agricultural extension services
(d) Farmers, farmers association and cooperatives	- Direct beneficiary of agricultural advisory services
Plant health information management	
(a) Department of Statistical and Agricultural Information (DSIA)	- Collection, analysis, and dissemination of agricultural statistical information
(b) Communication unit of Ministry of Environment, Agriculture and Livestock (MINEAGRIE)	- Posting media updates and engaging the public on agricultural matters
Diagnostic services	
(a) ISABU	- Soil analyses and plant health problems diagnostics
(b) DPV	- Diagnosis of plant health problems on samples brought by farmers
(c) Farmers, farmers association, and cooperatives	- Monitoring and reporting of plant health problems
Research and technology development	
(a) ISABU	- Main agricultural research agency in Burundi - Application of the research findings, technologies, and innovations
(b) Office National de Contrôle et de Certification des Semences (ONCCS)	- Research
(c) Institutions of higher learning (e.g., universities)	- Research
Agricultural input supply	
(a) DPV	- Quality assurance of agricultural inputs
(b) ISABU	- Seed variety development
(c) ONCCS	- Seed certification
(d) Collectif des Compagnies et Coopératives de Production des Semences du Burundi (COPROSEBU)	- National seed trade association - Seed production and processing - Seed production - Linking farmers with other actors in the agricultural input supply chain
(e) Agri-input supply and other private sector players	- Providing farmers with farm inputs (pesticides, seeds, fertilizer, and tools)
Policy, regulation, and control	
(a) Cabinet of MINEAGRIE	- Policy formulation - Plant protection regulation - Enforcement of sanitary and phytosanitary measures
(b) DPV	- Liaison office for International Plant Protection Convention (IPPC)
(c) Confédération des Associations des Producteurs Agricoles pour le Développement (CAPAD)	- Lobbying and advocacy for pro-farmer policies.
Crop production	
(a) MINEAGRIE	- Policy formulation, coordination, and mobilization of funds
(b) General Directorate of Agriculture	- Provision of the basics for the population's food needs - Adoption of new farming technologies, - Suppliers of agricultural credit and inputs to their members
(c) Farmers associations and cooperatives	- Seed production and processing
Agricultural training	
(a) Institutions of higher learning	- Education, training
(b) Ministry of Education	- Education, training
(c) CAPAD	- Training and provision of extension services to farmers

Currently, the primary government body responsible for Burundi plant health is MINEAGRIE. Other ministries also have agencies with responsibilities related to plant health, such as the Institute of Statistics and Economic Studies of Burundi (ISTEEBU) in the Ministry of Finance, Budget, and Economic Planification (MFBPE); technical colleges and universities in the Ministry of Education and Vocational and Professional Training; and the Ministry of National Solidarity, Human Rights, and Gender. Besides state actors, non-state actors are involved in the PHS of Burundi, including development partners, international agencies, NGOs, civil society, academia, and the private sector (Table 3).

Of the 20 stakeholders identified to be helpful to engage when it comes to various PHS functions, 13 were categorized as being critical (Table 4)—stakeholders with high interest and a significant influence: agricultural input suppliers; BPEAE; Cabinet of MINEAGRIE; Communication unit of MINEAGRIE; COPROSEBU; General Directorate of Agriculture; DGMAVAE; DPV; DSIA; farmers, farmers associations, and cooperatives; institutions of higher learning; ISABU; and Ministry of Education.

Table 4. Critical to engage stakeholders in the plant health system (PHS) of Burundi.

Stakeholder	PHS Function							
	AIS	AT	CP	DS	FAS	PHIM	PRC	RTD
(1) Agricultural input suppliers	x							
(2) Bureau Provinciale de L'Environnement, de l'Agriculture et de l'Elevage (BPEAE)				x	x			
(3) Cabinet of Ministry of Environment, Agriculture and Livestock (MINEAGRIE)							x	
(4) Communication unit of MINEAGRIE						x		
(5) Le Collectif des Compagnies et Coopératives de Production des Semences du Burundi (COPROSEBU)	x							
(6) General Directorate of Mobilization for Self-Development and Agricultural Extension				x	x			
(7) Directorate of Plant Protection	x			x	x		x	x
(8) Department of Statistical and Agricultural Information (DSIA)						x		
(9) Farmers, Farmers associations, and cooperatives			x	x				
(10) Institutions of higher learning		x						
(11) Institut des Sciences Agronomiques du Burundi (ISABU)				x	x			x
(12) Ministry of Education and Vocational and Professional Training		x						
(13) General Directorate of Agriculture				x				

AIS = Agricultural input supply; AT = Agricultural training; CP = Crop production; DS = Diagnostic services; FAS = Farmer advisory (extension) services; PHIM = Plant health information management; PRC = Policy, regulation, and control; and RTD = Research and technology development; and Key: x = involvement in a PHS function.

3.3.1. Agricultural Input Supply

Agricultural input suppliers play a significant role in the distribution system, linking farmers to products, particularly in areas not easily accessible. The input segment is concentrated by stakeholders in the private sector (agricultural input suppliers), public sector (DPV), and civil society (COPROSEBU and CAPAD) (Figure 3a). ISABU mainly controls the country's formal market for seed production. In terms of quantity and quality, the seed supply is considered inadequate for the requirements to increase yields and improve nutritional content and drought and pest resistance in the agricultural sector [30]. ONCCS is responsible for regulating the industry, improving the quality of seeds sold within the country, and facilitating the importation of foreign seeds to satisfy the local requirements for improved seed varieties [31].

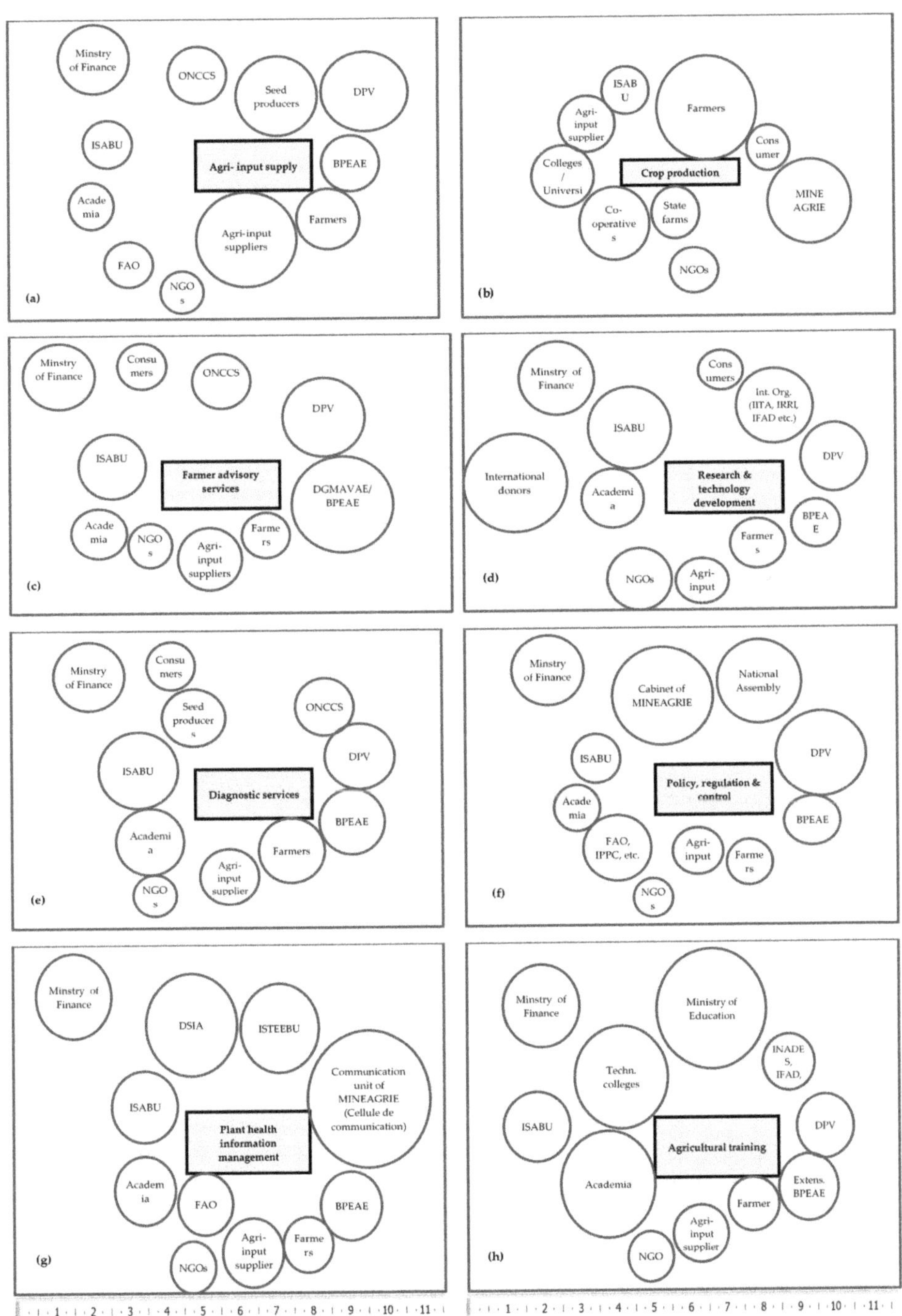

Figure 3. Schematic arrangement of interest (proximate to function) and influence (size of circle) of stakeholders in a plant health system's (PHS) functions. A ruler is included for scale.

Though available, mineral fertilizers and pest control products tend to be expensive due to inadequate competition and high costs associated with transportation. In addition, the available large pack sizes tend to be inappropriate for smallholders with limited needs due to their relatively low land areas and who are compelled to carry the products back to their farm by bicycle or on foot [8,30].

3.3.2. Crop Production

Agricultural production is mainly handled by smallholder farmers in Burundi (Figure 3b). Many are subsistence farmers, and many others are smallholder commercial farmers [6]. Smallholder farmers typically produce a varied range of commodities in limited quantities to manage risks for household food supplies and complement this with production for cash. These crops grown by smallholder farmers include roots and tubers, fresh fruit and vegetables, pulses, and cereals. Other smallholders may produce palm fruit. Production is generally anchored on traditional farming—manual, hand-hoe powered, minimal use of inputs, and limited irrigation and water management techniques.

There are only a few medium size producers (primarily producing fresh fruit and vegetables and cereals) and large commercial farms, mainly for export commodities like coffee, palm fruit, and vegetables [30]. Cooperatives are relatively less developed, but the government highly encourages their establishment, and their numbers have increased over time. State farms hardly exist. Nevertheless, Burundi can produce most of its food needs internally.

In 2020, roots and tubers, fresh fruit and vegetables, pulses, and cereals accounted for approximately 92% of the total crop production in Burundi [32]. Roots and tubers accounted for the largest share at 51% of production value, followed by fresh fruit and vegetables (22%), pulses (6%), and cereal production (6%) [32].

3.3.3. Farmer Advisory Services

Agricultural technical assistance is provided principally by BPEAE's extension workers, as well as through an assortment of programs by donor agencies and development partners and a small number of private sector contract farming schemes (Figure 3c). By 2020, Burundi had 1030 extension officers, of which BPEAE employed 996, while the remaining 34 were used by seed producers [31]. The ratio of extension officers to agricultural households stood at 1:3929 in 2017.

As noted above, the deployment of private extension service providers has primarily been low. Notwithstanding, a previous report observed that their performance might have been better compared to that of their counterparts in the public sector [30].

3.3.4. Research and Technology Development

Various public, international, and private organizations and institutions of higher learning undertake research in the agricultural sector (Figure 3d). Research and development are mainly focused on crop production and food technologies. In terms of crops, among the most researched crops include vegetables, rice, potatoes, and fruits.

Actors in the public sector include ISABU, the Agronomic and Zootechnique Research Institute (IRAZ), and the National Centre on Food Technologies (CNTA). The country's leading research center is ISABU accounting for most of Burundi's research capacity and investments. IRAZ supports food security through animal science and agriculture research, while CNTA focuses on developing and adapting processing and storage technologies [27]. Institutions of higher learning, such as the University of Burundi and the University of Ngozi, have departments specifically dedicated to research. Private sector investment in research and development is minimal. Besides local institutions, various international organizations, including IITA and IRRI, undertake applied research in Burundi.

3.3.5. Diagnostic Services

Diagnostics is mainly implemented by ISABU and the universities and certification bodies such as ONCCS (Figure 3e). Moreover, laboratories and institutions outside Burundi play a role. As for field diagnostics, the extension workers under BPEAE and farmers are key players and, to a small extent, the agricultural input supplier.

3.3.6. Policy, Regulation, and Control

The MINEAGRIE is responsible for developing rules relating to agriculture, as well as preparing agricultural policies. When it comes to the formulation of regulations, the cabinet of MINEAGRIE is critical in the decision-making process (Figure 3f). On the other hand, when it comes to agricultural policies, the national assembly plays an integral role. Some regulations are also adopted from international protocols, guidance, and agreements. Therefore, for example, FAO and IPPC influence regulations. DPV usually handles forecasting and warning systems under MINEAGRIE.

3.3.7. Plant Health Information Management

Agricultural data management seems less emphasized in Burundi, and linkages may need improvements. This low emphasis is, despite the existence of a department within MINEAGRIE (DSIA), meant to spearhead initiatives relating to agricultural data, and coupled with this is the existence of ISTEEBU (Institut de Statistiques et d'études économiques du Burundi) (Figure 3g). Some NGOs, such as AUXFIN, are capturing agricultural and farmer data and may be able to implement monitoring and quality assurance activities.

3.3.8. Agricultural Training

Agricultural training in Burundi has been chiefly driven by development partners through collaboration with MINEAGRIE's programs and involving a few private sector players who provide technical support. The partners who have supported agricultural training in the past include the World Bank and the French and Belgian cooperation agencies, the EU, IFAD, and FAO, as well as a raft of international and local NGOs (Figure 3h). Agricultural training programs have concentrated mainly on technical, administrative, and literacy issues to support the sector's development.

Formal training in agriculture is provided at high school and undergraduate levels. The agricultural training and educational institutions include the Faculty of Agronomy at the University of Ngozi and the Faculty of Agronomy and Bio-Engineering (FABI) at the University of Burundi; and the Institut Technique Agricole du Burundi (ITAB).

3.4. Strengths, Weaknesses, and Opportunities of the Functions of the Plant Health System of Burundi

Through discussions with stakeholders, the researchers identified the strengths and weaknesses of the PHS of Burundi. Additionally, the investigators undertook a comprehensive analysis of relevant secondary literature.

Across the PHS functions of Burundi, some significant weaknesses identified include low productivity, lack of adequate skills to serve farmers effectively, and inadequate capacity to diagnose pests.

Researchers have identified low agricultural productivity as a critical impediment to food security and fiscal development in post-conflict Burundi [7,8,33]. With the increasing population [34], poor yields have meant that the amount of food produced does not match the population, resulting in a constant food shortfall in the country.

The twin challenges of unpredictable smallholder agricultural production and the reality that most agricultural land is presently under production require that further growth emanates from product and process upgrades that aim for sustainable intensification [30]. These upgrades can include bench terraces on sloping land for enhanced hillside agriculture [35], adoption of minimum tillage and other improved land preparation methods [36], low-cost technologies on micro-fertilization [37,38], and small-scale irrigation schemes [39].

Additional upgrades include the safe use and handling of pesticides and the prioritization and promotion of integrated pest management techniques [40,41].

Adopting the above upgrades at the farm level requires strategic support and investment by government and development partners to stimulate demand and uptake by smallholders. Technical and financial support by government and development partners is necessary, particularly for upgrades whose returns on investment in the short term are negative.

Lack of adequate skills. As the agricultural sector of Burundi adapts to the upgrades above, development skills capacities for farmers and farmers' advisors on, among others, effective and practical methods of managing plant health problems is required.

In the short term, to complement formal education programs, the country can consult international experts from organizations such as CABI, IITA, FAO, and IFAD about developing suitable training programs. Besides developing the training programs and curricula, the country should seek opportunities for institutionalizing this training at the ITAB and university levels and in other institutions of higher learning [42–44].

Inadequate capacity to diagnose pests. Crop pests diminish the quality and yield of agricultural production [45]. They reduce food security at various levels (household, national and global) and are attributed to considerable economic losses, particularly in food-deficit regions with rapid population growth and often with emerging or re-emerging pests [46].

A scarcity of diagnostic service for plant health problems remains a critical contributing factor to plant health burden in resource-poor settings. Opportunities exist in Burundi for:

- Enhancing the capacity of agricultural extension workers to increase field-level detection of plant health problems.
- Supplementing the existing laboratory infrastructure with additional equipment and skilled personnel to enable the diagnosis of a broad spectrum of plant health problems
- Sharing and integrating plant health information sources in support of early warning systems
- Promoting innovations in surveillance and control that do not require and/or can lessen the physical presence of plant health practitioners

Notwithstanding the constraints above, the country has numerous crucial factors favoring the development of an enhanced PHS. For instance, the country is endowed with a good climate and geographical conditions that favor agricultural production, with several cropping seasons. In addition, the country enjoys competitive labor costs and is strategically placed to export in the region through Lake Tanganyika. However, the country must confront its fundamental weaknesses to take advantage of these opportunities. An analysis of the country's strengths and weaknesses is presented in Table 5.

Table 5. Strengths and weaknesses of each plant health system function of Burundi (n = 68 participants involved in three workshops).

Strengths	Weaknesses
Agricultural input supply	
- Availability of raw materials - Minimal or no cost in seed production in the informal sector - Availability of seed in the informal sector is assured	- Weak demand from producers - Lack of technical knowledge - Lack of financial resources - Market distortions from subsidy program - Scarcity of locally available organic fertilizer and mulch - Inadequate competition - High transportation cost - Minimal use of external inputs - Poor quality seed

Table 5. *Cont.*

Strengths	Weaknesses
Crop production	
- Good climate for crop production - Available arable land - Comparatively low labour costs - Significant involvement of women in production - Improved crop varieties/breeds	- Production based on traditional farming methods - Pests and diseases - Cooperatives and contract farming schemes are not well organized - Limited access to finance - Unclear land tenure system coupled with scarcity of land and an increasing population - Poor rural transportation infrastructure - Soil degradation - Majority of land that is cultivatable is already under production
Research and technology development	
- Existence of a national agricultural research system - Experience in collaborative research	- Shortage of human resource - Limited budget
Farmer advisory services	
- Decentralized and well-defined extension system - Available workforce of extension agents - Presence of farmers organizations, NGOs, and projects providing alternative extension service providers	- Many extension agents lack the requisite skills to effectively serve farmers either due to lack of practical skills or limited experience in the sector - Government extension staff are often poorly motivated due to low wages - Budgetary constraints limiting the hiring of more extension workers - Low literacy rates among farmers - Reluctance by trained agronomists to live or work in the rural areas
Plant health information management	
- Existence of institutions mandated to manage plant health information	- Lack of data science approach
Diagnostic services	
- Knowledge sharing platforms - Strong collaboration with experts outside the country	- Limited capacity to identify and isolate pests - Lack of equipment and infrastructure - inadequate capacity to prevent, detect and respond to new and emerging plant health problems
Policy, regulation and control	
- Extensive agricultural strategies detailing the vision and priority interventions	- Bureaucracy (e.g., overly centralized systems for planning and management) - Poor interdependencies between PHS and program investments - Inadequate regulation of agricultural inputs sector, improper industry practices - Underdeveloped sanitary and phytosanitary system
Agricultural training	
- Existence of agricultural vocational training institutes for training and producing middle-level skilled extension personnel	- Fast growing enrolment and inadequate facilities, equipment, and teaching staff - Training programs for universities are broad and focus on theoretical knowledge and lack training in applied and practical skills - Lack of resources - Discrepancy between the needs of the private sector and the skills provided by formal education schools - Lack of a clear mechanism for institutionalizing knowledge transfer and provision of adequate scale for reaching the fragmented production base. - Women not targeted in agricultural trainings

In terms of crucial needs (gaps) that the proposed project can realistically address, within the available time and resources, the participants, through consensus, identified the following:

- Management of linkages among stakeholders in plant health
- Training qualified staff in field diagnosis principles and giving farmers practical and effective plant health management recommendations. Also relevant are other training, including IPM/biocontrol, and the development of extension materials, e.g., pest management decision guides (PMDGs) and fact sheets for farmers.
- Operation of plant clinics and associated activities such as delivery of advice to farmers through complementary extension campaigns (such as plant health rallies and mass extension campaigns).
- Training of plant health practitioners in data management—processes involved in the collection and processing of plant clinic data.
- Setting up of ICT-based systems to support information sharing.
- Monitoring, evaluation, and learning to assess and document outcomes and im-pact of project interventions.
- Prioritization and addressing some of the vital gender-based constraints in agri-culture

In light of the identified needs (gaps), seven broad interventions are proposed to help rapidly strengthen the PHS of Burundi:

1. Promotion of a comprehensive approach to plant health governance. Plant health policies, operational processes, planning, and budgeting ought to be linked to the attainment of the relevant Sustainable Development Goals (SDGs), specifically SDG 2 (End hunger, achieve food security and improved nutrition, and promote sustainable agriculture). Means of firming accountability and mechanisms for engaging stake-holders, including alternative sectors, communities, external partners, civil society, private sector players, and academia, should be explored.
2. Provision of adequate, skilled, and well-distributed plant health practitioners. Staff cadres need to be streamlined following the requirements for essential services, and this needs to be mirrored in staffing standards, norms, conditions, and certification. Training curricula and programs need to be responsive to new and emerging priorities. The government and development partners should channel investments towards training programs (both pre-service and in-service) to ensure the personnel mirror both present and future plant health needs.
3. Building of an efficient agricultural advisory service delivery mechanism. The agri-cultural advisory service delivery mechanism needs to be rationalized at all levels (national and local) to reflect the country's priorities.
4. Provision of sufficient plant health equipment and infrastructure. The concerned entities should establish operational procedures and standards for critical plant health infrastructure. They should also develop medium-term and long-term plans for investment (including their maintenance and disposal) of plant health equipment and expansion of fixed infrastructure.
5. Provision of good-quality, affordable critical pest control products, diagnostics, and other agricultural inputs. To be implemented through a well-regulated procurement and supply system. The government must revise regulations and policies to enhance the capacity for judicious use of pest control products and other agricultural inputs, including during emergencies. The national plant protection organization should im-prove surveillance systems for monitoring new and emerging plant health problems, adverse effects, pesticide quality, and resistance.
6. Strengthening of plant health information systems and surveillance. The stakeholders should establish data coordination apparatuses to interlace information systems for research, surveys, surveillance, and critical statistics to support integration and lessen fragmentation. The capacity for data analysis and sharing should be prioritized, especially at the local level. Additionally, the stakeholders should scale up innovative

approaches for data collection and use. Finally, the stakeholders should prioritize the engagement of the research community for the generation and usage of evidence emanating from research for decision-making.

7. Provision of sustainable financing for plant health. The government should do this by establishing mechanisms for mobilizing additional sustainable domestic resources while sourcing more external resources for plant health. The country should strengthen accountability systems, public financial management, and institutional arrangement.

4. Conclusions

Functional plant health systems have been integral in mitigating the movement of non-native organisms, pests, and diseases within and between countries through trade. In some countries, such as Cuba [47], Venezuela [48], and Burundi, practitioners know little about the plant health system and other support systems. Efforts are underway in many regions to strengthen plant health systems through collaboration, including regulatory frameworks [49].

The accumulated experience of agroecological initiatives in thousands of small- and medium-scale farms in developing countries [47,48] is essential in defining national policies to support sustainable agriculture. In this regard, a vital aspect of the plant health systems is their pest and disease monitoring system that covers the most economically important crops in the country [49], providing pest thresholds to farmers and the importance of identifying possible interrelationships with biophysical components such as climate, soil, or agronomic management [48].

To facilitate the development of future trade partnerships between countries, agencies need to understand the organizational structure and diagnostic capacity of the plant health system of these countries, identify potential synergies between country systems, and identify steps toward cooperation. The results of this study fill this critical gap by presenting a descriptive analysis of the plant health system in Burundi. An understanding of plant health systems will be crucial for the regional economic and environmental stability of a relationship between countries.

Author Contributions: Conceptualization and methodology, W.N.O., S.T., C.N. and I.M.; investigation, W.N.O., S.T., C.N. and P.N.; data curation and visualization, S.T.; writing—original draft preparation, W.N.O.; writing—review and editing, S.T. and I.M., Funding and supervision, J.V. All authors have read and agreed to the published version of the manuscript.

Funding: CABI is an international intergovernmental organization, and we gratefully acknowledge the core financial support from our member countries (and lead agencies) including Australia (Australian Centre for International Agricultural Research), Canada (Agriculture and Agri-Food Canada), China (Chinese Ministry of Agriculture and Rural Affairs), the Netherlands (Directorate-General for International Cooperation), Switzerland (Swiss Agency for Development and Cooperation) and the United Kingdom (Foreign, Commonwealth and Development Office). This assessment was facilitated by CABI's Plantwise project in Burundi supported by the Embassy of the Kingdom of the Netherlands in Bujumbura, Burundi, contract number 4000004168 and Nuffic, grant number OKP-TMT + .20/00062. The assessment was also supported in-kind by the Institut des Sciences Agronomiques du Burundi (ISABU) and the Directorate of Plant Protection (DPV) of MINEAGRIE of Burundi as part of the Plantwise Project.

Institutional Review Board Statement: Not applicable.

Informed Consent Statement: Not applicable.

Data Availability Statement: Not applicable.

Acknowledgments: The authors gratefully acknowledge the contributions of all participants in the stakeholder workshops. We also acknowledge institutional support from ISABU and DPV, without which it would not have been possible to hold the workshops. The efforts of the project team members comprising staff from ISABU: Willy Irakoze, Emmerence Ndishimiyimana, and Espérance Habindavyi; and CABI: Mary Bundi, Bethel Terefe, Peter Karanja, Manfred Grossrieder, David Onyango, Duncan Chacha, Linda Likoko, Fredrick Mbugua, and Norah Kathuku are duly acknowledged.

Conflicts of Interest: The authors declare no conflict of interest.

References

1. Nijimbere, G.; Lizana, C.R. Assessment of Soil Erosion of Burundi Using Remote Sensing and GIS by RUSLE Model. *RUDN J. Ecol. Life Saf.* **2019**, *27*, 17–28. [CrossRef]
2. The World Bank Burundi. Available online: https://data.worldbank.org/country/BI (accessed on 26 March 2021).
3. Katungi, E.; Nduwarigira, E.; Ntukamazina, N.; Niragira, S.; Mutua, M.; Kalemera, S.; Onyango, P.; Nchanji, E.; Fungo, R.; Birachi, E.; et al. *Food Security and Common Bean Productivity: Impacts of Improved Technology Adoption among Smallholder Farmers in Burundi*; Pan-Africa Bean Research Alliance (PABRA): Nairobi, Kenya, 2020.
4. Ndagijimana, M.; Kessler, A.; van Asseldonk, M. Understanding Farmers' Investments in Sustainable Land Management in Burundi: A Case-Study in the Provinces of Gitega and Muyinga. *Land Degrad. Dev.* **2019**, *30*, 417–425. [CrossRef]
5. Aboyitungiye, J.B.; Prasetyani, D. Is Agriculture an Engine of Economic Reconstruction and Development in the Case of the Republic of Burundi? In Proceedings of the 8th International Conference on Sustainable Agriculture and Environment, Surakarta, Indonesia, 15 November 2021; Volume 905.
6. IFAD. *Republic of Burundi: Country Strategic Opportunities Programme (2022–2027)*; International Fund for Agricultural Development: Rome, Italy, 2022.
7. Okonya, J.S.; Ocimati, W.; Nduwayezu, A.; Kantungeko, D.; Niko, N.; Blomme, G.; Legg, J.P.; Kroschel, J. Farmer Reported Pest and Disease Impacts on Root, Tuber, and Banana Crops and Livelihoods in Rwanda and Burundi. *Sustainability* **2019**, *11*, 4296. [CrossRef]
8. Nyairo, R.; Machimura, T.; Matsui, T. A Combined Analysis of Sociological and Farm Management Factors Affecting Household Livelihood Vulnerability to Climate Change in Rural Burundi. *Sustainability* **2020**, *12*, 4296. [CrossRef]
9. Kim, J.-B.; Habimana, J.d.D.; Kim, S.-H.; Bae, D.-H. Assessment of Climate Change Impacts on the Hydroclimatic Response in Burundi Based on Cmip6 Esms. *Sustainability* **2021**, *13*, 2037. [CrossRef]
10. Mpozi, B.B.; Mizero, M.; Egesa, A.O.; Nguezet, P.M.D.; Vanlauwe, B.; Ndimanya, P.; Lebailly, P. Land Access in the Development of Horticultural Crops in East Africa. A Case Study of Passion Fruit in Burundi, Kenya, and Rwanda. *Sustainability* **2020**, *12*, 3041. [CrossRef]
11. Bisekwa, E.; Njogu, P.M.; Kufa-Obso, T. Wet Coffee Processing Discharges Affecting Quality of River Water at Kayanza Ecological Zone, Burundi. *Open J. Appl. Sci.* **2021**, *11*, 707–721. [CrossRef]
12. Bitama, P.C.; Lebailly, P.; Ndimanya, P.; Sabuhungu, E.G.; Burny, P. The Contribution of Microfinance to the Resilience Strategies of Smallholder Tea Farmers in Burundi. *AGROFOR Int. J.* **2019**, *4*, 33–41. [CrossRef]
13. Ndagijimana, M. Coping with Risk and Climate Change in Farming: Exploring an Index-Based Crop Insurance in Burundi. Ph.D. Thesis, Wageningen University and Research: Wageningen, The Netherlands, 2021.
14. Kessler, A.; Reemst, L.; Beun, M.; Slingerland, E.; Pol, L.; de Winne, R. Mobilizing Farmers to Stop Land Degradation: A Different Discourse from Burundi. *Land Degrad. Dev.* **2021**, *32*, 3403–3414. [CrossRef]
15. Othoche, B. Exploring Technologies for Sustainable Transboundary Water Resource Management in the Era of Climate Change: A Case for the Nile River Basin Riparian States. In *Nile and Grand Ethiopian Renaissance Dam*; Springer: Berlin/Heidelberg, Germany, 2021; pp. 181–193.
16. Gerard, A.; Lopez, M.C.; Clay, D.C.; Ortega, D.L. Farmer Cooperatives, Gender and Side-Selling Behavior in Burundi's Coffee Sector. *J. Agribus. Dev. Emerg. Econ.* **2021**, *11*, 490–505. [CrossRef]
17. Curtis, M. *Improving African Agriculture Spending: Budget Analysis of Burundi, Ghana, Zambia, Kenya and Sierra Leone*; Curtis Research: Oxford, UK, 2013.
18. Kinuthia, R.; Kiptot, E.; Nkurunzinza, C. *The Extension System in Burundi: Kayanza Province, Muruta Commune*; Australian Centre for International Agricultural Research: Canberra, ACT, Australia, 2016.
19. CABI. *Plantwise in Burundi Annual Report 2021*; Centre for Agriculture and Bioscience International: Wallingford, UK, 2022.
20. Otieno, W.; Ochilo, W.; Migiro, L.; Jenner, W.; Kuhlmann, U. Tools for Pest and Disease Management by Stakeholders: A Case Study on Plantwise. In *The Sustainable Intensification of Smallholder Farming Systems*; Burleigh Dodds Science Publishing: Cambridge, UK, 2020; pp. 151–174.
21. Raum, S. A Framework for Integrating Systematic Stakeholder Analysis in Ecosystem Services Research: Stakeholder Mapping for Forest Ecosystem Services in the UK. *Ecosyst. Serv.* **2018**, *29*, 170–184. [CrossRef]
22. Namazzi, G.; Kiwanuka, S.N.; Peter, W.; John, B.; Olico, O.; Allen, K.A.; Hyder, A.A.; Ekirapa, K.E. Stakeholder Analysis for a Maternal and Newborn Health Project in Eastern Uganda. *BMC Pregnancy Childbirth* **2013**, *13*, 58. [CrossRef] [PubMed]
23. Cooper, H.; Hedges, L.V.; Valentine, J.C. *The Handbook of Research Synthesis and Meta-Analysis*; Russell Sage Foundation: New York, NY, USA, 2009; ISBN 9780871541635.
24. Iacono, T.; Keeffe, M.; Kenny, A.; McKinstry, C. A Document Review of Exclusionary Practices in the Context of Australian School Education Policy. *J. Policy Pract. Intellect. Disabil.* **2019**, *16*, 264–272. [CrossRef]
25. Tubiello, F.N.; Salvatore, M.; Rossi, S.; Ferrara, A.; Fitton, N.; Smith, P. The FAOSTAT Database of Greenhouse Gas Emissions from Agriculture. *Environ. Res. Lett.* **2013**, *8*, 015009. [CrossRef]
26. Williams, F.; Ali, I.; Danielsen, S.; Alawy, A.; Romney, D. *A Step by Step Guide for Conducting a Stakeholder Analysis and Context Review*; Centre for Agriculture and Bioscience International: Wallingford, UK, 2015.

27. Danielsen, S.; Matsiko, F.; Mutebi, E.; Karubanga, G. *Second Generation Plant Health Clinics in Uganda: Measuring Clinic Performance from a Plant Health System Perspective (2010–2011)*; Centre for Agriculture and Bioscience International: Wallingford, UK, 2012.

28. WHO. *Everybody's Business: Strengthening Health Systems to Improve Health Outcomes*; WHO: Geneva, Switzerland, 2007; p. 44.

29. Hyder, A.; Syed, S.; Puvanachandra, P.; Bloom, G.; Sundaram, S.; Mahmood, S.; Iqbal, M.; Hongwen, Z.; Ravichandran, N.; Oladepo, O.; et al. Stakeholder Analysis for Health Research: Case Studies from Low- and Middle-Income Countries. *Public Health* **2010**, *124*, 159–166. [CrossRef]

30. Bamber, P.; Abdulsamad, A.; Gereffi, G. *Burundi in the Agribusiness Global Value Chain: Skills for Private Sector Development*; Duke University: Durham, NC, USA, 2014.

31. Mabaya, E.; Bararyenya, A.; Vyizigiro, E.; Waithaka, M.; Mugoya, M.; Kanyenji, G.; Tihanyi, K. Burundi 2021 Country Study—The African Seed Access Index; 2021. Available online: https://ageconsearch.umn.edu/record/317017/?ln=en (accessed on 26 May 2022).

32. FAO. FAOSTAT Database on Agriculture. Available online: https://www.fao.org/faostat/en/#data (accessed on 1 March 2022).

33. Schneiderbauer, S.; Baunach, D.; Pedoth, L.; Renner, K.; Fritzsche, K.; Bollin, C.; Pregnolato, M.; Zebisch, M.; Liersch, S.; López, M.; et al. Spatial-Explicit Climate Change Vulnerability Assessments Based on Impact Chains. Findings from a Case Study in Burundi. *Sustainability* **2020**, *12*, 6354. [CrossRef]

34. Nzokirishaka, A.; Itua, I. Determinants of Unmet Need for Family Planning among Married Women of Reproductive Age in Burundi: A Cross-Sectional Study. *Contracept. Reprod. Med.* **2018**, *3*, 11. [CrossRef]

35. Wolka, K.; Mulder, J.; Biazin, B. Effects of Soil and Water Conservation Techniques on Crop Yield, Runoff and Soil Loss in Sub-Saharan Africa: A Review. *Agric. Water Manag.* **2018**, *207*, 67–79. [CrossRef]

36. Osewe, M.; Mwungu, C.M.; Liu, A. Does Minimum Tillage Improve Smallholder Farmers' Welfare? Evidence from Southern Tanzania. *Land* **2020**, *9*, 513. [CrossRef]

37. Traore, K.; Traore, B.; Synnevåg, G.; Aune, J.B. Intensification of Sorghum Production in Flood Recession Agriculture in Yelimane, Western Mali. *Agronomy* **2020**, *10*, 726. [CrossRef]

38. Coulibaly, A.; Woumou, K.; Aune, J.B. Sustainable Intensification of Sorghum and Pearl Millet Production by Seed Priming, Seed Treatment and Fertilizer Microdosing under Different Rainfall Regimes in Mali. *Agronomy* **2019**, *9*, 664. [CrossRef]

39. Gidey, G. Impact of Small Scale Irrigation Development on Farmers' Livelihood Improvement in Ethiopia: A Review. *J. Resour. Dev. Manag.* **2020**, *62*, 10–18. [CrossRef]

40. Ochilo, W.N.; Otipa, M.; Oronje, M.; Chege, F.; Lingeera, E.K.; Lusenaka, E.; Okonjo, E.O. Pest Management Practices Prescribed by Frontline Extension Workers in the Smallholder Agricultural Subsector of Kenya. *J. Integr. Pest. Manag.* **2018**, *9*, 15. [CrossRef]

41. Okonya, J.S.; Petsakos, A.; Suarez, V.; Nduwayezu, A.; Kantungeko, D.; Blomme, G.; Legg, J.P.; Kroschel, J. Pesticide Use Practices in Root, Tuber, and Banana Crops by Smallholder Farmers in Rwanda and Burundi. *Int. J. Environ. Res. Public Health* **2019**, *16*, 400. [CrossRef]

42. Singh Malik, R. Educational Challenges in 21st Century and Sustainable Development. *J. Sustain. Dev. Educ. Res.* **2018**, *2*, 9–20. [CrossRef]

43. Raidimi, E.N.; Kabiti, H.M. A Review of the Role of Agricultural Extension and Training in Achieving Sustainable Food Security: A Case of South Africa. *S. Afr. J. Agric. Ext. (SAJAE)* **2019**, *47*, 120–130. [CrossRef]

44. Muilerman, S.; Wigboldus, S.; Leeuwis, C. Scaling and Institutionalization within Agricultural Innovation Systems: The Case of Cocoa Farmer Field Schools in Cameroon. *Int. J. Agric. Sustain.* **2018**, *16*, 167–186. [CrossRef]

45. Bebber, D.P.; Field, E.; Gui, H.; Mortimer, P.; Holmes, T.; Gurr, S.J. Many Unreported Crop Pests and Pathogens Are Probably Already Present. *Glob. Chang. Biol.* **2019**, *25*, 2703–2713. [CrossRef]

46. Savary, S.; Willocquet, L.; Pethybridge, S.J.; Esker, P.; McRoberts, N.; Nelson, A. The Global Burden of Pathogens and Pests on Major Food Crops. *Nat. Ecol. Evol.* **2019**, *3*, 430–439. [CrossRef]

47. Gomez, D.F.; Adams, D.C.; Cossio, R.E.; Carton De Grammont, P.; Messina, W.A.; Royce, F.S.; Galindo-Gonzalez, S.; Hulcr, J.; Muiño, B.L.; Vá Zquez, L.L. Peering into the Cuba Phytosanitary Black Box: An Institutional and Policy Analysis. *PLOS ONE* **2020**, *15*, e0239808. [CrossRef] [PubMed]

48. Olivares, B.O.; Rey, J.C.; Lobo, D.; Navas-Cortés, J.A.; Gómez, J.A.; Landa, B.B. Fusarium Wilt of Bananas: A Review of Agro-Environmental Factors in the Venezuelan Production System Affecting Its Development. *Agronomy* **2021**, *11*, 986. [CrossRef]

49. Martínez-Solórzano, G.; Mesoamericana, J.R.-B.-A. Bananas (Musa AAA): Importance, production and trade in COVID-19 times. *Agron. Mesoam.* **2021**, *32*, 1034–1046. [CrossRef]

 *sustainability*

Article

Effects of Combined Application of Salicylic Acid and Proline on the Defense Response of Potato Tubers to Newly Emerging Soft Rot Bacteria (*Lelliottia amnigena*) Infection

Richard Osei [1], Chengde Yang [1,2,*], Lijuan Wei [1], Mengjun Jin [1] and Solomon Boamah [1]

[1] College of Plant Protection, Gansu Agricultural University, Lanzhou 730070, China; sammytoga31@gmail.com (R.O.); weilijuan12345@126.com (L.W.); jmj1678541254@163.com (M.J.); solomon4408boamah@gmail.com (S.B.)

[2] Biocontrol Engineering Laboratory of Crop Diseases and Pests of Gansu Province, Lanzhou 730070, China

* Correspondence: yangcd@gsau.edu.cn

Abstract: Potato soft rot, caused by the pathogenic bacterium *Lelliottia amnigena* (*Enterobacter amnigenus*), is a serious and widespread disease affecting global potato production. Both salicylic acid (SA) and proline (Pro) play important roles in enhancing potato tuber resistance to soft rot. However, the combined effects of SA and Pro on defense responses of potato tubers to *L. amnigena* infection remain unknown. Hence, the combined effects of SA and Pro in controlling newly emerging potato soft rot bacteria were investigated. Sterilized healthy potato tubers were pretreated with 1.5 mM SA and 2.0 mM Pro 24 h before an inoculation of 0.3 mL of *L. amnigena* suspension (3.69×10^7 CFU mL^{-1}). Rotting was noticed on the surfaces of the hole where the *L. amnigena* suspension was inoculated. Application of SA and Pro with *L. amnigena* lowered the activity of pectinase, protease, pectin lyase, and cellulase by 64.3, 77.8, 66.4 and 84.1%, and decreased malondialdehyde and hydrogen peroxide contents by 77.2% and 83.8%, respectively, compared to the control. The activities of NADPH oxidase, superoxide dismutase, peroxide, catalase, polyphenol oxidase, phenylalanine ammonia-lyase, cinnamyl alcohol dehydrogenase, 4-coumaryl-CoA ligase and cinnamate-4-hydroxylase were increased in the potato tubers with combined treatments by 91.4, 92.4, 91.8, 93.5, 94.9, 91.3, 96.2, 94.7 and 97.7%, respectively, compared to untreated stressed tubers. Six defense-related genes, pathogenesis-related protein, tyrosine-protein kinase, Chitinase-like protein, phenylalanine ammonia-lyase, pathogenesis-related homeodomain protein, and serine protease inhibitor, were induced in SA + Pro treatment when compared with individual application of SA or Pro. This study indicates that the combined treatment of 1.5 mM SA and 2.0 mM Pro had a synergistic effect in controlling potato soft rot caused by a newly emerging bacterium.

Keywords: antioxidant enzymes; extracellular enzymes; pathogen; reactive oxygen species; systemic acquired resistance

Citation: Osei, R.; Yang, C.; Wei, L.; Jin, M.; Boamah, S. Effects of Combined Application of Salicylic Acid and Proline on the Defense Response of Potato Tubers to Newly Emerging Soft Rot Bacteria (*Lelliottia amnigena*) Infection. *Sustainability* **2022**, *14*, 8870. https://doi.org/10.3390/su14148870

Academic Editors: Barlin Orlando Olivares Campos, Miguel Araya-Alman, Edgloris E. Marys and Sean Clark

Received: 4 June 2022
Accepted: 12 July 2022
Published: 20 July 2022

1. Introduction

According to Lim et al. [1], potato is the fourth most important food crop and staple food in the world. Potato tubers are in most cases stored for 3–6 months before being processed and/or consumed by humans. The average annual loss due to potato soft rot is estimated to be between 6 and 25%, with up to 60% of tubers in some cases being damaged [2,3]. One of the most common potato diseases that reduce tuber quality during storage is soft rot caused by *Pectobacterium carotovorum* subsp. *Carotovorum* [4]. Postharvest diseases in fruits and vegetables are caused by bacteria during storage. Throughout the growing season and during storage, members of the families *Enterobacteriaceae* and *Pectobacteriaceae* are the primary cause of potato soft rot [4]. These pathogens cause infection either during the preharvest stage in the field or after harvesting during storage and transportation. Soft rot bacteria degrade pectate molecules, which bind plant cells together, eventually causing the plant structure to fall apart. Potato wounds or damage are the primary sources of soft rot bacteria invasion. This usually happens during harvesting and grading, allowing bacteria to invade the tuber [5]. When this is combined with water on the tuber's surface, the bacteria can overcome the tuber's natural defenses and initiate tuber rot. Soft rot pathogens produce many enzymes capable of degrading the plant cell wall other than saprophytic pectolytic bacteria. These enzymes include pectinases, cellulases, proteases, pectin lyase, and xylanases, each with its own set of properties [6]. Due to its ability to synthesize a broader variety of isoenzymes faster and in greater amounts than pectolytic saprophytic bacteria, *L. amnigena* enters tissues more easily and causes infection [7,8].

Chemical control is an important method of controlling plant diseases. However, because of their negative effects on humans and the environment, synthetic bactericides are not the preferred method for controlling plant pathogens [9]. As a result, an eco-friendly alternative strategy for bacterial soft rot management must be developed. Increased natural defense system of plants is one of the promising environmentally friendly methods for postharvest disease control [10,11]. The activation of plant defense systems that help to delay the spread of various pathogens can help to protect the plant from bacterial pathogens [11]. Induced disease resistance in plants is a viable option for preventing the invasion of bacterial pathogens and an appealing disease control strategy [12].

Salicylic acid (SA) and proline (Pro), which induce natural resistance in plants to bacterial infections and can provide year-round protection, can be considered a promising alternative to the use of synthetic bactericides [13]. SA is essential for plant growth, development, and defense responses, and can also be used to inhibit microbial growth [14]. Plant resistance to pathogens is induced by SA application via mechanisms such as oxidative burst, cell wall reinforcement, and gene expression regulation [13]. SA is important in adaptable interactions, in which a small amount of SA controls the expression of a collection of defense-related genes, resulting in a defense-like response [15]. Several studies [16,17] show that applying SA to tomatoes can improve their resistance to *Ralstonia solanacearum* in the greenhouse or the field. According to previous studies, SA is an essential elicitor that triggers plant resistance to pathogens such as bacteria and fungi [18,19]. SA has been identified to mediate resistance in a variety of plant–pathogen interactions. Depending on the pathogen, SA can prevent pathogen proliferation and cell-to-cell or long-distance pathogen migration [20]. As a result, SA has been demonstrated to be a potential compound for inhibiting postharvest fungal pathogens and thus improving fruit postharvest quality. For example, SA treatments effectively controlled postharvest damage in the cases of *Colletotrichum gloeosporioides* on mango [21], *Penicillium expansum* on sweet cherry [22] and peach [23], *Botrytis cinerea* on peach [24], and *Monilinia* spp. on sweet cherry [22], apricot [25], and nectarine [26].

Pro is a water-soluble amino acid that is essential and multifunctional in plants, and accumulates in high contents under stress [27,28]. Pro as a proteinogenic amino acid naturally increases in response to biotic and abiotic stress by increasing Pro synthesis or decreasing Pro degradation [27]. Pro is essential for maintaining osmotic balance, preserving the structure of enzymes and membranes in key proteins, protecting photosynthetic products, and scavenging free radicals [29,30]. Pro application improves plant resistance to both biotic and abiotic stresses according to many studies [31,32]. Pro is a widely studied and used osmoprotectant under stress conditions, and it has been found to mitigate the effect of stress in plants [31,33–35]. Many studies on the effects of Pro on various crops under oxidative stress have been carried out. For example, Pro application in rose, rice seedlings, chickpeas, and citrus improved stress tolerance by increasing the activity of antioxidant enzymes, decreasing membrane lipid peroxidation, and retaining ascorbic acid content as non-enzymatic components of the antioxidant system [36]. As far as we know, many reports have been published evaluating the impact of SA and Pro applied as single treatments on crops growing under biotic and abiotic stress environments. However, there is no study on SA and Pro treatment combinations to ameliorate the effects of *L. amnigena* on potato tubers. We hypothesized that SA and Pro can control potato soft rot bacteria, *L.*

amnigena due to their antibacterial properties. As a result, this study aimed to evaluate the combined effects of SA and Pro in controlling potato soft rot caused by a newly emerging bacterium (*L. amnigena*).

2. Materials and Methods

2.1. Source of Materials

Potato tubers, var. Atlantic, were obtained from the production field in Lanzhou, Gansu Province, China. *Lelliottia amnigena* (PC3) was obtained from the Plant Pathology Laboratory, Gansu Agricultural University, Lanzhou, China [37]. *Lelliottia amnigena* was cultured on a nutrient agar (NA) medium (3.0 g peptone, 4.0 g glucose, 9.0 g agar, 1.5 g beef extract, and 500 mL water) in Petri dishes for 2 d at 28 °C. The bacterial inoculum was prepared from the 2-d-old cultured bacteria according to the method of Ben-David and Davidson [38]. The bacterial inoculum (3.69×10^7 CFU mL^{-1}) was quantified and stored at 4 °C. SA and Pro were purchased from Sangon Biotech Company Limited, Shanghai in China. 1.5 mM SA was chosen based on our previous study [39], and 2.0 mM Pro concentration was prepared following the method described by Perveen and Nazir [40], with a few modifications.

2.2. Experimental Design

The two independent experiments were arranged in a randomized complete block design using a factorial experiment with three replications of each treatment. The treatments were: (i) sterilized distilled water (negative control potato tubers treated with water without *L. amnigena* and SA and Pro treatments); (ii) *L. amnigena* (PC3) (positive control potato tubers treated with *L. amnigena* without SA and Pro treatments); (iii) *L. amnigena* + 1.5 mM SA; (iv) *L. amnigena* + 2.0 mM Pro; and (v) *L. amnigena* + 1.5 mM SA + 2.0 mM Pro. SA and Pro were purchased from Sangon Biotech Co., Ltd., Shanghai and Sigma ($C_5H_9NO_2$, CAS No. 147-85-3) Shanghai, China, respectively.

2.3. Effect of SA and Pro on Extracellular Enzyme Production by L. amnigena

The *L. amnigena* culture was inoculated into 250 mL of bacterial liquid (1.5 g peptone, 2.0 g glucose, 0.75 g beef extract, and 250 mL water) with and without SA and Pro, and kept for 2 d at 37 °C in an Honour Instrument Shaker Machine (HNYC-202T, Guangdong, China). One mL for each test of cultured bacterial liquid was taken and centrifuged at $10,000 \times g$ for 10 min at 4 °C. Supernatants were collected to perform the production levels of protease, pectinase, pectin lyase, and cellulase, following protocol assay kits (Solarbio Science and Technology Co., Ltd., Beijing, China). Pectinase, protease, pectin lyase, and cellulose were measured at 540, 450, 235 and 540 nm, respectively, using a spectrophotometer (EPOCH2 Plate Reader, BioTek, Santa Clara, CA, USA). The activities were expressed as U mL^{-1}. Each activity was carried out three times.

2.4. Effect of SA and Pro on Potato Soft Rot

The combined effect of SA and Pro on the severity of potato soft rot was investigated in this experiment. The experiment was carried out in a laboratory. Uniform healthy potato tubers were sterilized for 1 min with 75% ethanol before being washed three times in distilled water. The tubers were dried at room temperature. Using a sterile cork borer (BML505-15 mm, Wuhan Servicebio Technology Co., Ltd., Wuhan, China), a hole (about 5 mm in diameter and 10 mm in depth) was made in the center of each sterilized tuber. The tubers were pretreated with 1.5 mM SA and 2.0 mM Pro by pipetting them into the holes of the tubers using a 1000 µL Filtered Pipette Tip (ISO9001:2015, Wuhan Servicebio Technology Co., Ltd., Wuhan, China) 1 d before inoculation with the *L. amnigena* inoculum. A total of 0.3 mL of the *L. amnigena* inoculum was pipetted into the holes of the pretreated and untreated tubers with a 1000 µL Filtered Pipette Tip, and water as a control (CK). The inoculated tubers were placed in sterilized sealed plastic containers packed with sterilized moist cotton (moderate) and kept for 7 d at room temperature. The experiment was repeated with three replicates and for each experiment; 15 tubers were used for all the treatments.

2.5. Disease Assessment

The disease index (DI) of the tubers was measured up to 7 d after inoculation with PC3 inoculum. The disease index was determined using a method described by Scherf et al. [41], with a 5 degrees scale (0–4), where 0 = no disease, 1 = trace to 25% of the tubers were rotted, 2 = 26%–50% of the tubers were rotted, 3 = 51%–75% of the tubers were rotted, and 4 = 76%–100% of the tubers were rotted. The DI was determined using the formula DI (%) = [$\sum$ (number of diseased tubers $\times$ disease index)/(total number of tubers investigated $\times$ highest disease index)] $\times$ 100.

2.6. Sampling

Tissues were collected from inoculated tubers according to Zhang et al. [42] after 7 d of treatment with SA and Pro using a sterilized sharp blade. Tissues of untreated tubers were collected. Liquid nitrogen (Henan Boss Liquid Nitrogen Container Co., Ltd., Dongtai, China, 78.0% by volume, 75.5% by weight) was used to freeze the collected tissues, ground with a pestle and mortar into a powdery form, and then stored at $-80\,°C$ until use.

2.7. Malondialdehyde and Hydrogen Peroxide Content in Potato Tuber

The malondialdehyde (MDA) content was determined following the assay kit provided (BC0025, Solarbio Science and Technology Co., Ltd., Beijing, China). In brief, 0.1 g of frozen potato tuber was ground in liquid nitrogen (Henan Boss Liquid Nitrogen Container Co., Ltd., Dongtai, China, 78.03% by volume, 75.5% by weight). One mL of extract solution was added and centrifuged at $8000\times g$ for 10 min at $4\,°C$, and the supernatant was collected. Regents were added according to the manufacturer's instructions. MDA was measured at 600, 532 and 450 nm, and expressed as $\mu mol\ kg^{-1}$ FW. This was repeated three times. The hydrogen peroxide (H_2O_2) content was estimated following the protocol of the assay kit provided (BC3595, Solarbio Science and Technology Co., Ltd., Beijing, China). In brief, 0.1 g of frozen potato tuber was crushed in liquid nitrogen and placed on an ice bath in 1 mL of acetone. One mL of the extract solution was added and centrifuged at $8000\times g$ for 10 min at $4\,°C$, and the supernatant was collected. Other regents were added according to the manufacturer's instructions. The H_2O_2 content was measured at 415 nm and expressed as $\mu mol\ kg^{-1}$ FW. All the results are expressed on a fresh weight (FW).

2.8. Assay of Some Enzymatic Activities

NADPH oxidase (NOX; EC 1.6.3.1) activity was measured following the protocol of the assay kit provided (BC0630, Solarbio Science and Technology Co., Ltd., Beijing, China). 0.1 g of the frozen potato tuber was ground in liquid nitrogen. 1 mL of extract solution was added to 0.1 g of frozen powder of potato tuber and centrifuged at $600\times g$ for 5 min at $4\,°C$. The supernatant was transferred to another centrifuge tube and centrifuged at $11,000\times g$ for 10 min at $4\,°C$. The various reagents were added as instructed by the manufacturer. The absorbance (OD) value was determined at 600 nm, then NOX activity was expressed as $U\ g\ kg^{-1}$ FW. Peroxidase (POD; EC 1.11.1.7) activity was measured using the instructions of the assay kit provided (BC0090, Solarbio Science and Technology Co., Ltd., Beijing, China). 0.1 g of the frozen potato tuber was ground in liquid nitrogen. One mL of the extract solution was added to 0.1 g of frozen powder of potato tuber and centrifuged at $8000\times g$ for 10 min at $4\,°C$, and the supernatant was collected. Reagents were added as instructed. The OD value was measured at 470 nm using a spectrophotometer and POD activity was expressed as $U\ g\ kg^{-1}$ FW.

Catalase (CAT; EC 1.11.1.6) activity was analyzed following the instructions of the assay kit provided (BC0200, Solarbio Science and Technology Co., Ltd., Beijing, China). 0.1 g of the frozen potato tuber was ground in liquid nitrogen. One mL of the extract solution was added to 0.1 g of frozen powder of potato tuber and centrifuged at $8000\times g$

for 10 min at 4 °C, and the supernatant was collected. Reagents were added as instructed. The OD value was performed spectrophotometrically at 240 nm and CAT activity was expressed as U g kg^{-1} FW. Superoxide dismutase (SOD; EC 1.15.1.1) activity was analyzed using the instructions of the assay kit provided (BC0170, Solarbio Science and Technology Co., Ltd., Beijing, China). 0.1 g of the frozen potato tuber was ground in liquid nitrogen and 1 mL of the extract solution was added, then centrifuged at 8000× g for 10 min at 4 °C, and the supernatant was collected. Reagents were added as instructed by the manufacturer. The OD value of SOD was measured at 560 nm and SOD activity was expressed as U g kg^{-1} FW. Polyphenol oxide (PPO; EC 1.14.81.1) activity was analyzed according to the protocol of the assay kit provided (BC0195, Solarbio Science and Technology Co., Ltd., Beijing, China). 0.1 g of the frozen potato tuber was ground in liquid nitrogen. One mL of the extract solution was added and centrifuged at 8000× g for 10 min at 4 °C, and the supernatant was collected. Reagents were added as instructed by the manufacturer. The OD value of PPO was measured at 420 nm and PPO activity was expressed as U g kg^{-1} FW.

Phenylalanine ammonia-lyase (PAL; EC 4.3.1.5) activity was determined using the protocol of the assay kit provided (BC0210, Solarbio Science and Technology Co., Ltd., Beijing, China). 0.1 g of the frozen potato tuber was ground in liquid nitrogen, 1 mL of the extract solution was added and centrifuged at 8000× g for 10 min at 4 °C, and the supernatant was collected. The OD of PAL was measured spectrophotometrically at 290 nm, then expressed as U g kg^{-1} FW. Cinnamyl alcohol dehydrogenase (CAD; EC 1.1.1.195) activity was determined according to the instructions of the assay kit provided (BC4170, Solarbio Science and Technology Co., Ltd., Beijing, China). One mL of extract solution was added to 0.1 g of frozen powder of potato tuber, then centrifuged at 10,000× g for 10 min at 4 °C, and the supernatant was collected. Other reagents were added as instructed by the manufacturer. The OD was determined at 340 nm and CAD activity was expressed as U g kg^{-1} FW. 4-coumaryl-CoA ligase (4CL; EC 6.2.1.12) activity was assayed following the instructions of the assay kit provided (BC4220, Solarbio Science and Technology Co., Ltd., Beijing, China). One mL of the extract solution was added to 0.1 g of frozen powder of potato tuber, then centrifuged at 8000× g for 10 min at 4 °C, and the supernatant was collected. Other reagents were added as instructed by the manufacturer. The OD was measured at 333 nm and 4CL activity was expressed as U g kg^{-1} FW. Cinnamate-4-hydroxylase (C4H; EC 1.14.13.11) activity was determined using a kit provided (BC4080, Solarbio Science and Technology Co., Ltd., Beijing, China). One mL of the extract solution was added to 0.1 g of frozen powder of potato tuber, then centrifuged at 12,000× g for 15 min at 4 °C, and the supernatant was collected. Regents were added as instructed The OD was measured at 340 nm and expressed as U g kg^{-1} FW.

2.9. Quantitative Real-Time (qRT) PCR Analysis

Quantitative RT–PCR (ABI 7500, Applied Biosystems, Foster City, CA, USA) was conducted to evaluate gene expression (*NOXB*, *PAL*, *CAD*, *4CL*, and *C4H*) and plant defense-related genes [pathogenesis-related protein (*PR1*), tyrosine-protein kinase (*PR2*), phenylalanine ammonia-lyase (*PAL*), Chitinase-like protein (*CTL1*), pathogenesis-related homeodomain protein (*PRH3*), and serine protease inhibitor (*SPI1*)] in the potato tubers subjected to SA and Pro treatments under *L. amnigena* stress. A PureLink® RNA Mini Kit (Tiangen Biotechnology, Beijing, China) was used to extract total RNA. A Nano-Drop spectrophotometer was used to measure the quantity and quality of isolated RNA at absorbances of 230 and 260 nm. The A$_{260}$/A$_{280}$ ratio showed that the RNA was not contaminated with proteins. The Revert AidTM First Strand cDNA Synthesis Kit was used for first-strand cDNA synthesis (Tiangen Biotechnology, Beijing, China). The 23 μL reaction of cDNA contained 0.5 μL of RNA, 2 μL oligo (dT), 4 μL 5 × M-MLV buffer, 1 μL dNTPs, 0.5 μL RNasin, 1 μL M-MLV, and 14 μL of ddH$_2$O. The primer sequences and NCBI gene IDs are presented in Table 1. qRT–PCR was determined using 2 × SYBR Green qPCR Master Mix (Shanghai LZ Biotech Co., Ltd., Shanghai, China). The 20 μL reaction mixture contained 1 μL of each primer, 1 μL of cDNA, 0.4 μL of ROX reference dye, 10 μL 2 × SYBR

Green qPCR Master Mix, and 6.6 µL of ddH$_2$O, and was determined. Using potato actin, relative expression levels were calculated using the formula $2^{-\Delta\Delta CT}$ [43]. For each gene, three biological replicates were used.

Table 1. Gene description, primers sequences, and NCBI gene ID for the genes used for the qRT–PCR.

Gene Symbol	Description	Primer Sequence (5′-3′)	Gene ID	Activity
Act	Actin	F: ACAATGCTTGCACGTTTCCTC R: TTAGCTGGGACCATTGCCTG	102605823	Antioxidant
NOXB	NADPH oxidase	F: CATTGCTTCTTCAGGCTCCG R: CCACAAAGCCATCACCCAAA	111509039	Antioxidant
PAL	Phenylalanine ammonia-lyase	F: GAGGAGTATAGGAAGCCGG R: CTCATCCCTTCCATCACCCA	102596017	Antioxidant
CAD	Cinnamyl alcohol dehydrogenase	F: GGCCTGATGATGTGCAAGTC R: CCAACAAGCAATCCAACTCCA	102584791	Antioxidant
4CL	4-counmaryl-CoA ligase	F: GCCCTGAATTTGTGTTTGCG R: CCTTCACTTTCCCCGCAAAA	102596056	Antioxidant
C4H	Cinnamate-4-hdroxylase	F: AGTCTGAGGCTGCTAGTGT R: GAGTCTGAGGCTGCTAGTGT	817599	Antioxidant
PR 1	Pathogenesis-related protein	F: GCCAATCCAGGCTGTAGCA R: AGTGGGGAAGAAGAATGTGGAC	102580826	Plant defense
PR 2	Tyrosine-protein kinase	F: ACCGCCTTCGAGAACTAGAG R: CCACAAACTTGCCATATCACCA	111517981	Plant defense
PRH3	Pathogenesis-related homeodomain protein	F: GCAAAGGGGAAGCTGGGTAA R: TGTTACTTTCAGCTGCATCCTCT	102596310	Plant defense
CTL1	Chitinase-like protein 1	F: ATTACGGTCGTGGTGCCTTG R: ATCTGCAACTGCTTTCCGTG	102595303	Plant defense
PAL	Phenylalanine ammonia-lyase	F: TGGTGGTGCCCTTCAAAAAG R: CGTAGCTTGTTATGTCATGATGAT	102596017	Plant defense
SPI1	Serine protease inhibitor-1	F: TAGGTGGCCAGAACTGGTTG R: TGTGTTAGCGATTGTCCTTCGA	823839	Plant defense
Act	Actin	F: ACAATGCTTGCACGTTTCCTC R: TTAGCTGGGACCATTGCCTG	102593148	Plant defense

2.10. Statistical Analysis

The data were subject to one-way ANOVA using the SPSS package (SPSS V16.0; SPSS, Inc., Chicago, IL, USA). Treatment effects were determined using Duncan's multiple range test and significant results were expressed at $p < 0.05$.

3. Results

3.1. Effect of SA and Pro on Extracellular Enzyme Production by L. amnigena

The results of our study show that SA and Pro affected the synthesis of pectinase, protease, pectin lyase, and cellulase, which are virulence factors in *L. amnigena*. The application of SA and *L. amnigena* (PC3) (SA + PC3) lowered the production of pectinase, protease, pectin lyase, and cellulase by 55.6, 73.1, 55.1, and 62.5%, respectively, compared to the control (Figure 1). In addition, co-cultured Pro and PC3 (Pro + PC3) decreased pectinase, protease, pectin lyase, and cellulase by 40.7, 64.5, 53.3, and 34.1%, respectively. However, the combined SA and Pro with PC3 (SA + Pro + PC3) reduced pectinase, protease, pectin lyase, and cellulase synthesis by 64.3, 77.8, 66.4, and 84.1%, respectively, compared to the control (Figure 1).

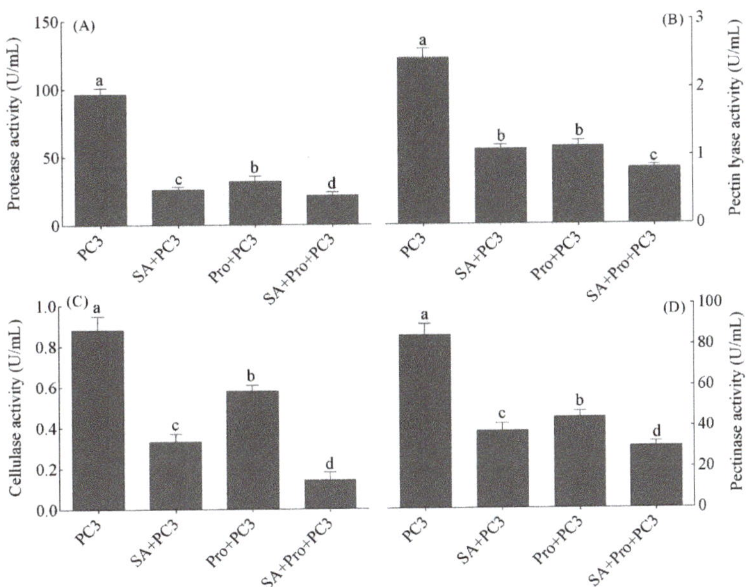

Figure 1. Effect of salicylic acid (SA) and proline (Pro) on Protease (**A**), Pectin lyase (**B**), Cellulase (**C**), and Pectinase (**D**) production by *L. amnigena*. Data are presented as mean ± standard error (SE) of two independent experiments performed in three replicates. Means with the same lowercase letters are not significantly different at $p < 0.05$ according to Duncan's multiple range test. PC3–*L. amnigena*.

3.2. Disease Assessment

The results show that SA or Pro and their combined treatments reduced the disease index of potato soft rot. However, *L. amnigena*-treated tubers experienced a high incidence of soft rot compared to potato tubers treated with SA and Pro (Figure 2). The results show that applied SA reduced the disease index by 67.9% after 6 d of treatment. In addition, applied Pro reduced the disease index by 64.6% after 6 d of treatment. However, the combined application of SA and Pro reduced the disease index by 72.5% after 6 d of treatment. The combination of SA and Pro treatment provided better disease control than either SA or Pro application alone (Figure 2).

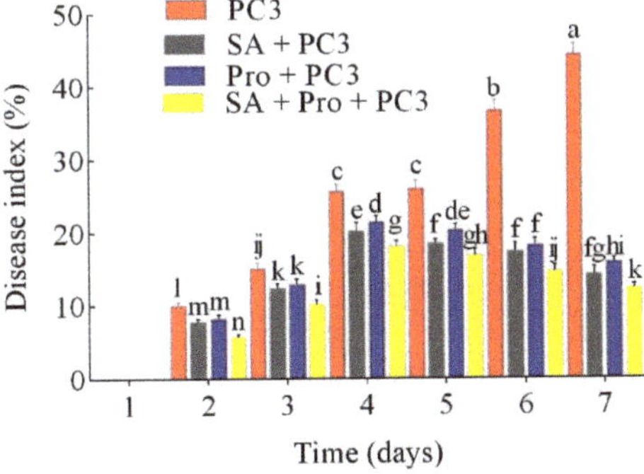

Figure 2. Effect of salicylic acid (SA) and proline (Pro) on the disease index of tubers inoculated with *L. amnigena* (PC3). Data are presented as mean ± standard error (SE), based on two independent experiments with three replicates. Means with the same lowercase letters are not significantly different at $p < 0.05$ according to Duncan's multiple range test.

3.3. Effects of SA and Pro on MDA and H$_2$O$_2$ Content in Potato Tubers Inoculated with L. amnigena

Our results in Figure 3 show that *L. amnigena* stress significantly increased MDA and H$_2$O$_2$ contents in potato tubers by 50.0 and 40.7%, respectively, compared to the control. In Figure 3, the beneficial impact of SA or proline or SA + Pro on decreasing oxidative stress and MDA and H$_2$O$_2$ is observed. These treatments resulted in a significant decrease in MDA and H$_2$O$_2$, and the best treatment was SA + Pro (77.2 and 83.8%), followed by SA then Pro (Figure 3).

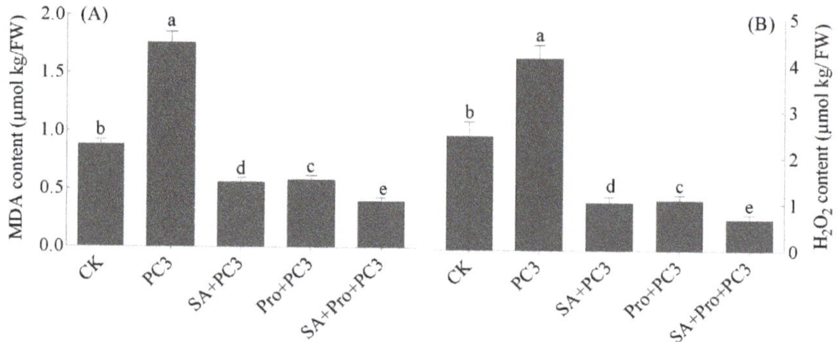

Figure 3. Effect of salicylic acid (SA) and proline (Pro) on the MDA (**A**) and H$_2$O$_2$ (**B**) content of potato tubers under *L. amnigena* (PC3) stress, where CK represents the control treatment with distilled water. Data are presented as mean ± standard error (SE), based on two independent experiments with three replicates. Means with the same lowercase letters are not significantly different at $p < 0.05$ according to Duncan's multiple range test.

3.4. Effects of SA and Pro on NOX, SOD, POD, PPO, and CAT Activity in Potato Tubers Inoculated with L. amnigena

The activities of NOX, SOD, POD, CAT, and PPO as stress indicators were higher in potato tubers exposed to *L. amnigena* than in controls. The data presented in Figure 4A–E show that antioxidant enzymes NOX, SOD, POD, PPO, and CAT activity significantly increased by 69.3, 80.9, 78.0, 86.0, and 83.0%, respectively, in potato tubers under *L. amnigena* stress treated with SA, respectively, compared with untreated stressed tubers. In addition, Pro-treated tubers increased NOX, SOD, POD, CAT, and PPO activity by 62.1, 69.8, 70.4, 66.9, and 73.9%, respectively, compared with untreated stressed tubers. However, the best results of NOX, SOD, POD, CAT, and PPO activity were recorded with SA + Pro treatment compared with untreated stressed tubers (Figure 4A–E).

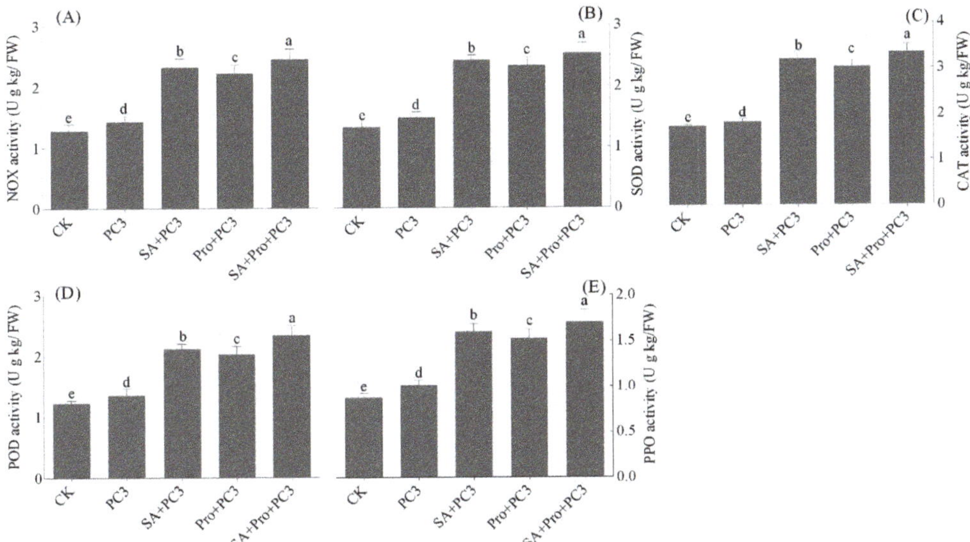

Figure 4. Effect of salicylic acid (SA) and proline (Pro) on NOX (**A**), SOD (**B**), CAT (**C**), POD (**D**), and PPO (**E**) activities in potato under *L. amnigena* (PC3) stress, where CK represents the control treatment with distilled water. Data are presented as mean ± standard error (SE), based on two independent experiments with three replicates. Means with the same lowercase letters are not significantly different at $p < 0.05$ according to Duncan's multiple range test.

3.5. SA and Pro Treatment Increased PAL, CAD, 4CL, and C4H Activity in Potaato Tubers Inoculated with L. amanigena

In the current study, the results show that PAL, CAD, 4CL, and C4H activities increased significantly by exposure to *L. amnigena* stress (Figure 5). However, the application of SA led to an increase in PAL, CAD, 4CL, and C4H activities by 77.7, 73.6, 79.1, and 76.0%, respectively, compared with untreated stressed tubers. Additionally, application of Pro increased PAL, CAD, 4CL, and C4H activities by 70.4, 60.4, 72.1, and 64.9%, respectively, as compared to the control. Similarly, the activities of PAL, CAD, 4CL, and C4H were higher in the SA + Pro treatments than in the two single treatments (Figure 5).

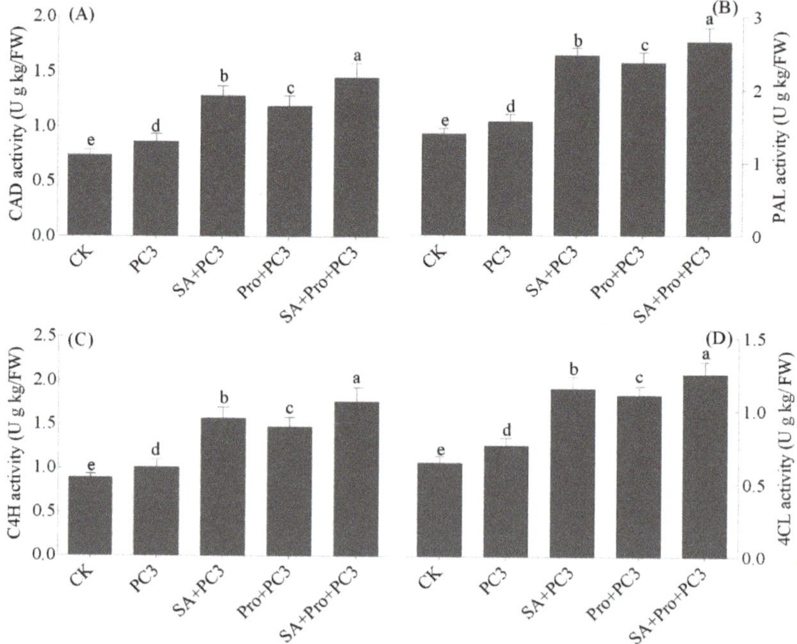

Figure 5. Effect of salicylic acid (SA) and proline (Pro) on CAD (**A**), PAL (**B**), C4H (**C**), and 4CL (**D**) activities in potato tubers under *L. amnigena* (PC3) stress, where CK represents the control treatment with distilled water. Data are presented as mean ± standard error (SE), based on two independent experiments with three replicates. Means with the same lowercase letters are not significantly different at $p < 0.05$ according to Duncan's multiple range test.

3.6. SA and Pro Treatment Up-Regulated NOX, PAL, CAD, 4CL, and C4H Genes in Potato Tubers Inoculated with L. amnigena

When compared to untreated stressed tubers, the NOX, PAL, CAD, 4CL, and C4H transcripts were significantly induced in the SA, Pro, and SA + Pro treatments (Figure 6A–E). SA treatment increased the expression of NOX, PAL, CAD, 4CL, and C4H genes 2.7, 2.8, 1.8, 1.8, and 1.6-fold, respectively, compared with untreated stressed tubers. Similarly, Pro-treated tubers elevated the expression of NOX, PAL, CAD, 4CL, and C4H genes 2.5, 2.6, 1.7, 1.7, and 1.4-fold, respectively, compared with untreated stressed tubers. However, in the SA + Pro treatments, the transcriptional levels of NOX, PAL, CAD, 4CL, and C4H genes were significantly higher (3.1, 3.1, 2.2, 2.5, and 2.1-fold), respectively, compared to the control (Figure 6A–E).

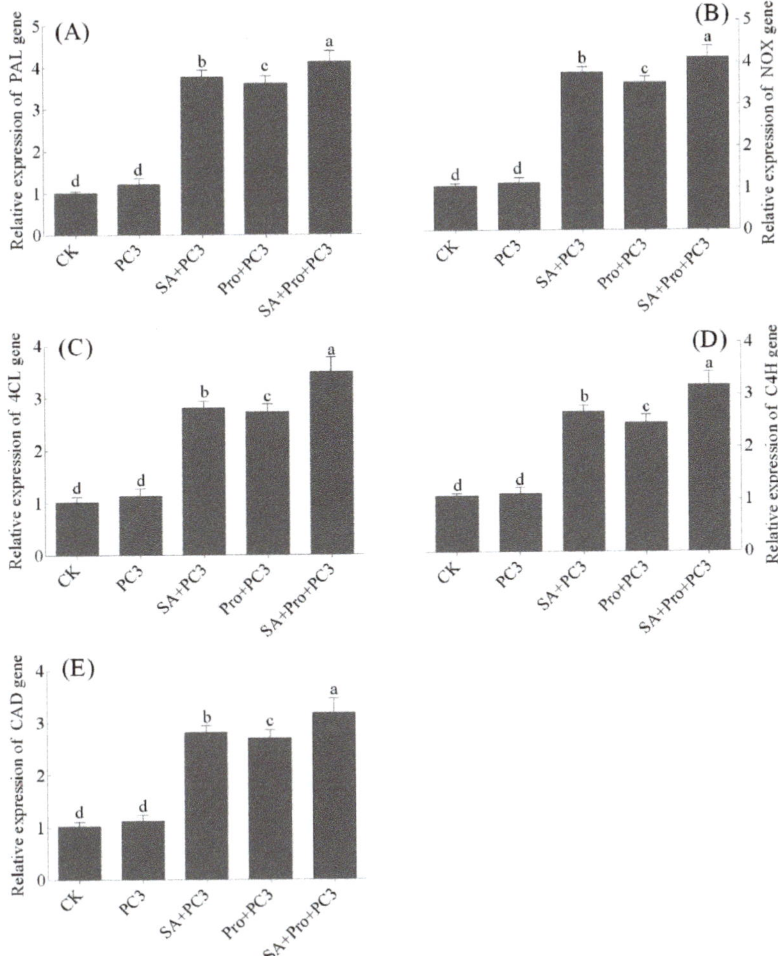

Figure 6. Effect of salicylic acid (SA) and proline (Pro) on the gene expression of PAL (**A**), NOX (**B**), 4CL (**C**), C4H (**D**), and CAD (**E**) in potato tubers under *L. amnigena* (PC3) stress, where CK represents the control treatment with distilled water. Data are presented as mean ± standard error (SE), based on two independent experiments with three replicates. Means with the same lowercase letters are not significantly different at $p < 0.05$ according to Duncan's multiple range test.

3.7. Effects of Combined SA and Pro Treatment on the Expression Levels of Plant Defense-Related Genes in Potato Tubers Inoculated with L. amnigena

SA, Pro, and SA + Pro treatments affected plant defense-related genes (PR1, PR2, CTL1, PAL, PRH3, and SPI1) (Figure 7A–F). The results show that PR1, PR1, CTL1, PAL, PRH3, and SPI1 genes were elevated by exposure to *L. amnigena* stress (Figure 7A–F). However, treatment of *L. amnigena*-infected tubers with SA increased the transcription levels of PR1, PR2, CTL1, PAL, PRH3, and SPI1 2.8, 2.0, 2.1, 2.1, 2.5, and 2.7-fold, respectively, compared to the untreated stressed tubers. Pro treatment also increased PR1, PR2, CTL1, PAL, PRH3, and SPI1 2.5, 1.9, 2.0, 2.0, 2.4, and 1.5-fold, respectively, compared to *L. amnigena* alone (Figure 7). However, treatment of combined SA and Pro further increased PR1, PR2, CTL1, PAL, PRH3, and SPI1 3.1, 2.5, 2.6, 2.4, 3.0, and 2.9-fold, respectively, compared to the control.

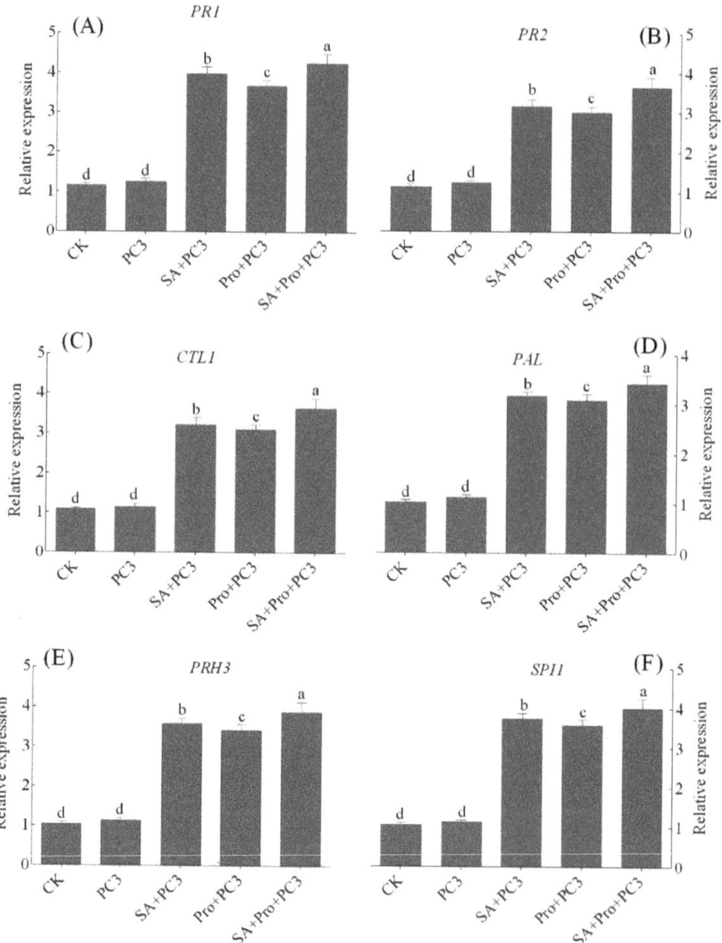

Figure 7. Effect of salicylic acid (SA) and proline (Pro) on the relative expression of pathogenesis-related protein (PR1) (**A**), tyrosine-protein kinase (PR2) (**B**), Chitinase-like protein (CTL1) (**C**), phenylalanine ammonia-lyase (PAL) (**D**), pathogenesis-related homeodomain protein (PRH3) (**E**), and serine protease inhibitor (SPI1) (**F**) in potato tubers under *L. amnigena* (PC3) stress, where CK represents the control treatment with distilled water. Data are presented as mean ± standard error (SE), based on two independent experiments with three replicates. Means with the same lowercase letters are not significantly different at $p < 0.05$ according to Duncan's multiple range test.

4. Discussion

Extracellular enzymes can be used as a primary mechanism by pathogens, including bacteria, to develop plant diseases. The activity of the extracellular enzymes permits the pathogen to penetrate host tissues by breaking down the host cells' defensive outer layers. Extracellular enzymes secreted by microorganisms play an important role in disease progression. These degrade plant materials into smaller particles that pathogens can easily absorb and use for growth and development. In this current study, pectinase, protease, pectin lyase, and cellulase were produced by *L. amnigena*, aiding it to cause potato soft rot. Application of SA or Pro alone to *L. amnigena* reduced the synthesis of pectinase, protease, pectin lyase, and cellulase. However, combined SA and Pro further reduced the production of these enzymes. This means that, while using SA and Pro alone can limit extracellular enzyme production by *L. amnigena*, combining the two is the most effective. Similarly, Bandara et al. [44] discovered that applied SA had an impact on the secretion of protease and elastase by *Pseudomonas aeruginosa*, which is involved in the development of microbial keratitis. Plant pathogens have evolved complex penetration, invasion, and colonization tactics to disable susceptible hosts' plant defense mechanisms and cause disease [45]. The pathogen normally infects the plant by producing components that are utilized to manipulate plant tissue to get physical access to the tissues and to draw nutrients [46]. These factors allow the plant both to induce disease and to advance the infection into the tissue's core, resulting in a worsening of the disease situation. Pathogens attack plants in different ways. One mode necessitates that the plants and living plant tissues cooperate. Within the invaded plant tissues, the pathogen produces penetration structures and a network of flagella [47]. This mass of cell wall structures generated in the plant's intercellular gaps extracts nutrients from the plant.

The reduction in disease severity caused by SA and Pro could be the result of the antibacterial consequence of SA and Pro, which reduces *L. amnigena* production of extracellular enzymes. The results show that the treatment of SA and Pro significantly reduced potato soft rot induced by *L. amnigena*; we suggested that combined SA and Pro treatment in the present study has synergistic effects on the control of potato soft rot caused by *L. amnigena*. Our results could also be attributed to either a direct toxicity effect of SA and Pro on *L. amnigena* growth or an indirect plant defense-related effect by inducing resistance in infected tissues. Although the molecular mechanism of SA and Pro-induced *L. amnigena* resistance is unknown, it appears to include inhibition of pathogen-secreted extracellular enzymes. SA was found to be effective in generating localized acquired tolerance to *Pectobacterium carotovora* subsp. *carotovora* infection [48]. Lastochkina et al. [49] previously discovered that SA can boost potato resistance to *F. oxysporum* and *Phytophthora infestans* postharvest disease. Eshgpour et al. [50] also reported that applied SA reduced infection of *Pectobacterium carotovorum* causing potato soft rot in vitro studies, which are similar to the current study's findings. Our findings are comparable with those of Bawa et al. [51], who found that applying SA to soybean seedlings decreased disease severity and induced resistance to *Fusarium solani*. According to Li and Zou [52], foliar spraying tomato plants with SA at a dosage of 2.0 mM resulted in a substantial reduction in disease severity. According to Yao and Tian [22], the roles of SA in reducing brown rot may be due to the direct toxicity of SA on fungal mycelia and/or an indirect plant defense-related effect by activation of some defense enzymes that play an important role, such as: (i) breaking down the fungus cell wall (such as chitinase and -1,3-glucanase); (ii) saving the plant cell wall; or (iii) increasing antioxidants (such as PAL or POD). According to Cecchini et al. [53], Pro is a defense compound contributing to hypersensitive response and tolerance to diseases. Pro treatments have thus been shown in several investigations to reduce the negative impacts of environmental pressures such as pathogen infections [54]. Ben et al. [55] reported that Pro application induces plant resistance to various pathogens through mechanisms such as oxidative scavenger and gene expression regulation. Qian et al. [56] also found that applied Pro mitigated superficial scald incidence and index in pear fruit.

A well-known side impact of stress is a buildup of oxidatives (reactive oxygen species, ROS) [57]. To minimize ROS damage effectively, plants have evolved scavenging systems like antioxidants. SOD, POD, PAL, and CAT, which are essential enzyme systems for ROS scavenging mechanisms, are essential metrics used for measuring plant resistance to stress. ROS scavenging mechanisms, are mediated by antioxidant enzymes and are the first edge of defense against stress. To reduce ROS generation and interference, as well as to alleviate the negative impacts of stress on plant growth and development, effective antioxidant capacity is required [57]. Protein denaturation occurs in plant cells due to oxidative stress, and significant amounts of MDA and H_2O_2 accumulate, which may act as primary stress mediators and triggers of plant defense systems [58]. While reactive oxygen species (ROS) can help improve plant tissue resistance to pathogen infections, high accumulations of ROS also lead to lipid peroxidation and loss of plant organ membrane virtue. In this study, potato tubers responded to *L. amnigena* treatment by producing more oxidants such as MDA and H_2O_2. SA reduced oxidant content in tubers inoculated with *L. amnigena* by acting as an antibiotic against bacteria pathogens, which is consistent with the study of Mishra and Baek [59]. Bawa et al. [60] found a decrease in H_2O_2 and MDA levels in soybean plants treated with SA, compared to untreated soybean plants under biotic stress. Studies by Estaji [61] and Sayyari et al. [62] discovered that applying SA reduced oxidants such as MDA and H_2O_2 content in pepper and purslane, respectively, under drought stress. According to Naeem et al. [63], SA at 0.5 mM may be administered to tomato plants under saline conditions up to 90 mM, significantly alleviating the harmful effect of salt stress. Iqbal et al. [64] demonstrated that SA supplementation reduced the negative effects of salt stress on wheat cultivar development. The results show that the content of MDA and H_2O_2 in Pro-treated tubers compared to untreated inoculated tubers indicates that proline has efficacy in alleviation of *L. amnigena*-induced oxidative damage in potato tubers. This finding agrees with the findings of Hayat et al. [65], who discovered that applied Pro decreased MDA and H_2O_2 contents in pigeon peas subjected to cadmium stress. Pro application enhances plant resistance to biotic stress [66]. However, the combined therapy of SA and Pro with *L. amnigena*-infected tubers significantly reduced MDA and H_2O_2 content. De Carvalho et al. [67] discovered that applied Pro reduced the negative impacts of ROS as a result of increased SOD, POD, and CAT. The results of the current study collaborate with those of Abdelaal et al. [32], who found that combined SA and Pro lowered MDA and H_2O_2 levels in barley plants under drought stress. Sanchez-Rodriguez et al. [68] demonstrated that tomato plants treated with Pro reduced MDA and H_2O_2 levels, while increasing the activity of the antioxidant enzymes.

In a variety of plants, the role of antioxidant enzymes in plant defense against pathogen stress has been studied [69]. The current study found that either SA or Pro alone increased the activity of NOX, SOD, POD, CAT, and PPO in tubers subjected to *L. amnigena* stress. However, combined treatment of SA and Pro with *L. amnigena*-infected tubers further increased the activity of NOX, SOD, POD, CAT, and PPO, compared to the controls. The role of SA in the antioxidative system is widely assumed to be that of a signal molecule. This result suggests that SA treatment of *L. amnigena*-inoculated tubers may induce resistance because it activates the plant defense system by increasing some enzymes (NOX, SOD, POD, CAT, and PPO). Previous research has shown that, under stress, SA can maintain the activity of antioxidant enzymes to some extent, and it may also help to limit the impact of oxidative processes associated with disease development and spread, implying that SA may play an important antioxidant role in oxidative processes associated with plant defense responses [70,71]. Furthermore, SA-induced protein synthesis could be used to activate antioxidants [72]. The findings support the findings of Ma et al. [73], who found that during salt stress, SA raised the antioxidant enzyme activity in *Dianthus superbus*. Similar results were obtained in a previous work on *Portulaca oleracea* L., in which SA increased antioxidant enzyme activity, lowering ROS concentrations under diverse environmental conditions [74]. By adhering to hydrogen bonds, Pro can increase protein stability and safeguard membrane integrity [75]. Pro may also protect cells by enhancing water absorption capacity and promoting enzyme activity [76]. Pro, in addition to being an osmolyte, is a powerful antioxidative defense molecule, a metal chelator, a protein stabilizer, a ROS scavenger, and an inhibitor of programmed cell death [77,78]. Exogenous Pro has been shown in several studies to improve plant stress tolerance [7,79]. This type of treatment may activate stress avoidance systems by increasing stress tolerance and enhancing reactivity to stress triggers during a later stressful condition [54]. Pro contributes to the plant's stress response by enhancing antioxidant enzymes [65]. Pro was found to increase the enzyme's activity associated with the ascorbate–glutathione revolution [80], implying a role in improving cell antioxidant capacity. Abdelhamid et al. [81] discovered that applied Pro raised SOD, CAT, and POD in *Phaseolus vulgaris* L. plants subjected to salt stress. Furthermore, Tabssum et al. [82] reported that applied Pro at 50 mM increased SOD and CAT activity in rice under salt conditions. Ghaffari et al. [83] reported that

Pro-mediated changes in antioxidant enzymatic activities and the physiology of sugar beet under drought stress. Pro regulates the activity of SOD, CAT, and POX enzymes in plant cells, as well as their involvement in metabolic response development in response to environmental factors, according to Abdallah and El-Bassiouny [84]. The results of the present study are also in agreement with those of Abdelaal et al. [32], who discovered that combining SA and Pro increased drought resistance in barley plants through modulation of antioxidant activities. Based on the findings, it is proposed that the reduction of oxidative stress by activated antioxidant enzymatic systems may contribute to the synergistic effects of SA and Pro in the control of *L. amnigena*-caused potato soft rot.

Phenylpropanoid metabolism aids in the reduction of potato soft rot by providing substrates for the synthesis of phenolic and monolignin with antibacterial properties [85]. PAL is an important enzyme that initiates this metabolism by deaminating L-phenylalanine to trans-cinnamic acid [86]. In the current study, SA and Pro treatments significantly increased the production of phenylpropanoids such as PAL, CAD, 4CL, and C4H in potato tubers inoculated with *L. amnigena*. This could be due to an increase in the expression levels of phenylpropanoid-related genes following treatment with SA or Pro, as previous research has shown that different types of elicitors increase the expression levels of phenylpropanoid-related genes and increase phenolic compound accumulation. Thus, SA and Pro treatment may promote the phenylpropanoid pathway by increasing enzyme activity and the synthesis of phenol compounds and lignin at the tuber's rotting site. SA may affect the phenylpropanoid pathway by inducing key enzymes such as PAL, CAD, 4CL, and C4H, resulting in phenolic compound accumulation. PAL, CAD, 4CL, and C4H also act as plant defense mechanisms, and the activities of these phenylpropanoids help the potato tubers to reduce the excess reactive oxygen species which is the source of oxidative stress during pathogen infection [87]. Plants can successfully induce PAL, CAD, 4CL, and C4H activity when infected with microorganisms such as bacteria pathogens [20,88]. Our results confirm this; PAL, CAD, 4CL, and C4H activities increased in *L. amnigena*-infected tubers treated with SA and Pro.

The important role of SA and Pro in protecting plants from stress may be due to their capability to increase the gene expression of *PR* proteins [89]. In this current study, potato tubers infected with *L. amnigena* increased relative expression of *NOX, PAL, CAD, 4CL,* and *C4H* compared with the control. However, combined treatment of SA and Pro with *L. amnigena*-infected tubers significantly up-regulated *NOX, PAL, CAD, 4CL,* and *C4H* genes. Lavrova et al. [90] reported that applied SA promotes gene expression in tomato plants infected with pathogens. These findings also support those of El-Esawi et al. [91], who found that *bZIP62, DREB2, ERF3,* and *OLPb* were increased in rosemary plants treated with SA, implying that applied SA-modulated genes enhanced the rosemary plants' resistance to the salinity condition. SA reduces oxidative damage caused by *L. amnigena* stress by up-regulating antioxidant defense mechanisms, making potato tubers more resistant to this type of biotic stress [92]. The Pro is a radical scavenger, in addition to being a suitable osmolyte [93]. As a result, Pro served as both an osmolyte molecule and an antioxidant [94]. Pro also serves as an antioxidative defense molecule and a signaling molecule during stress, in addition to being an effective osmolyte [93].

Plant defenses defined by pathogen gene-for-gene recognition by plants containing resistance genes have also been connected to the stimulation of SA-dependent R-genes. Higher expression of PR proteins has been linked to activated resistance attained with SA in earlier research [95]. In addition, SA-primed stressed-infected plants had 1.7, 2.9, 2.1, 2.5, and 2-fold *IAA27, MPK1, GPX,* chitinase, and 1,3-glucanase, respectively, than non-primed susceptibility inoculation controls. Since these genes are critical for developing resistance during the host–pathogen interaction, the high level of *IAA27, MPK1, GPX,* chitinase, and 1,3-glucanase correlates with disease protection studies. The increased stimulation of defense gene processes in potato tubers treated with SA or Pro alone suggests the molecular mechanisms governing both SA and Pro to alleviate *L. amnigena* infection in tubers. Combined treatment of SA and Pro with *L. amnigena*-infected tubers induced the up-regulation of *PR1, PR2, CTL1, PAL, PRH3,* and *SPI1*, compared to the controls, which led to the most important effector genes for systemic acquired resistance (SAR) mediated by SA and Pro [96,97]. Chen et al. [98] discovered that applied Pro at 0.5 mM activated PR gene expression. The discovery that Pro raises the *OxyR* gene implies that Pro metabolism raises ROS scavengers. *OxyR* reacts with H_2O_2 to form a disulfide bond between *Cys199* and *Cys208*, resulting in *OxyR* regulon transcriptional activation [99].

5. Conclusions

The use of SA and Pro could help to reduce potato soft rot caused by *L. amnigena*. It could be attributed to the stimulation of antioxidant enzymes and the alleviation of oxidative damage caused by pathogenic infection, which stabilized intracellular redox homeostasis. SA and Pro treatment also increased the activities of *PR1, PR2, CTL1, PAL, PRH3,* and *SPI1* in the tuber, which may slow the growth of pathogenic bacteria (*L. amnigena*). These findings suggest that antioxidants and plant defense-related genes, particularly in potato tubers, are important in the defense response to pathogen infection.

Author Contributions: Conceptualization, R.O. and C.Y.; methodology, R.O., S.B., L.W. and M.J.; software, R.O. and S.B.; validation, R.O., C.Y. and L.W.; formal analysis, R.O., C.Y. and M.J.; investigation, R.O.; resources, C.Y.; data curation, R.O. and S.B.; writing—original draft preparation, R.O.; writing—review and editing, C.Y.; visualization, R.O. and S.B.; supervision, C.Y.; project administration, C.Y.; funding acquisition, C.Y. All authors have read and agreed to the published version of the manuscript.

Funding: The Science and Technology Innovation Fund of Gansu Agricultural University (No. GAU-XKJS-2018–148) and the Project of National Potato Industry Technology System (No. CARS-10-P18).

Institutional Review Board Statement: Not applicable.

Informed Consent Statement: Not applicable.

Data Availability Statement: Not applicable.

Acknowledgments: We thank our supervisor, C. Yang and K.G. Santo who provided insight and expertise that greatly assisted the research.

Conflicts of Interest: The authors declare no conflict of interest.

References

1. Lim, S.; Chisholm, K.; Coffin, R.H.; Peters, R.D.; Al-Mughrabi, K.I.; Wang-Pruski, G.; Pinto, D.M. Protein profiling in potato (*Solanum tuberosum* L.) leaf tissues by differential centrifugation. *J. Proteome Res.* **2012**, *11*, 2594–2601. [CrossRef] [PubMed]
2. Sun, X.; Li, Y.; Bi, Y.; Liu, J.; Yin, Y. Investigation analysis on potato disease during storage in northwest China. *Chin. Potato J.* **2009**, *23*, 364–365.
3. Bojanowski, A.; Avis, T.J.; Pelletier, S.; Tweddell, R.J. Management of potato dry rot. *Postharvest Biol. Technol.* **2013**, *84*, 99–109. [CrossRef]
4. Czajkowski, R.; Perombelon, M.C.; van Veen, J.A.; van der Wolf, J.M. Control of blackleg and tuber soft rot of potato caused by *Pectobacterium* and *Dickeya* species: A review. *J. Plant Pathol.* **2011**, *60*, 999–1013. [CrossRef]
5. Dees, M.W.; Lebecka, R.; Perminow, J.I.S.; Czajkowski, R.; Motyka, A.; Zoledowska, S.; Śliwka, J.; Lojkowska, E.; Brurberg, M.B. Characterization of *Dickeya* and *Pectobacterium* strains obtained from diseased potato plants in different climatic conditions of Norway and Poland. *Eur. J. Plant Pathol.* **2017**, *148*, 839–851. [CrossRef]
6. Duarte, V.; De Boer, S.; Ward, L.d.; De Oliveira, A. Characterization of atypical *Erwinia carotovora* strains causing blackleg of potato in Brazil. *J. Appl. Microbiol.* **2004**, *96*, 535–545. [CrossRef]
7. Orsini, F.; Pennisi, G.; Mancarella, S.; Al Nayef, M.; Sanoubar, R.; Nicola, S.; Gianquinto, G. Hydroponic lettuce yields are improved under salt stress by utilizing white plastic film and exogenous applications of proline. *Sci. Hortic.* **2018**, *233*, 283–293. [CrossRef]
8. Bateman, D. Plant cell wall hydrolysis by pathogens. In *Biochemical Aspects of Plant-Parasite Relationships*; Friend, J., Threlfall, D.R., Eds.; Academic Press: London, UK, 2012; pp. 79–103.
9. Jess, S.; Kildea, S.; Moody, A.; Rennick, G.; Murchie, A.K.; Cooke, L.R. European Union policy on pesticides: Implications for agriculture in Ireland. *Pest Manag. Sci.* **2014**, *70*, 1646–1654. [CrossRef]
10. Torres, R.; Vilanova, L.; Usall, J.; Teixidó, N. Insights into Fruit Defense Mechanisms Against the Main Postharvest Pathogens of Apples and Oranges. In *Postharvest Pathology*; Springer: Berlin/Heidelberg, Germany, 2021; pp. 21–40. [CrossRef]
11. Tian, S.; Torres, R.; Ballester, A.; Li, B.; Vilanova, L.; González-Candelas, L. Molecular aspects in pathogen-fruit interactions: Virulence and resistance. *Postharvest Biol. Technol.* **2016**, *122*, 11–21. [CrossRef]

12. Droby, S.; Vinokur, V.; Weiss, B.; Cohen, L.; Daus, A.; Goldschmidt, E.; Porat, R. Induction of resistance to *Penicillium digitatum* in grapefruit by the yeast biocontrol agent *Candida oleophila*. *Phytopathology* **2002**, *92*, 393–399. [CrossRef]

13. Tripathi, D.; Raikhy, G.; Kumar, D. Chemical elicitors of systemic acquired resistance—Salicylic acid and its functional analogs. *Curr. Plant Biol.* **2019**, *17*, 48–59. [CrossRef]

14. Lowe-Power, T.M.; Jacobs, J.M.; Ailloud, F.; Fochs, B.; Prior, P.; Allen, C. Degradation of the plant defense signal salicylic acid protects *Ralstonia solanacearum* from toxicity and enhances virulence on tobacco. *MBio* **2016**, *7*, e00656-16. [CrossRef]

15. Zhang, J.J.; Wang, Y.K.; Zhou, J.H.; Xie, F.; Guo, Q.N.; Lu, F.F.; Jin, S.F.; Zhu, H.M.; Yang, H. Reduced phytotoxicity of propazine on wheat, maize and rapeseed by salicylic acid. *Ecotoxicol. Environ. Saf.* **2018**, *162*, 42–50. [CrossRef] [PubMed]

16. Afroz, A.; Khan, M.R.; Ahsan, N.; Komatsu, S. Comparative proteomic analysis of bacterial wilt susceptible and resistant tomato cultivars. *Peptides* **2009**, *30*, 1600–1607. [CrossRef] [PubMed]

17. Narasimhamurthy, K.; Soumya, K.; Udayashankar, A.; Srinivas, C.; Niranjana, S. Elicitation of innate immunity in tomato by salicylic acid and Amomum nilgiricum against *Ralstonia solanacearum*. *Biocatal. Agric. Biotechnol.* **2019**, *22*, 101414. [CrossRef]

18. Lefevere, H.; Bauters, L.; Gheysen, G. Salicylic acid biosynthesis in plants. *Front. Plant Sci.* **2020**, *11*, 338. [CrossRef] [PubMed]

19. Wallis, C.M.; Galarneau, E.R. Phenolic Compound induction in plant-microbe and plant-insect interactions: A meta-analysis. *Front. Plant Sci.* **2020**, *11*, 2034. [CrossRef]

20. Poveda, J. Use of plant-defense hormones against pathogen-diseases of postharvest fresh produce. *Physiol. Mol. Plant Pathol.* **2020**, *111*, 101521. [CrossRef]

21. Zeng, K.; Cao, J.; Jiang, W. Enhancing disease resistance in harvested mango (*Mangifera indica* L. cv.'Matisu') fruit by salicylic acid. *J. Sci. Food Agric.* **2006**, *86*, 694–698. [CrossRef]

22. Yao, H.; Tian, S. Effects of pre-and post-harvest application of salicylic acid or methyl jasmonate on inducing disease resistance of sweet cherry fruit in storage. *Postharvest Biol. Technol.* **2005**, *35*, 253–262. [CrossRef]

23. Yang, Z.; Cao, S.; Cai, Y.; Zheng, Y. Combination of salicylic acid and ultrasound to control postharvest blue mold caused by *Penicillium expansum* in peach fruit. *Innov. Food Sci. Emerg. Technol.* **2011**, *12*, 310–314. [CrossRef]

24. Zhang, H.; Ma, L.; Wang, L.; Jiang, S.; Dong, Y.; Zheng, X. Biocontrol of gray mold decay in peach fruit by integration of antagonistic yeast with salicylic acid and their effects on postharvest quality parameters. *Biol. Control* **2008**, *47*, 60–65. [CrossRef]

25. Ezzat, A.; Szabó, S.; Szabó, Z.; Hegedűs, A.; Berényi, D.; Holb, I.J. Temporal patterns and inter-correlations among physical and antioxidant attributes and enzyme activities of apricot fruit inoculated with *Monilinia laxa* under salicylic acid and methyl jasmonate treatments under shelf-life conditions. *Fungi* **2021**, *7*, 341. [CrossRef]

26. Lyousfi, N.; Lahlali, R.; Letrib, C.; Belabess, Z.; Ouaabou, R.; Ennahli, S.; Blenzar, A.; Barka, E.A. Improving the biocontrol potential of bacterial antagonists with salicylic acid against brown rot disease and impact on nectarine fruits quality. *Agronomy* **2021**, *11*, 209. [CrossRef]

27. Mattioli, R.; Palombi, N.; Funck, D.; Trovato, M. Proline accumulation in pollen grains as potential target for improved yield stability under salt stress. *Front. Plant Sci.* **2020**, 1699. [CrossRef] [PubMed]

28. Ami, K.; Planchais, S.; Cabassa, C.; Guivarc'h, A.; Véry, A.-A.; Khelifi, M.; Djebbar, R.; Abrous-Belbachir, O.; Carol, P. Different proline responses of two *Algerian durum* wheat cultivars to in vitro salt stress. *Acta Physiol. Plant.* **2020**, *42*, 21. [CrossRef]

29. Trovato, M.; Forlani, G.; Signorelli, S.; Funck, D. Proline metabolism and its functions in development and stress tolerance. In *Osmoprotectant-Mediated Abiotic Stress Tolerance in Plants*; Springer: Berlin/Heidelberg, Germany, 2019; pp. 41–72.

30. Sofy, M.R.; Seleiman, M.F.; Alhammad, B.A.; Alharbi, B.M.; Mohamed, H.I. Minimizing adverse effects of pb on maize plants by combined treatment with jasmonic, salicylic acids and proline. *Agronomy* **2020**, *10*, 699. [CrossRef]

31. Tonhati, R.; Mello, S.C.; Momesso, P.; Pedroso, R.M. L-proline alleviates heat stress of tomato plants grown under protected environment. *Sci. Hortic.* **2020**, *268*, 109370. [CrossRef]

32. Zali, A.G.; Ehsanzadeh, P. Exogenous proline improves osmoregulation, physiological functions, essential oil, and seed yield of fennel. *Ind. Crops Prod.* **2018**, *111*, 133–140. [CrossRef]

33. Merwad, A.-R.M.; Desoky, E.-S.M.; Rady, M.M. Response of water deficit-stressed Vigna unguiculata performances to silicon, proline or methionine foliar application. *Sci. Hortic.* **2018**, *228*, 132–144. [CrossRef]

34. Hanif, S.; Saleem, M.F.; Sarwar, M.; Irshad, M.; Shakoor, A.; Wahid, M.A.; Khan, H.Z. Biochemically triggered heat and drought stress tolerance in rice by proline application. *J. Plant Growth Regul.* **2021**, *40*, 305–312. [CrossRef]

35. El-Beltagi, H.S.; Mohamed, H.I.; Sofy, M.R. Role of ascorbic acid, glutathione and proline applied as singly or in sequence combination in improving chickpea plant through physiological change and antioxidant defense under different levels of irrigation intervals. *Molecules* **2020**, *25*, 1702. [CrossRef] [PubMed]

36. Mohammadrezakhani, S.; Hajilou, J.; Rezanejad, F.; Zaare-Nahandi, F. Assessment of exogenous application of proline on antioxidant compounds in three Citrus species under low temperature stress. *J. Plant Interact.* **2019**, *14*, 347–358. [CrossRef]

37. Osei, R.; Yang, C.; Cui, L.; Ma, T.; Li, Z.; Boamah, S. Isolation, identification, and pathogenicity of *Lelliottia amnigena* causing soft rot of potato tuber in China. *Microb. Pathog.* **2022**, *164*, 105441. [CrossRef]

38. Ben-David, A.; Davidson, C.E. Estimation method for serial dilution experiments. *J. Microbiol. Methods* **2014**, *107*, 214–221. [CrossRef]

39. Osei, R.; Yang, C.; Cui, L.; Wei, L.; Jin, M.; Boamah, S. Salicylic acid effect on the mechanism of *Lelliottia amnigena* causing potato soft rot. *Folia Hortic.* **2021**, *33*, 376–389. [CrossRef]

40. Perveen, S.; Nazir, M. Proline treatment induces salt stress tolerance in maize (*Zea Mays* L. CV. Safaid Afgoi). *Pak. J. Bot.* **2018**, *50*, 1265–1271.

41. Scherf, J.M.; Milling, A.; Allen, C. Moderate temperature fluctuations rapidly reduce the viability of Ralstonia solanacearum race 3, biovar 2, in infected geranium, tomato, and potato plants. *Appl. Environ. Microbiol.* **2010**, *76*, 7061–7067. [CrossRef]

42. Zhang, X.; Zong, Y.; Li, Z.; Yang, R.; Li, Z.; Bi, Y.; Prusky, D. Postharvest *Pichia guilliermondii* treatment promotes wound healing of apple fruits. *Postharvest Biol. Technol.* **2020**, *167*, 111228. [CrossRef]

43. Livak, K.J.; Schmittgen, T.D. Analysis of relative gene expression data using real-time quantitative PCR and the $2^{-\Delta\Delta CT}$ method. *Methods* **2001**, *25*, 402–408. [CrossRef]

44. Bandara, M.B.; Zhu, H.; Sankaridurg, P.R.; Willcox, M.D. Salicylic acid reduces the production of several potential virulence factors of *Pseudomonas aeruginosa* associated with microbial keratitis. *Investig. Ophthalmol. Vis. Sci.* **2006**, *47*, 4453–4460. [CrossRef] [PubMed]

45. Gibson, D.M.; King, B.C.; Hayes, M.L.; Bergstrom, G.C. Plant pathogens as a source of diverse enzymes for lignocellulose digestion. *Curr. Opin. Microbiol.* **2011**, *14*, 264–270. [CrossRef] [PubMed]

46. Kim, H.-S.; Ma, B.; Perna, N.T.; Charkowski, A.O. Phylogeny and virulence of naturally occurring type III secretion system-deficient *Pectobacterium* strains. *Appl. Environ. Microbiol.* **2009**, *75*, 4539–4549. [CrossRef]

47. Stępień, Ł. *Fusarium*: Mycotoxins, Taxonomy, Pathogenicity. *Microorganisms* **2020**, *8*, 1404. [CrossRef] [PubMed]

48. López, M.M.; López-López, M.J.; Martí, R.; Zamora, J.; López-Sanchez, J.; Beltra, R. Effect of acetylsalicylic acid on soft rot produced by *Erwinia carotovora* subsp. *carotovora* in potato tubers under greenhouse conditions. *Potato Res.* **2001**, *44*, 197–206. [CrossRef]

49. Lastochkina, O.; Baymiev, A.; Shayahmetova, A.; Garshina, D.; Koryakov, I.; Shpirnaya, I.; Pusenkova, L.; Mardanshin, I.d.; Kasnak, C.; Palamutoglu, R. Effects of endophytic *Bacillus subtilis* and salicylic acid on postharvest diseases (*Phytophthora infestans, Fusarium oxysporum*) development in stored potato tubers. *Plants* **2020**, *9*, 76. [CrossRef]

50. Eshghpour, E.; Shahryari, F.; Ghasemi, A. Evaluation of salicylic acid effect on the reduction of *Pectobacterium carotovorum* damage in potato tubers. In Proceedings of the 22nd Iranian Plant Protection Congress, Karaj, Iran, 27–30 August 2016.

51. Bawa, G.; Feng, L.; Yan, L.; Du, Y.; Shang, J.; Sun, X.; Wang, X.; Yu, L.; Liu, C.; Yang, W. Pre-treatment of salicylic acid enhances resistance of soybean seedlings to *Fusarium solani*. *Plant Mol. Biol.* **2019**, *101*, 315–323. [CrossRef]

52. Li, L.; Zou, Y. Induction of disease resistance by salicylic acid and calcium ion against *Botrytis cinerea* in tomato (*Lycopersicon esculentum*). *Emir. J. Food Agric.* **2017**, *29*, 78–82. [CrossRef]

53. Cecchini, N.M.; Monteoliva, M.I.; Alvarez, M.E. Proline dehydrogenase contributes to pathogen defense in Arabidopsis. *Plant Physiol.* **2011**, *155*, 1947–1959. [CrossRef]

54. Ashraf, M.; Foolad, M.R. Roles of glycine betaine and proline in improving plant abiotic stress resistance. *Environ. Exp. Bot.* **2007**, *59*, 206–216. [CrossRef]

55. Ben Ahmed, C.; Ben Rouina, B.; Sensoy, S.; Boukhriss, M.; Ben Abdullah, F. Exogenous proline effects on photosynthetic performance and antioxidant defense system of young olive tree. *J. Agric. Food Chem.* **2010**, *58*, 4216–4222. [CrossRef] [PubMed]

56. Qian, M.; Wang, L.; Zhang, S.; Sun, L.; Luo, W.; Posny, D.; Xu, S.; Tang, C.; Ma, M.; Zhang, C. Investigation of proline in superficial scald development during low temperature storage of 'Dangshansuli'pear fruit. *Postharvest Biol. Technol.* **2021**, *181*, 111643. [CrossRef]

57. Hasanuzzaman, M.; Bhuyan, M.; Zulfiqar, F.; Raza, A.; Mohsin, S.M.; Mahmud, J.A.; Fujita, M.; Fotopoulos, V. Reactive oxygen species and antioxidant defense in plants under abiotic stress: Revisiting the crucial role of a universal defense regulator. *Antioxidants* **2020**, *9*, 681. [CrossRef] [PubMed]

58. Kolupaev, Y.E.; Yastreb, T. Physiological functions of nonenzymatic antioxidants of plants. *Bull. Kharkiv Natl. Agrar. Univ. Ser. Biol.* **2015**, *2*, 6–25.

59. Mishra, A.K.; Baek, K.-H. Salicylic Acid Biosynthesis and Metabolism: A Divergent Pathway for Plants and Bacteria. *Biomolecules* **2021**, *11*, 705. [CrossRef]

60. Munsif, F.; Shah, T.; Arif, M.; Jehangir, M.; Afridi, M.Z.; Ahmad, I.; Jan, B.L.; Alansi, S. Combined effect of salicylic acid and potassium mitigates drought stress through the modulation of physio-biochemical attributes and key antioxidants in wheat. *Saudi J. Biol. Sci.* **2022**, *29*, 103294. [CrossRef]

61. Estaji, A. Impact of exogenous applications of salicylic acid on the tolerance to drought stress in pepper (*Capsicum Annuum* L.) plants. *Res. Sq.* **2020**, *6*, 1–17. [CrossRef]

62. Sayyari, M.; Ghavami, M.; Ghanbari, F.; Kordi, S. Assessment of salicylic acid impacts on growth rate and some physiological parameters of lettuce plants under drought stress conditions. *Int. J. Agric. Crop Sci.* **2013**, *5*, 1951–1957.

63. Naeem, M.; Basit, A.; Ahmad, I. Effect of Salicylic Acid and Salinity Stress on the Performance of Tomato Plants. *Gesunde Pflanz.* **2020**, *72*, 393–402. [CrossRef]

64. Iqbal, M.; Ashraf, M.; Jamil, A.; Ur-Rehman, S. Does seed priming induce changes in the levels of some endogenous plant hormones in hexaploid wheat plants under salt stress? *J. Integr. Plant Biol.* **2006**, *48*, 181–189. [CrossRef]

65. Hayat, K.; Khan, J.; Khan, A.; Ullah, S.; Ali, S.; Fu, Y. Ameliorative effects of exogenous proline on photosynthetic attributes, nutrients uptake, and oxidative stresses under cadmium in Pigeon Pea (*Cajanus cajan* L.). *Plants* **2021**, *10*, 796. [CrossRef] [PubMed]

66. Kumar, V.; Yadav, S.K. Proline and betaine provide protection to antioxidant and methylglyoxal detoxification systems during cold stress in *Camellia sinensis* (L.) O. Kuntze. *Acta Physiol. Plant.* **2009**, *31*, 261–269. [CrossRef]

67. de Carvalho, K.; de Campos, M.K.F.; Domingues, D.S.; Pereira, L.F.P.; Vieira, L.G.E. The accumulation of endogenous proline induces changes in gene expression of several antioxidant enzymes in leaves of transgenic Swingle citrumelo. *Mol. Biol. Rep.* **2013**, *40*, 3269–3279. [CrossRef]

68. Sánchez-Rodríguez, E.; del Mar Rubio-Wilhelmi, M.; Blasco, B.; Leyva, R.; Romero, L.; Ruiz, J.M. Antioxidant response resides in the shoot in reciprocal grafts of drought-tolerant and drought-sensitive cultivars in tomato under water stress. *Plant Sci.* **2012**, *188*, 89–96. [CrossRef]

69. Rivas-San Vicente, M.; Plasencia, J. Salicylic acid beyond defence: Its role in plant growth and development. *J. Exp. Bot.* **2011**, *62*, 3321–3338. [CrossRef]

70. Liu, J.; Li, L.; Yuan, F.; Chen, M. Exogenous salicylic acid improves the germination of *Limonium bicolor* seeds under salt stress. *Plant Signal. Behav.* **2019**, *14*, e1644595. [CrossRef] [PubMed]

71. Martínez-Esplá, A.; Serrano, M.; Valero, D.; Martínez-Romero, D.; Castillo, S.; Zapata, P.J. Enhancement of antioxidant systems and storability of two plum cultivars by preharvest treatments with salicylates. *Int. J. Mol. Sci.* **2017**, *18*, 1911. [CrossRef]

72. Kováčik, J.; Klejdus, B.; Hedbavny, J.; Bačkor, M. Effect of copper and salicylic acid on phenolic metabolites and free amino acids in *Scenedesmus quadricauda* (Chlorophyceae). *Plant Sci.* **2010**, *178*, 307–311. [CrossRef]

73. Ma, X.; Zheng, J.; Zhang, X.; Hu, Q.; Qian, R. Salicylic acid alleviates the adverse effects of salt stress on *Dianthus superbus* (Caryophyllaceae) by activating photosynthesis, protecting morphological structure, and enhancing the antioxidant system. *Front. Plant Sci.* **2017**, *8*, 600. [CrossRef]

74. Saheri, F.; Barzin, G.; Pishkar, L.; Boojar, M.M.A.; Babaeekhou, L. Foliar spray of salicylic acid induces physiological and biochemical changes in purslane (*Portulaca oleracea* L.) under drought stress. *Biologia* **2020**, *75*, 2189–2200. [CrossRef]

75. Hossain, M.A.; Kumar, V.; Burritt, D.J.; Fujita, M.; Mäkelä, P. Osmoprotectant-mediated abiotic stress tolerance in plants. In *Proline Metabolism and Its Functions in Development and Stress Tolerance*; Psikol. Çalışmaları; Springer Nature: Cham, Switzerland, 2019; pp. 41–72. [CrossRef]

76. Burritt, D.J. Proline and the cryopreservation of plant tissues: Functions and practical applications. *Curr. Front. Cryopreserv.* **2012**, *14*, 415–426. [CrossRef]

77. Adejumo, S.A.; Oniosun, B.; Akpoilih, O.A.; Adeseko, A.; Arowo, D.O. Anatomical changes, osmolytes accumulation and distribution in the native plants growing on Pb-contaminated sites. *Environ. Geochem. Health* **2021**, *43*, 1537–1549. [CrossRef] [PubMed]

78. Dar, M.I.; Naikoo, M.I.; Rehman, F.; Naushin, F.; Khan, F.A. Proline accumulation in plants: Roles in stress tolerance and plant development. In *Osmolytes and Plants Acclimation to Changing Environment: Emerging Omics Technologies*; Springer: Berlin/Heidelberg, Germany, 2016; pp. 155–166. [CrossRef]

79. Priya, M.; Sharma, L.; Singh, I.; Bains, T.; Siddique, K.H.; Bindumadhava, H.; Nair, R.M.; Nayyar, H. Securing reproductive function in mungbean grown under high temperature environment with exogenous application of proline. *Plant Physiol. Biochem.* **2019**, *140*, 136–150. [CrossRef] [PubMed]

80. Hoque, M.A.; Banu, M.N.A.; Okuma, E.; Amako, K.; Nakamura, Y.; Shimoishi, Y.; Murata, Y. Exogenous proline and glycine-betaine increase NaCl-induced ascorbate–glutathione cycle enzyme activities, and proline improves salt tolerance more than glycinebetaine in tobacco Bright Yellow-2 suspension-cultured cells. *J. Plant Physiol.* **2007**, *164*, 1457–1468. [CrossRef] [PubMed]

81. Abdelhamid, M.T.; Rady, M.M.; Osman, A.S.; Abdalla, M.A. Exogenous application of proline alleviates salt-induced oxidative stress in *Phaseolus vulgaris* L. plants. *J. Hortic. Sci. Biotechnol.* **2013**, *88*, 439–446. [CrossRef]

82. Tabssum, F.; uz Zaman, Q.; Chen, Y.; Riaz, U.; Ashraf, W.; Aslam, A.; Ehsan, N.; Nawaz, R.; Aziz, H. Exogenous application of proline improved salt tolerance in rice through modulation of antioxidant activities. *Pak. J. Agric. Res.* **2019**, *32*, 140. [CrossRef]

83. Ghaffari, H.; Tadayon, M.R.; Nadeem, M.; Cheema, M.; Razmjoo, J. Proline-mediated changes in antioxidant enzymatic activities and the physiology of sugar beet under drought stress. *Acta Physiol. Plant.* **2019**, *41*, 23. [CrossRef]

84. Abdallah, M.M.; El-Bassiouny, H. Impact of exogenous proline or tyrosine on growth, some biochemical aspects and yield components of quinoa plant grown in sandy soil. *Int. J. Pharm. Technol.* **2016**, *9*, 12–23.

85. Yang, R.; Han, Y.; Han, Z.; Ackah, S.; Li, Z.; Bi, Y.; Yang, Q.; Prusky, D. Hot water dipping stimulated wound healing of potato tubers. *Postharvest Biol. Technol.* **2020**, *167*, 111245. [CrossRef]

86. Wang, C.; Chen, L.; Peng, C.; Shang, X.; Lv, X.; Sun, J.; Li, C.; Wei, L.; Liu, X. Postharvest benzothiazole treatment enhances healing in mechanically damaged sweet potato by activating the phenylpropanoid metabolism. *J. Sci. Food Agric.* **2020**, *100*, 3394–3400. [CrossRef]

87. Rais, A.; Jabeen, Z.; Shair, F.; Hafeez, F.Y.; Hassan, M.N. *Bacillus* spp., a bio-control agent enhances the activity of antioxidant defense enzymes in rice against Pyricularia oryzae. *PLoS ONE* **2017**, *12*, e0187412. [CrossRef] [PubMed]

88. Yadav, V.; Wang, Z.; Wei, C.; Amo, A.; Ahmed, B.; Yang, X.; Zhang, X. Phenylpropanoid pathway engineering: An emerging approach towards plant defense. *Pathogens* **2020**, *9*, 312. [CrossRef] [PubMed]

89. Kidokoro, S.; Maruyama, K.; Nakashima, K.; Imura, Y.; Narusaka, Y.; Shinwari, Z.K.; Osakabe, Y.; Fujita, Y.; Mizoi, J.; Shinozaki, K. The phytochrome-interacting factor PIF7 negatively regulates DREB1 expression under circadian control in Arabidopsis. *Plant Physiol.* **2009**, *151*, 2046–2057. [CrossRef] [PubMed]

90. Lavrova, V.; Udalova, Z.V.; Matveeva, E.; Khasanov, F.; Zinovieva, S. Mi-1 gene expression in tomato plants under root-knot nematode invasion and treatment with salicylic acid. *Dokl. Biochem. Biophys.* **2016**, *471*, 413–416. [CrossRef]

91. El-Esawi, M.A.; Elansary, H.O.; El-Shanhorey, N.A.; Abdel-Hamid, A.M.; Ali, H.M.; Elshikh, M.S. Salicylic acid-regulated antioxidant mechanisms and gene expression enhance rosemary performance under saline conditions. *Front. Physiol.* **2017**, *8*, 716. [CrossRef]

92. Mutlu, S.; Atıcı, Ö.; Nalbantoğlu, B.; Mete, E. Exogenous salicylic acid alleviates cold damage by regulating antioxidative system in two barley (*Hordeum vulgare* L.) cultivars. *Front. Life Sci.* **2016**, *9*, 99–109. [CrossRef]

93. Hayat, S.; Hayat, Q.; Alyemeni, M.N.; Wani, A.S.; Pichtel, J.; Ahmad, A. Role of proline under changing environments: A review. *Plant Signal. Behav.* **2012**, *7*, 1456–1466. [CrossRef]

94. Sperdouli, I.; Moustakas, M. Interaction of proline, sugars, and anthocyanins during photosynthetic acclimation of *Arabidopsis thaliana* to drought stress. *J. Plant Physiol.* **2012**, *169*, 577–585. [CrossRef]

95. Mahesh, H.; Murali, M.; Pal, M.A.C.; Melvin, P.; Sharada, M. Salicylic acid seed priming instigates defense mechanism by inducing PR-Proteins in *Solanum melongena* L. upon infection with *Verticillium dahliae Kleb. Plant Physiol. Biochem.* **2017**, *117*, 12–23. [CrossRef]

96. Sekhon, P.; Sangha, M. Influence of different SAR elicitors on induction and expression of *PR*-proteins in Potato and Muskmelon against Oomycete pathogens. *Indian Phytopathol.* **2019**, *72*, 43–51. [CrossRef]

97. Ghanbari, S.; Fakheri, B.A.; Mahdinezhad, N.; Khedri, R. Systemic acquired resistance. *J. New Biol. Rep.* **2015**, *23*, 105–110.

98. Chen, J.; Zhang, Y.; Wang, C.; Lü, W.; Jin, J.B.; Hua, X. Proline induces calcium-mediated oxidative burst and salicylic acid signaling. *Amino Acid* **2011**, *40*, 1473–1484. [CrossRef] [PubMed]

99. Zheng, M.; Aslund, F.; Storz, G. Activation of the OxyR transcription factor by reversible disulfide bond formation. *Science* **1998**, *279*, 1718–1721. [CrossRef] [PubMed]

MDPI

St. Alban-Anlage 66

4052 Basel

Switzerland

www.mdpi.com

Sustainability Editorial Office

E-mail: sustainability@mdpi.com

www.mdpi.com/journal/sustainability